Virtual Reality in Health and Rehabilitation

Rehabilitation Science in Practice Series

Series Editors:

Marcia J. Scherer
Institute for Matching Person & Technology, Webster, New York, USA

Dave J. Muller
Visiting Professor Rehabilitation Psychology, University of Suffolk, UK

Everyday Technologies in Healthcare
Christopher M. Hayre, Dave J. Muller, Marcia J. Scherer

**Enhancing Healthcare and Rehabilitation: The Impact
of Qualitative Research**
Christopher M. Hayre, Dave J. Muller

**Neurological Rehabilitation: Spasticity and Contractures
in Clinical Practice and Research**
Anand D. Pandyan, Hermie J. Hermens, Bernard A. Conway

Quality of Life Technology Handbook
Richard Schulz

**Computer Systems Experiences of Users with and Without Disabilities:
An Evaluation Guide for Professionals**
Simone Borsci, Masaaki Kurosu, Stefano Federici, Maria Laura Mele

Assistive Technology Assessment Handbook - 2nd Edition
Stefano Federici, Marcia J. Scherer

Ambient Assisted Living
Nuno M. Garcia, Joel Jose P.C. Rodrigues

For more information about this series, please visit: https://www.crcpress.com/
Rehabilitation-Science-in-Practice-Series/book-series/CRCPRESERIN

Virtual Reality in Health and Rehabilitation

Edited by

Christopher M. Hayre,
Dave J. Muller and Marcia J. Scherer

CRC Press
Taylor & Francis Group
Boca Raton London New York

CRC Press is an imprint of the
Taylor & Francis Group, an **informa** business

CRC Press
Boca Raton and London
First edition published 2021
by CRC Press
6000 Broken Sound Parkway NW, Suite 300, Boca Raton, FL 33487-2742
and by CRC Press
2 Park Square, Milton Park, Abingdon, Oxon, OX14 4RN

ISBN: 978-0-367-36771-8 (hbk)
ISBN: 978-0-429-35136-5 (ebk)

Typeset in Times LT Std
by KnowledgeWorks Global Ltd.

Dedication

Dr. Christopher M. Hayre would like to dedicate this book to his wife Charlotte, to whom this would not be possible. He would also like to dedicate this book to his daughters Ayva, Evelynn, and Ellena. Love to all.

Professor Dave J. Muller would like to dedicate this to his extended family with love, Pam, Emily, Pete, Lucy, Simon, Tasha, Harlie, Luke, Toby, Edie, Kaya, and Freya.

Professor Marcia J. Scherer would like to dedicate this to her husband and colleagues from whom she continues to learn so much. Further, she dedicates this to the users and providers of today and tomorrow in their quest to employ technologies wisely and in ways that add quality to life.

The editors would also like to dedicate this book to a recent colleague and contributor of this book, Professor Pierre-Alain Joseph, who recently passed away. Our thoughts are with his family, friends, and colleagues.

Collectively, the editors congratulate one another for completing another worthwhile and enjoyable collaborative project.

Contents

SECTION I Introductory Perspective

SECTION II Virtual Reality in Neurological Rehabilitation

SECTION III   Virtual Reality for Health Education

SECTION IV   Gamification and Virtual Reality in Contemporary Contexts

SECTION V   Emerging Perspectives and Applications of Virtual Reality in Practice

Acknowledgements

The editors would like to thank all contributing authors for sharing their innovative work in the field of virtual reality. This is a growing area of interest for practitioners, and your commitment to this book reflects the multifaceted use of virtual reality in the contemporary settings. On behalf of the editors, it has been a great pleasure working with you all in bringing together this collection of academic work. The work demonstrates not only the existing virtues of virtual reality in practice but also highlights developmental opportunities within the field. This has been a thoroughly enjoyable project as we have been able to collaborate with colleagues from various parts of the world. Thank you.

Editors

Dr. Christopher M. Hayre is currently a Senior Lecturer in Diagnostic Radiography at Charles Sturt University, Australia. He is also a Senior Fellow for the School of Health and Sport Sciences at the University of Suffolk. He has published over 50 outputs, involving commissioned texts, book chapters, journal articles, and conference outputs. In 2016, he founded the Journal of Social Science & Allied Health Professions and remains Editor-in-Chief. He is currently involved in a number of projects and has research interests in virtual reality, personalized medicine, and artificial intelligence and how they will impact on the medical imaging profession and wider healthcare community. In 2019 he founded his own book series and remains Editor, with CRC Press.

Professor Dave J. Muller is currently Editor of the CRC series with Professor Marcia Scherer on Rehabilitation Science in Practice. He was founder Editor of the Journal Aphasiology and is currently Editor in Chief of the Journal Disability and Rehabilitation. He has published over 40 refereed papers and has been involved either as Series Editor, Editor or author of over 50 books. He is a visiting Professor at the University of Suffolk, United Kingdom.

Professor Marcia J. Scherer is a rehabilitation psychologist and founding President of the Institute for Matching Person & Technology. She is also Professor of Physical Medicine and Rehabilitation, University of Rochester Medical Center where she received both her Ph.D. and MPH degrees. She is a past member of the National Advisory Board on Medical Rehabilitation Research, National Institutes of Health, and is Editor of the journal Disability and Rehabilitation: Assistive Technology. She is Co-Editor of the book series for CRC Press, Rehabilitation Science in Practice Series. Dr. Scherer is Fellow of the American Psychological Association, American Congress of Rehabilitation Medicine, and the Rehabilitation Engineering and Assistive Technology Society of North America (RESNA). Dr. Scherer has authored, edited, or co-edited nine book titles and has published over 80 articles in peer-reviewed journals, 50 published proceedings papers, and 30 book chapters on disability and technology. Her research has been cited more than 5000 times by others.

Foreword

Over the last 25 years, intrepid researchers and clinicians have pursued the dream of using virtual reality (VR) as a tool to advance clinical assessment, intervention, and scientific research. This dream was inspired by the many lines of reasoning that make the intuitive case for why VR is well matched to address the challenges for providing effective clinical strategies for mental health, rehabilitation, and general medical purposes.

At its core, VR technology, along with other related simulation-based formats (e.g., augmented/mixed reality), offers new capabilities that simply did not exist with the traditional methods that were available back in the bygone days of the 20th century! Since the mid-1990s, VR-based testing, training, teaching, and treatment approaches have been developed that would be difficult, if not impossible, to deliver without leveraging the power of modern computing, 3D graphics, body tracking and novel user interfaces, gaming/narrative principles, robotics, big data analytics, and artificial intelligence. Such VR enabling technologies allow for the creation of highly realistic (and sometimes surreal), interactive, engaging, and systematically controllable stimulus environments that users can be immersed in and interact with, for human performance measurement and training, as well as clinical assessment and intervention. This "Ultimate Skinner Box" perspective makes VR well matched to the experimental- and clinical targets that comprise the area sometimes referred to as, "Virtual Rehabilitation".

At the same time, the costs and complexity for developing clinical VR applications have gone down as the sheer capacity of the technology has concomitantly gone up! A complex six degree of freedom tracked VR headset and hand controllers that might have cost tens of thousands of dollars in 2000, now checks in at well under 1,000 dollars. Exciting clinical opportunities enabled by this trend loom on the horizon with the recent developments in "Standalone" VR headsets (e.g., Oculus Quest, Pico Neo, Vive Focus, etc.). Such VR display systems do not require a tethered computer, with all the graphic and interaction processing taking place onboard the device. 5G cloud-computing promises to advance this capability by offloading much of the real-time simulation processing to remote servers that will offer up large libraries of clinical VR scenarios designed to meet the varied needs of diverse clinical populations. Moreover, in addition to the dramatic advancements in the audio-visual and 3D user interface tracking fidelity of these new systems, a ready group of capable and passionate VR programmers is now waiting in the wings, many having learned VR coding in high school computer science courses. This bodes well for the continued growth and opportunities that VR will enable in the future of clinical healthcare and rehabilitation.

However, as encouraging as all this sounds, VR as a clinical tool cannot evolve and add value simply based on intuition, technical prowess, and optimistic projections. The hard and sometimes cold reality of scientific inquiry and evaluation must be an essential part of the mix. That is where the current volume, *Virtual Reality in Health and Rehabilitation*, enters the picture to provide very current and detailed accountings of where we stand with the science of virtual rehabilitation. As I read this book, I was impressed not only with the honest and critical reviews of the state of the science in each clinical domain but also in the informed rational analysis of what factors may have contributed to the results reported in the outcome studies reviewed. When viewed across the many clinical conditions where VR has been applied, an encouraging picture for the future emerges.

Most of the research reviewed in this book falls into what I call the "Phase 1 Period" of Virtual Rehabilitation with outcomes that are often at least equivalent to standard practice, often with recommendations that VR could be a useful "supplement" to standard practice. That is impressive for a nascent field that has often been insufficiently funded and where the early research was often challenged by technologic limitations. The technology required to deliver high fidelity rehabilitation has been a moving target over the years where researchers have had to put focus on adapting systems to the incremental updates and occasional failures of technology that routinely confounded system access and usability, let alone documented usefulness. Researchers in this area also commonly had two significantly daunting development requirements—garner enough resources (both financial and technical capability) to build a quality VR system and then acquire sufficient funding to run a decent clinical trial. I believe that the significant weight of these dual challenges are in large part responsible for the common critique that appears in various reviews and meta-analyses in this literature—that the evidence is limited by "lack of controlled comparisons, small sample sizes, absence of long term follow-ups, and the need for more comprehensive outcome assessments." That "wish-list," while essential for good science, is a tall order for VR clinicians and researchers. Such scientists are tasked with creating innovative, theoretically informed, complex, and usable systems, managing hardware/software distribution and system maintenance issues, recruiting for and implementing the standards needed for quality clinical trials, and of course the necessary and additional job of fund raising to translate an aspirational vision into real hardware applications that may sometimes challenge established paradigms and generate professional or funder resistance. In spite of these demands, Virtual Rehabilitation professionals have been inspired to rise to those challenges and the chapters in this book provide optimism that the field is well poised to move forward with Virtual Rehabilitation's "Phase 2 Period."

In this next phase, within the backdrop of the growing public awareness/appreciation of VR's capabilities and a growing army of young, passionate, and well-trained VR developers, the field will be empowered to leverage what Virtual Rehabilitation science has revealed thus far. Informed by that knowledge base, this next phase will be further fueled by the availability of higher quality, lower cost, and familiar VR technology solutions that are now being honed on the proving ground of the accelerating commercial gaming and entertainment markets. These factors will converge to chart a path towards conducting the research needed to further drive applications and clinical strategies that have previously documented positive impact on the users that these systems were designed to benefit. The beginnings of such a trajectory are already being observed in the clinical VR exposure therapy and pain management R&D areas.

Within this next phase, we may also expect to see a sharper focus on research questions beyond the general outcome findings generated in large sample efficacy studies. For example, future well-powered research could serve to inform better prediction in advance, as to WHO will benefit from a VR approach as opposed to a more traditional strategy. Dismantling studies will be needed to more precisely specify what components of a VR intervention approach (e.g., sensory realism/fidelity, 3D user interface parameters, stimulus challenge pacing, reinforcement/feedback scheduling, addition of narrative/gaming to promote engagement/adherence, etc.) will serve to produce positive treatment outcomes. Lower cost personal VR systems will also enable clinician-supervised, independent user access to home-based VR systems that provide increased opportunities to maximize rehabilitation activity "doses" in engaging formats that motivate users to do the number of trials needed to produce desired outcomes. I believe that the knowledge conveyed in the chapters in *Virtual Reality in Health and Rehabilitation* will figure significantly in guiding the science for this next phase in the clinical application of VR, and thus, this impressive volume should be viewed as a "field manual" for the future of Virtual Rehabilitation!

Skip Rizzo, Ph.D.
University of Southern California
Institute for Creative Technologies

Section I

Introductory Perspective

1 Introducing Virtual Reality for Health and Rehabilitation

Christopher M. Hayre
Charles Sturt University
Wagga Wagga, Australia

Dave J. Muller
University of Suffolk
Ipswich, Suffolk, UK

Marcia J. Scherer
Institute for Matching Person and Technology
Webster, New York

CONTENTS

SETTING THE SCENE

The use of virtual reality (VR) is captured within this edited volume. It not only provides contributions from researchers transnationally but also importantly represents the far reaching application(s) of VR in an array of health and rehabilitation settings. This book is in part a celebration of the emerging and sustained applications of VR by researchers and practitioners alike, and henceforth, up and coming chapters reaffirm the utility of VR technology, which is anticipated to add value to readers transnationally.

The book begins by demonstrating the application of VR for neurological rehabilitation. This remains a large section of this book and continues to represent a large volume of VR use around the world (Section II). The book then draws from experienced educators that currently utilize VR technology in tertiary educational settings. We see the application of virtual simulation for the use of ultrasound, in response to a shortage in clinical supervision. In addition, we see the advantages of VR use for cancer treatment and interventional procedures, facilitating the training and education of diagnostic and therapeutic radiographers. Section IV combines the notion of gamification within the VR sphere, whereby gameplay and reward are interlinked to enhancing health outcomes. The final section of this book, then, captures the emerging clinical applications of VR, in particular, aphasia, orthopaedics, paediatrics and loss of vision and weight management.

It is anticipated this book will be of use to a number of audiences. First, for undergraduate students, the examples provide insight for learning and understanding about VR technology within contemporary health and rehabilitation environments. Second, classroom will find the content

useful for knowledge transfer, whilst also utilizing topics for class discussion. Lastly, it is anticipated that postgraduate students and early career researchers will find the content and references
helpful in order to define their own research objectives and importantly uncover unique areas of
VR research. This introductory chapter will now provide a brief overview of VR, identify its role in
contemporary practice, coincided with its role in a Coronavirus (COVID-19) world.

WHAT IS VIRTUAL REALITY?

The development of VR in recent years has evolved in light of technological advances associated
with enhanced computing power and advances in graphical and three dimension representations.
Devices that help enhance the experience are those that are closely linked to the human senses,
notably visual, auditory, haptic and olfactory. These sensory connections between humans and technology enable individuals to resemble experiences that may be part of the real world. VR, then, is
a concept that aims to simulate an experience, which is either similar or different to the real world
around us, but replicated within a digital sphere. As we write this chapter and bring together this
edited book volume, the unknown prospect of the COVID-19 pandemic has led to an immediate
physical disconnect and 'touch' with family, friends, colleagues and society. In response, whilst
our social distancing in society has led to an increase in 'virtual connectedness', it does bring into
question the relevance of VR becoming increasingly integrated and managed into our everyday
lives. Further, not only for the fulfilling clinical contexts but also for the purpose of remaining safe
and well, as individuals in society. In addition to the widely recognized concept of VR, augmented
reality (AR) is also generally accepted through an interactive experience of real-world environments
and is encompassed through objects that reside in the real world, using multiple sensory tools. The
role of AR has three central facets in order for it to be considered 'augmented'; (1) combines real
and virtual worlds; (2) provides real-time interaction and (3) accurately provides 3D registration of
virtual and real objects. In short, AR alters the perception of a real-world environment, whereas VR
completely replaces the participant's real-world environment with a simulated one.

VR has classically been utilized in a number of environments prior to emerging in the health
and rehabilitation context. Jaron Lanier, for instance, in the late 1980s built a digital model using a
graphic interface, a VR environment, which used external devices in order to facilitate the interaction with graphical elements of that environment. Today, VR is home to no discipline and whilst
first observed within the computer gaming industry and enhancing cinematic experiences, it now
plays a significant role in medical training (Ruthenbeck and Reynolds, 2015), aviation (Eschen
et al, 2018), training for the military (Bhagat, Liou and Chang, 2016) and healthcare in general. In
short, a VR system provides an interface between the individual and a computer, which can be supported by sensors and a viewer, usually fixed to the user's head. Furthermore, multisensory trackers
capture movements of a user and can now send quantifiable data for research/audit purposes with
real-time visualization (Moloney et al, 2018). The multifaceted uses within a multivariate professional space shares a common goal (and remains a central facet throughout this book): that of
enhancing the visual, auditory, haptic and/or olfactory experience of individuals in order to assist in
the healthcare delivery and rehabilitation.

THE ROLE OF VIRTUAL REALITY IN THE HEALTH
AND REHABILITATION ENVIRONMENT

VR has attracted attention as a means of managing health and rehabilitation in a wide range of
health fields in recent years, which is reflected within the literature. It is generally accepted that
VR provides patients and service users with sensory stimulation, and a more immersive environment, with real-time feedback during specific task-orientated actions reflecting both motor
learning and neuroplasticity. The benefits of VR in rehabilitation provide control over a particular
stimulus and hence deliver consistency to service users; it has the ability to vary the stimuli from

say a 'simple' to more 'complex situations', thus enabling professionals to 'test' users in a safe and controlled manner. The ability to quantifiably record (and perhaps reward) an individual's progress remains key in the role of VR use amongst individuals. For instance, VR therapy for stroke survivors has an added benefit by providing instant feedback to individuals with the level of difficulty of the therapy modified when appropriate (Singh et al, 2013). Individuals recovering from a stroke have reported that the use of VR was more engaging and stimulating for them, when compared with conventional therapy (Levin et al, 2012). It is evident that VR has the ability to deliver 'personalized care' to individuals, reflecting their needs and abilities due to the extent of their condition, thus arguably falling into the personalized medicine realm of healthcare delivery. The example of such a personalized medical approach resonates with patients with stroke and brain-related injuries, whereby the scope of injury and/or recovery will vary, depending on the situation.

In short, VR applications have wide ranging applications and potential in terms of neuroplasticity for individuals (Cheung et al, 2014). VR remains a promising tool for treating and managing patient experiences, coincided with cognitive and physical recovery (Shin and Kim, 2015), upper limb function (Goncalves et al, 2018), whilst improving memory and retention, regardless of the level of immersion (Mendes et al, 2012). Thus, whilst a clear focus still remains in rehabilitation, there are other areas which lead the way with VR use, which this book examines.

VR has the capability of generating the physical world analogs via simulated settings, of which patients and/or service users interact (Pausch, 1993). There is an argument that the conventional one-to-one application of either training or rehabilitation (without the use of VR) is seen to be monotonous or even boring to patients and as such many struggle to integrate it into their daily lives (Feng et al, 2019). The counter argument favours VR technology, whereby favoured games encourage motivation, which lead to reward upon completion of a particular exercise. This interconnection with gamification via reward mechanisms aims to help improve patient outcomes, which can be supported with choice of music and exquisite pictures of family, friends or idyllic scenery in order to help instill motivation and engagement (Dias et al, 2019).

Other examples of VR use have been associated with The Virtual Reality Exposure Therapy (VRET) application, helping prospective radiotherapists position and treat cancer patients using high levels of ionising radiation. Further, we are witnessing VR use in psychology, as a treatment for phobias. For instance, a recent study by Gujjar et al (2018) identified how VR facilitated dental phobias, alleviating anxiety in patients. Here, we can see that positive stimuli may help foster a unique health and rehabilitation environment in order to help reduce phobias and is being considered in other health settings. For instance, a study by Brown et al (2019) examined the application of VR as a tool to desensitize claustrophobic patients prior to attending magnetic resonance imaging (MRI) examinations. The authors of this study prospectively aim to use a mobile phone application and headset in order to provide a glimpse of an MRI experience (from within the bore of the magnet). By providing the patient with an initial virtual experience, it is anticipated that both hospital cancellations and abandonment of MRI examinations during imaging may reduce, leading to a more cost-effective service.

Another interesting area of VR use is the notion of anaesthesia. Because VR is now able to combine visual, auditory, haptic and olfactory senses, it is arguably able to distract individuals from painful processes. A study by Hayashi et al (2019) demonstrated that exercised-induced pain in healthy subjects showed VR immersion significantly reduced pain intensity by the participants. Further, a meta-analysis by Mallari et al (2019) examined the effectiveness of VR in reducing both acute and chronic pain. A good example of the application of VR to limit pain has been demonstrated by Maani et al (2011), whereby the use of immersive VR reduced pain of patients with combat-related bring injuries during severe burn wound debridement. In other examples, VR has demonstrated pain reduction in intravenous placement (Gold et al, 2005), dental work (Gujjar et al, 2018), neck pain (Chen et al, 2017), spinal cord injuries (Chi et al, 2019) and cancer treatment (Mohammad and Ahmad, 2019).

Whilst the aforementioned offers a unique paradigm shift in terms of healthcare delivery, VR has the potential to cause discomfort, nausea, disorientation, dizziness, eyestrain and fatigue when used (Davis et al, 2014). This is generally described as cybersickness and there are two common theories that aim to describe its occurrence. First, the sensory conflict theory assigns the occurrence of symptoms to discrepancies between the perceived motion of the virtual optic flow and the motion of the participant detected by their vestibular, proprioceptive and somatosensory systems (Keshavarz et al, 2015). For instance, it has been reported that stroke survivors may experience issues processing information from the above-mentioned systems, thus making them susceptible to cybersickness (Massetti et al, 2018). Second, the postural instability theory identifies postural instability as the cause for symptoms and suggests that they are due to postural instability in unfamiliar situations in order to maintain balance (Lim et al, 2018). Again, stroke survivors might be more prone to develop cybersickness as they could already experience postural instability (Weech et al, 2019). Cybersickness symptoms might rapidly fade but could even persist for days afterwards (Lackner, 2014), increasing a potential risk for falls, which could preclude the use of immersive VR.

Whilst we are observing the unique application of VR within health and rehabilitation settings, it is important to remain mindful of other associated risks, such as cybersickness, which may cause additional stress and potentially lead to additional risks for patients.

VIRTUAL REALITY IN A POST COVID-19 WORLD

In a post COVID-19 world, what will the application of VR look like for patient use? Is it plausible to suggest that a combination of VR, telerehabilitation and self-management will be required in order to enhance the outcomes of patients? Russell (2007) comments on the shorter history associated with telerehabilitation, when compared to other fields, in particular medical imaging. This has arguably been predominately because of the requirement to 'physically touch' in rehabilitation, which is used to guide, direct and facilitate movement of patients. The conceptual shift in assessment and treatment practices, however, for telerehabilitation services is considered a possibility (Russell, 2007). Further, we hasten to add that in relation for our immediate need to change due to COVID-19, the integration of VR and telerehabilitation should be further expedited to ensure remote rehabilitation, facilitated with quantitative data collection in future years.

Historically, telemedicine remains an excellent tool that aims to reach and treat individuals isolated as a result of physical impairment, or geographical locality. It has notably been used in previous pandemics, for instance, the treatment of Ebola (Ohannessian, 2015). Yet, the COVID-19 pandemic has been far reaching when compared to Ebola and thus impacted communities around the world. On reflection, as healthcare practitioners we should now begin to think how telerehabilitation, VR and self-management could be integrated in order to ensure the sound delivery of rehabilitation, if another outbreak was to occur.

As academics, educators and scholars, the authors of this chapter have had to think critically about their own academic practices of education and scholarly delivery via remote technology. On reflection, the first author's ability to communicate with students and peers via teleconferencing tools, the examination of quantifiable metrics to assess student engagement with content, and the delivery of final assessments via online methods has required adaption from all within academic institutions. As we continue to reflect throughout and beyond this pandemic, further thought, optimism and opportunism will be necessary in order into consider the self-management of others and how VR can play an integral part. Whilst many factors need to be considered, such as encryption, confidentiality, ergonomic safety and digital networking, the technology currently exists for an online VR platform that can engage patients, practitioners, academics and students alike. Lastly, the aforementioned combined with the rise of artificial intelligence and machine learning, it may begin to help quantify, predict and utilize technology in a way that promotes the health, safety and well-being of others.

SUMMARY

This introductory chapter sought to set the scene for this book. We began by providing an overview of the forthcoming content, which we anticipate readers will find useful. We then reflected upon the notion of VR and how it is being incorporated in the health and rehabilitation space. The authors have identified that although VR has been around for several decades, we are firmly becoming accustomed to its virtues in which it may hold for patients, service users and students alike. Further, the COVID-19 pandemic, which is currently sweeping our planet, should enable us to think critically, offer reflection and examine the importance of sustaining the delivery of digital care, whereby tools such as VR, telemedicine, AI and self-management may become the everyday norms within society in order to ensure health and rehabilitation practices continue and react if/when social isolation again becomes a requirement.

REFERENCES

Bhagat, K.K., Liou, W., and Chang, C. (2016) A cost-effective interactive 3D virtual reality system applied to military live firing training. *Virtual Reality*. 20, pp. 127–140.

Brown, R.K.J., Petty, S., O'Malley, S., Stojanovska, J., Davenport, M.S., Kazerooni, E.A., and Fessahazion, D. (2018) Virtual reality tool simulates MRI experience. *Tomography*. 4 (3), pp. 95–98.

Chen, K.B., Sesto, M.E., Ponto, K., Leonard, J., Mason, A., Vanderheiden, G., Williams, J., and Radwin, R.G. (2017) Use of virtual reality feedback for patients with chronic neck pain and kinesiophobia. *IEEE Trans Neural System Rehabilitation Engineering*. 25 (8), pp. 1240–1248.

Cheung, K.L., Tunik, E., Adamovich, S.V., and Boyd, L.A. (2014) Neuroplasticity and virtual reality. *Virtual Reality for Physical and Motor Rehabilitation*, pp. 5–24.

Chi, B., Chau, B., Yeo, E., and Ta, P. (2019) Virtual reality for spinal cord injury-associated neuropathic pain: Systematic review. *Annals of Physical and Rehabilitation Medicine*. 62 (1), pp. 49–57.

Davis, S., Nesbitt, K., and Nalivaiko, E. (2014) A systematic review of cybersickness. *Association for Computing Machinery*. December 2014. 1, pp. 1–9.

Dias, P., Silva, R., Amorium, P., Lains, J., Roque, E., Pereira, I.S.F, Santos., B.S., and Potel, M. (2019) Using virtual reality to increase motivation in poststrike rehabilitation. *IEEE Computer Graphics Applications*. 39 (1), pp. 64–70.

Eschen, H., Kotter, T., Rodeck, R., Harnisch, M., and Schuppstuhl, T. (2018) Augmented and virtual reality for inspection and maintenance processes in the aviation inductry. *Prodedia Manufacturing*. 19 (1), pp. 156–163.

Feng, H., Cuiyun, L., Liu, J., Wang, L., Ma, J., Li, G., Gan, L., Shang, X., and Wu, Z., (2019) Virtual reality rehabilitation versus conventional physical therapy for improving balance and gait in Parkinson's disease patients: A randomized controlled trial. *Medical Science Monitor*. 25, pp. 4186–4192.

Gold J.I., Kant A., Kim S., and Rizzo A. (2005) Virtual anesthesia: The use of virtual reality for pain distraction during acute medical interventions. *Seminars in Anesthesia, Perioperative Medicine and Pain*. 24, pp. 203–210.

Goncalves, M.G., Piva, M.F.L., Marques, C.L.S., da Costa, R.D.M., Bazan, R., Luvizutto, G.J., Gomes, L.E., and Betting, G. (2018) Effects of virtual reality therapy on upper limb function after stroke and the role of neuroimaging as a predictor of a better response. *Arquivos de Neuro-Psiquiatria*. 76 (10), pp. 654–662.

Gujjar, K.R., Wijk, A.V., Sharma, R., and Jongh, A.D. (2018) Virtual reality exposure therapy for the treatment of dental phobia: A controlled feasibility study. *Behavioral Cognitive Psychotherapy*. 46 (3), pp. 367–373.

Hayashi, K., Aono, S., Shiro, Y., and Ushida, T. (2019) Effects of virtual reality-based exercise imagery on pain in healthy individuals. *BioMed Research International*. Apr 17. 1, pp. 1–9.

Keshavarz, B., Riecke, B.E., Hettinger, L.J., and Campos, J.L. (2015) Vection and visually induced motion sickness: How are they related? *Frontiers in Psychology*. 20 April, [Online]. Available at: https://www.frontiersin.org/articles/10.3389/fpsyg.2015.00472/full (Accessed: 24/06/2020).

Levin, M.F., Snir, O., Liebermann, D.G., Weingarden, H., and Weiss, P.L. (2012) Virtual reality versus conventional treatment of reaching ability in chronic stroke: Clinical feasibility study. *Neurology and Therapy*. 1 (1), p. 3.

Lackner, J.R. (2014) Motion sickness: More than nausea and vomiting. *Experimental Brain Research*. 232 (8), pp. 2493–2510.

Lim, Y., Kim, J., Lee, H., and Kim, S. (2018) Postural instability induced by visual motion stimuli in patients with vestibular migraine. *Frontiers in Neurology*. 9 (433), pp. 1–8.

Maani, C.V., Hoffman, H.G., Morrow, M., Maiers, A., Gaylord, K., McGhee, L.L., and DeSocio, PA. (2011) Virtual reality pain control during burn wound debridement of combat-related burn injuries using robot-like arm mounted VR goggles. *The Journal of Trauma*. 71 (1), pp. S125–30.

Mallari, B., Spaeth, E.K., Goh, H., and Boyd, B.S. (2019) Virtual reality as an analgesic for acute and chronic pain in adults: A systematic review and meta-analysis. *Journal of Pain Research*. 12, pp. 2053–2085.

Massetti, T., da Silva, T.D., Crocetta, T.B., Guarnieri, R., de Freitas, B.L., Lopes, P.B., Watson, S., Tonks, J., and Monterio, C.B. (2018) The clinical utility of virtual reality in neurorehabilitation: A systematic review. *Journal of Central Nervous System Disease*. 10, pp. 1–18.

Mendes, F.A., Pompeu, J.E., Lobo, A.M., da Silva, K.G., Oliveira, T.P., Zomignani, A.P., and Piemonte, M.E.P. (2012) Motor learning, retention and transfer after virtual-reality-based training in Parkinson's disease – effect of motor and cognitive demands of games: A longitudinal, controlled clinical study. *Physiotherapy*. 98 (2), pp. 217–223.

Mohammad, E.B., and Ahmad, M. (2019) Virtual reality as a distraction technique for pain and anxiety among patients with breast cancer: A randomized control trail. *Palliative Support Care*. 17 (1), pp. 29–34.

Moloney, J., Spehar, B., Globa, A., and Wang, R. (2018) The affordance of virtual reality to enable the sensory representation of multi-dimensional data for immersive analytics: From experience to insight. *Journal of Big Data*. 5 (53), pp. 1–19.

Ohannessian, R. (2015) Telemedicine: Potential applications in epidemic situations. *European Research in Telemedicine*. 4, pp. 95–98.

Pausch, R. (1993) The views of virtual reality: An overview. *In Computer*. 26 (2) pp. 79–80.

Russell, T.G. (2007) Physical rehabilitation using telemedicine. *Journal Telemedicine and Telecare*. 13 (5), pp. 217–220.

Ruthenbeck, G.S., and Reynolds, K.J. (2015) Virtual reality for medical training: The state-of-the-art. *Journal of Simulation*. 9 (1) pp. 16–26.

Shin, H., and Kim, K. (2015) Virtual reality for cognitive rehabilitation after brain injury: A systematic review. *Journal of Physical Therapy Science*. 27 (9), pp. 2999–3002.

Singh, D.K.A., Nordin, N.A.M., Aziz, N.A.A., Lim, B.K., and Soh, L.C. (2013) Effects of substituting a portion of standard physiotherapy time with virtual reality games among community-dwelling stroke survivors. *BMC Neurology*. 13 (1):199.

Weech, S., Kenny, S., and Barnett-Cowan, M. (2019) Presence and cybersickness in virtual reality are negatively related: A review. *Frontiers in Psychology*, 10 (158), pp. 1–19.

Section II

Virtual Reality in Neurological Rehabilitation

2 On the Possibility of Using Virtual Reality to Improve the Mobility of People with Parkinson's Disease

Amin Amini
Brunel University London
London, UK

CONTENTS

INTRODUCTION

Parkinson's is a neurological condition in which parts of the brain responsible for movements become incapacitated over time due to the abnormal dopamine equilibrium. Freezing of Gait (FOG) is one of the main Parkinson's disease (PD) symptoms that affects patients not only physically, but also psychologically as it prevents them from fulfilling simple tasks such as standing up or walking. Different auditory and visual cues have been proven to be very effective in improving the mobility of People with Parkinson's (PwP). Nonetheless, many of the available methods require user intervention or devices to be worn, charged, etc. to activate the cues.

PD, caused by the depletion of dopamine in the substantia nigra, is a degenerative neurological condition affecting the initiation and control of movements, particularly those related to walking (Yarnall *et al.* 2015, Dirkx *et al.* 2017). There are many physical symptoms associated with PD including akinesia, hypokinesia, and Bradykinesia (Young *et al.* 2016). An additional symptom is FOG, usually presenting in advanced stages of Parkinson's (Bloem *et al.* 2004, Okuma 2006, Pickering *et al.* 2007, Johnson *et al.* 2013). FOG is one of the most debilitating and least understood symptoms associated with Parkinson's. It is exacerbated by several factors including the need to walk through narrow spaces, turning as well as stressful situations (Okuma 2006, Beck *et al.* 2015).

FREEZING OF GAIT (FOG)

FOG is one of the most disabling symptoms in PD that affects its sufferers by impacting their gait performance and locomotion. FOG is an episodic phenomenon that prevents the initiation or continuation of a patient's locomotion and usually occurs in latter stages of PD where patients' muscles freeze in place as they are trying to move (Bloem *et al.* 2004, Okuma 2006, Donovan *et al.* 2011, Yarnall *et al.* 2015, Amini 2018).

FOG and associated incidents of falling often incapacitate PwP and, as such, can have a significant detrimental impact at both physical and psychological levels (Bloem *et al.* 2004). Consequently, the patient's quality of life decreases and health care and treatment expenditures increase substantially (Nutt *et al.* 2011). A research study conducted by the University of Rochester's Strong Memorial Hospital (University of Rochester 1999) showed that approximately 30% of PwP experience sudden, unexpected freezing episodes, thus highlighting the high level of dependency that many PwP have on physical or psychological strategies that may assist in alleviating FOG and help people start walking again.

POSSIBLE TREATMENTS

There is no proven therapy to eradicate the PD or slow down its progression. As a result, the focus of the medical therapy is on the treating or reducing the effect of its symptoms (Factor and Weiner 2007). There are different treatments available to improve PwP living standards and help deal with the symptoms including supportive therapies, medications, and surgery.

Supportive therapies focus towards pain relief using different methods including physiotherapy that relieves joint pain and muscle stiffness as well as exercises and occupational therapy that provide support for day-to-day activities of PwP and programmes that help them maintain their independence. Moreover, supportive therapies also cover dietary advice that would be beneficial to some extent for symptom relieve. Lastly, speech, and language therapy can also help PwP improving speech impairment caused by the disease or reduce the patient's swallowing difficulties (dysphagia), also related to PD (Parkinson's disease - Treatment 2016).

Medications are also beneficial in reducing the frequency or effect of PD's main symptoms including FOG and tremors. Nonetheless, there are usually possible short- and long-term side effects in these methods. The three types of the mainstream medication for PwP are (Parkinson's disease - Treatment 2016):

- **Levodopa**: Levodopa helps to the increase the dopamine production by the nerve cells; an agent for message transmission between brain parts and nerves responsible of controlling movement. Consequently, this would improve the patient's movement irregularities and locomotion (Factor and Weiner 2007, Galvez-Jimenez 2013).
- **Dopamine agonists**: These chemicals act as a substitute for the imbalanced dopamine level in the brain, which yields similar effect as levodopa. Dopamine agonists could have many side effects including hallucinations and confusion (Factor and Weiner 2007, Galvez-Jimenez 2013).
- **Monoamine oxidase-B inhibitors**: Monoamine oxidase-B (MOA-B) inhibitors aim at blocking the effect of an enzyme responsible of breaking down dopamine. As a result, the dopamine level would be increased. MOA-B can improve the PD symptoms and can be prescribed to be used alongside other medications such as dopamine agonists or levodopa (Factor and Weiner 2007, Galvez-Jimenez 2013).

Finally, a pulse generator can be surgically implanted into the subject's chest wall connected using wires to a specific part of the brain. This acts as a deep brain stimulation that produces a tiny electrical current that stimulates the brain in order to ease PD symptoms (Marks 2010).

SENSORY STIMULATION

Many studies suggest that auditory (Rubinstein *et al.* 2002a, Suteerawattananon *et al.* 2004, Rochester *et al.* 2005, Khan 2013) and visual cues (Azulay *et al.* 1999, 2006, Rubinstein *et al.*

2002, Suteerawattananon *et al.* 2004, Jiang and Norman 2006, Carrel 2007, Kaminsky *et al.* 2007, McAuley *et al.* 2009, Donovan *et al.* 2011, Dvorsky *et al.* 2011, Griffin *et al.* 2011, Lebold and Almeida 2011, Velik 2012, Velik *et al.* 2012) can improve PwP's gait performance, especially during FOG. Rubinstein et al. (2002) observed that in the presence of an external 'movement trigger' (i.e., a sensory cue), a patient's self-paced actions, such as walking, can be significantly improved, a phenomenon known as '*kinesia paradoxica*'.

THE EFFECT OF VISUAL CUE ON PD LOCOMOTION

Many previous studies have developed methods for monitoring FOG behaviours and intervening to improve motor symptoms with the use of external visual cues. Many studies utilised computer vision technologies to minimise the need for patients to wear measurement devices, which can be cumbersome and also have potential to alter a person's movement characteristics. Since the release of the Microsoft Kinect camera, several attempts have been made to use the Kinect sensor as a non-invasive approach for monitoring PD-related gait disorders. Many previous research studies have focussed on rehabilitation outcomes and experimental methods for monitoring patients' activities.

For instance, in Takač, et al. (2013), a home tracking system was developed using Microsoft Kinect sensors to help PwP who experience regular FOG. The research interconnected multiple Kinect sensors together to deliver a wider coverage of the testing environment. The model operated by collectively gathering data from multiple Kinect sensors into a central computer and storing them in a centralised database for further analysis and processing. The research employed a model based on the subject's histogram colour and height together with the known average movement delays between each camera. Nonetheless, as a Kinect camera produces a raw RGB (red, green, blue) data stream, analysing multiple Kinect colour data stream for the histogram of colour in real-time requires a very powerful processor and significant amount of computer memory. Moreover, the synchronisation between each camera feed would add extra computation for this approach.

Previous research has demonstrated that dynamic visual cues (such as laser lines projected on the floor) can deliver a profound improvement to walking characteristics in PwP (Rubinstein *et al.* 2002). Furthermore, strong evidence now exists suggesting that it is not only the presence of sensory information (or an external 'goal' for movement) that 'drives' improvements/*kinesia paradoxia*, but rather the presence of continuous and dynamic sensory information. This was first demonstrated by Azulay et al. (1999) who showed that the significant benefits to gait gained when walking on visual stepping targets were lost when patients walked on the same targets under conditions when the room was illuminated by strophic lighting, thus making the visual targets appear static. Similar observations have also been made in the auditory domain (Young *et al.* 2016).

In Zhao et al. (2013), in order to improve PwP's gait performance, a visual cue system was implemented based on a wearable system installed on subjects' shoes. This system employed laser pointers as visual cues fitted on a pair of modified shoes using a 3D printed caddy. The system consisted of pressure sensors that detect the stance phase of gait and trigger the laser pointers when a freeze occurs.

While effective and intuitive to use, the reliance on any attachable/wearable apparatus can be cumbersome and also required users to remember to attach appropriate devices, even around the house; where many people experience significant problems with FOG at times when they are not wearing their shoes. In another approach based on wearable devices (Lewis *et al.* 2000), the effect of a subject mounted light device (SMLD) projecting visual step length markers on the floor was evaluated. The study showed that an SMLD induced a statically significant improvement on subjects' gait performance. Nevertheless, it was suggested that the requirement of wearing SMLD might lead to practical difficulties both in terms of comfort and on the potential for the devices impacting on characteristics of patients' movements. In Velik et al. (2012) the entire SMLD visual cue system included a backpack consisting of a remotely controlled laptop (needed to be carried by the subjects). Although the SMLD method was employed, researchers added the 10 seconds on-demand option to the 'constantly on' visual cue casting.

Moreover, similar to the aforementioned technologies, the laser visual cues are always turned on, regardless of the subject's FOG status of gait performance. McAuley et al. (2009) and Kaminsky et al. (2007) proposed the use of Virtual Cueing Spectacles (VCS) that, similar to approaches that project targets on the floor; project virtual visual targets on to a user's spectacles. The use of VCS might eliminate major disadvantages introduced by SMLD (or other wearable approaches), but these systems still need to either be sensitive to a FOG onset or constantly turned on, even when not required. In Amini Maghsoud Bigy *et al.* (2015) and Amini *et al.* (2018), in order to eliminate the obtrusiveness of above-mentioned approaches, an automatic and dynamic visual cue system was developed to provide visual stimuli as green laser lines projected in front of the subject according to his location and orientation in a room. This was achieved by employing Microsoft Kinect v2 to detect FOG and localise the subject's location in a room without the need for any on-body attached sensors. Moreover, a set of pan/tilt servo motors was developed to cast laser lines produced pro-jection system accordingly. Although some of limitations of on-body sensors were addressed, the developed system was only limited to a room.

VIRTUAL AND AUGMENTED REALITY

With regard to virtual reality (VR) technology, there have been many studies conducted towards providing virtual/dynamic visual cues to PwP. This includes VR and its another variant augmented reality (AR). In AR, a subject can see her surroundings, but virtual objects are provided to her as an overlay. AR enables users to not only see real environment through a goggle, but also see virtual information on top of it. Although there are advantages of using AR over VR in these scenarios, the efficacy of AR for providing virtual/dynamic visual cues are still lower than other methods (Griffin *et al.* 2011). In Griffin et al. (2011), the effect of real and virtual visual cueing was compared, and it was concluded that real transverse lines casted on the floor are more impactful than the virtual counterparts. Nonetheless, using VCS eliminates the shortcomings in other techniques such as limi-tations in mobility, steadiness and symmetry. VCS also has the advantage of being capable to be used in outdoor when the patient is out and about.

In Wang *et al.* (2011), a study was conducted to compare a VR-based solution to physical reality in terms of patients' movement speed in reaching stationary and moving targets. It was concluded that using VR technology in improving the movement speed and reaction time of a subject had an equal result compared to physical reality. Although this study did not provide any visual stimuli to the patients, it paved the way for the possibility of using VR technology for such disorders and future studies. This study has successfully demonstrated that using VR-based technologies has a potential to be used as a tool for providing visual motion stimuli to improve movement speed in PwP. In Mirelman *et al.* (2013), a platform has been developed to improve the balance and gait in PwP using motor imagery (MI) and VR. Although the study had a small sample, it diversified its methodologies to compensate. During the study, several key areas about patients' assessments were taken into account including (1) "participant characteristics, including number of participants, age, disease duration, H & Y staging, and whether the study was conducted "on" or "off" medication; (2) method of delivery, design, intervention type, duration, and frequency; and (3) outcomes, includ-ing assessment, primary outcome measures, and results'. It was concluded that user feedback varied between MI and VR where the former had an internal and latter an external feedback. In VR, the feedback was derived from environment and exergames, whereas for MI, the users were required to see and feel their movements without apparent movements. In conclusion, the research explored the benefits of using the combination of IM and VR in PwP with promising results. In another attempt, Park et al. (2011) developed a VR-based treadmill control interface used for gait analysis and assess-ment of PwP during FOG episodes. The platform provided a safe and manageable walking platform based on VR bodyweight supported treadmill interface (BWSTI) to investigate and assess the irregularities in gait cycles associated with FOG under various conditions. The developed BWSTI allowed the user to control the speed of the treadmill not by relying on the interaction force between the robotic device and the subject's limbs, but by using VR environment as a virtual interface to adopt the treadmill speed based on the user while keeping her at the middle of the treadmill. This resulted in more stable and higher response system eliminating the need for a longer treadmill. The VR environment also provides a realistic walking condition to evoke FOG in a laboratory environment.

Moreover, Liao et al. (2015) developed a VR-based training system with the aim to improve the obstacle-crossing performance and dynamic balance in PwP. The research explored the effect and feasibility of VR-based exercise tools on dynamic balancing and obstacle crossing performance in PwP. Thirty-six PD diagnosed participants were involved in the trial that were split randomly into three groups. The participants received different exercises including VR-based and traditional exercises for 45 minutes, followed by 15 minutes of treadmill training in each session for a total of 12 sessions over 6 weeks. Moreover, the control group's participants received no exercise programme. It was concluded that the VR-based exercise group had the greatest improvement both in obstacle crossing velocity, dynamic balance, crossing stride length, sensory organization test (SOT), Parkinson's Disease Questionnaire (PDQ39), fall efficacy scale (FES-I), and timed up and go test (TUG) compared to the control group. The VR-based exercise group also showed greater improvement in movement velocity of limits-of-stability test when compared against the traditional exercise group. In another research (Mirelman *et al.* 2011), the use of VR for gait training and analysis in PwP was investigated. Moreover, the paper explored whether VR-based systems can induce motor learning in order to improve complex gait cycles as well as reducing the patient's fall risk associated with PD. Twenty PwP received regular VR-based intensive treadmill training sessions (18 sessions – 3 per week). The sessions included normal condition walking as well as dual task walking with obstacles. Moreover, functional performance and cognitive function were assessed. A virtual environment was developed to simulate a pathway while the subjects' gait cycles were captured and analysed using shoe-mounted IR-LEDs.

The results showed an improvement in VR-based dual task test with obstacles among the subjects. Moreover, there was also an improvement in cognitive function and functional performance.

Finally, Severiano et al. (2018) developed a system to assess the effectiveness of VR-based balance exercise games in PwP. Sixteen PwP participated in this study that involved sitting–rising assessment (the mobility and ability of subjects' lower limbs during sitting and rising actions). The system provided an immersive virtual environment to verify the effectiveness of balance exercises for PwP with promising results. It was concluded that Dizziness Handicap Inventory and Berg Balance Scale scores were improved after the introduction of VR-based exercises among the participants. Moreover, SRT and SF-36 showed a significant change in functional capacity after the rehabilitation programme. It was concluded that there was a statistically significant clinical improvement in participants in the final assessment after VR-based rehabilitation.

DISCUSSION

The aforementioned studies and researches on the possibility of using VR and AR technology in improving the mobility of PwP showed promising results. Moreover, although on-body attached solutions provide an effective gait analysis and improving mean, these systems are quite simple from an end user perspective by providing visual cues without requiring external data processing and FOG detection system. As a result, these systems provide a constantly-on visual cue to the subject's eyes. This study explores the possibility of VR/AR technology that can provide an unobtrusive facility to monitor and improve the mobility of PwP especially during a FOG episode using virtual visual cues.

These solutions target mainly patients with mid-cognitive impairments improving their independence and mitigating accidents risks, but also long-term risks of cognitive decline and dementia.

This study aims to explore the feasibility of the development of a novel integrated system that features an unobtrusive monitoring tool for fall and FOG incidents using VR/AR technology designed to improve patients' mobility during a FOG incident using automatic/dynamic visual cue projection. Using AR-based visual cue can help cast overlay lines to the viewer's eyes according to patients' orientation and position; the system will have the capability of delivering bespoke and tailored sensory information for each user in a manner that eliminates the limitation of the area of coverage. These solutions eliminate or minimise the limitations and issues with other AR technologies developed by other researches. Additionally, they would help decreasing costs involved in social care section of expenditures.

Additional costs of social care can be reduced significantly as such a technology would make Parkinson's patients less dependent on assistance and requiring less transformations to the home. Moreover, by being more mobile and requiring less assistance, some of the patients can maintain a professional activity reducing the risks of income loss for Stage 3-4 patients according to the Unified Parkinson's Disease Rating Scale. As the patients become more independent, the need for a healthcare provider would be lessened as the PD patients would be able to carry out tasks more easily thanks to the decrease of their FOG incidents both in frequency and amplitude. It is estimated that by implementing the proposed system, a patient's PD-related costs especially in social care section would be reduced significantly; this reduction would be even higher for PwP at their earlier phase of PD as they will be able to reduce the FOG symptoms by using the system.

VR/AR solutions have proven to have a promising result in providing dynamic visual cues to PwP in improving their mobility during FOG episodes. One could expect to see more mainstream solutions being developed for such needs. With the aid of 3D printing and Printed Circuit Board (PCB) miniaturisation, more ergonomic and smaller formfactor smart AR goggles could be produced resulting in the increase of the acceptance rate of patients in using them. These goggles could not only provide dynamic/virtual cues to the patients, but would also be capable of providing auditory cues as well. Moreover, these goggles have the potential to include more features; embedded sensors such as accelerometers and gyroscopes, coupled with smartphones' and smartwatches' sensors that could yield a robust fall detection alert system.

Finally, these solutions will provide reliable information for doctors and healthcare providers to examine PwP's daily activity as well as the number and the frequency of their FOG and fall incidents as well as the degradation of their cognitive and functional impairments. They can provide an overall view of a patient's progress that consequently can help towards their independence. Another benefit will be requiring fewer appointments with medical and paramedical staff for evaluating the patient health reducing further financial strains on healthcare sector.

REFERENCES

Amini, A., 2018. *Using 3D Sensing and Projecting Technology to Improve the Mobility of Parkinson's Disease Patients*. Brunel University London.

Amini, A., Banitsas, K., and Young, W.R., 2018. Kinect4FOG: Monitoring and improving mobility in people with Parkinson's using a novel system incorporating the Microsoft Kinect v2. *Disability and Rehabilitation: Assistive Technology*, 14 (6), 566–573.

Amini Maghsoud Bigy, A., Banitsas, K., Badii, A., and Cosmas, J., 2015. Recognition of postures and Freezing of Gait in Parkinson's disease patients using Microsoft Kinect sensor. *2015 7th International IEEE/ EMBS Conference on Neural Engineering (NER)*, 731–734.

Azulay, J., Mesure, S., Amblard, B., Blin, O., Sangla, I., and Pouget, J., 1999. Visual control of locomotion in Parkinson's disease. *Analysis*, 17, 111–120.

Azulay, J.P., Mesure, S., and Blin, O., 2006. Influence of visual cues on gait in Parkinson's disease: Contribution to attention or sensory dependence? *Journal of the Neurological Sciences*, 248 (1–2), 192–195.

Beck, E.N., Ehgoetz Martens, K.A., and Almeida, Q.J., 2015. Freezing of gait in Parkinson's disease: An overload problem? *Plos One*, 10 (12), e0144986.

Bloem, B.R., Hausdorff, J.M., Visser, J.E., and Giladi, N., 2004. Falls and freezing of Gait in Parkinson's disease: A review of two interconnected, episodic phenomena. *Movement Disorders*, 19 (8), 871–884.

Carrel, A.J., 2007. *The Effects of Cueing on Walking Stability in People with Parkinson's Disease.* Iowa State University.

Dirkx, M.F., den Ouden, H.E.M., Aarts, E., Timmer, M.H.M., Bloem, B.R., Toni, I., and Helmich, R.C., 2017. Dopamine controls Parkinson's tremor by inhibiting the cerebellar thalamus. *Brain*, 140 (3), 721–7734.

Donovan, S., Lim, C., Diaz, N., Browner, N., Rose, P., Sudarsky, L.R., Tarsy, D., Fahn, S., and Simon, D.K., 2011. Laserlight cues for gait freezing in Parkinson's disease: An open-label study. *Parkinsonism and Related Disorders*, 17 (4), 240–245.

Dvorsky, B.P., Elgelid, S., and Chau, C.W., 2011. The Effectiveness of utilizing a combination of external visual and auditory cues as a gait training strategy in a pharmaceutically untreated patient with Parkinson's disease: A case report. *Physical & Occupational Therapy in Geriatrics*, 29 (4), 320–326.

Factor, S.A. and Weiner, W.J., 2007. *Parkinson's Disease: Diagnosis and Clinical Management.* 2nd ed. Demos Medical Publishing.

Galvez-Jimenez, N., 2013. *Scientific Basis for the Treatment of Parkinson's Disease.* 2nd ed. CRC Press.

Griffin, H.J., Greenlaw, R., Limousin, P., Bhatia, K., Quinn, N.P., and Jahanshahi, M., 2011. The effect of real and virtual visual cues on walking in Parkinson's disease. *Journal of Neurology*, 258 (6), 991–1000.

Jiang, Y. and Norman, K.E., 2006. Effects of visual and auditory cues on gait initiation in people with Parkinson's disease. *Clinical Rehabilitation*, 20 (1), 36–45.

Johnson, L., James, I., Rodrigues, J., Stell, R., Thickbroom, G., and Mastaglia, F., 2013. Clinical and posturographic correlates of falling in Parkinson's disease. *Movement Disorders*, 28 (9), 1250–1256.

Kaminsky, T. a, Dudgeon, B.J., Billingsley, F.F., Mitchell, P.H., and Weghorst, S.J., 2007. Virtual cues and functional mobility of people with Parkinson's disease: A single-subject pilot study. *Journal of Rehabilitation Research and Development*, 44 (3), 437–448.

Khan, A.A., 2013. *Detecting Freezing of Gait in Parkinson's Disease for Automatic Application of Rhythmic Auditory Stimuli.* University of Reading.

Lebold, C.A. and Almeida, Q.J., 2011. An evaluation of mechanisms underlying the influence of step cues on gait in Parkinson's disease. *Journal of Clinical Neuroscience*, 18 (6), 798–802.

Lewis, G.N., Byblow, W.D., and Walt, S.E., 2000. Stride length regulation in Parkinson's disease: The use of extrinsic, visual cues. *Brain: A Journal of Neurology*, 123 (Pt 1), 2077–2090.

Liao, Y.Y., Yang, Y.R., Cheng, S.J., Wu, Y.R., Fuh, J.L., and Wang, R.Y., 2015. Virtual reality-based training to improve obstacle-crossing performance and dynamic balance in patients with Parkinson's disease. *Neurorehabilitation and Neural Repair*, 29 (7), 658–667.

Marks, W.J., 2010. *Deep Brain Stimulation Management.* Cambridge Medicine.

McAuley, J.H., Daly, P.M., and Curtis, C.R., 2009. A preliminary investigation of a novel design of visual cue glasses that aid gait in Parkinson's disease. *Clinical rehabilitation*, 23 (8), 687–695.

Mirelman, A., Maidan, I., and Deutsch, J.E., 2013. Virtual reality and motor imagery: Promising tools for assessment and therapy in Parkinson's disease. *Movement Disorders*, 28 (11), 1597–1608.

Mirelman, A., Maidan, I., Herman, T., Deutsch, J.E., Giladi, N., and Hausdorff, J.M., 2011. Virtual reality for gait training: Can it induce motor learning to enhance complex walking and reduce fall risk in patients with Parkinson's disease? *Journals of Gerontology - Series A Biological Sciences and Medical Sciences*, 66 A (2), 234–240.

Nutt, J.G., Bloem, B.R., Giladi, N., Hallett, M., Horak, F.B., and Nieuwboer, A., 2011. Freezing of gait: Moving forward on a mysterious clinical phenomenon. *The Lancet Neurology*, 10 (8), 734–744.

Okuma, Y., 2006. Freezing of gait in Parkinson's disease. *Journal of Neurology*, 253 (SUPPL. 7), 27–32.

Park, H.S., Yoon, J.W., Kim, J., Iseki, K., and Hallett, M., 2011. Development of a VR-based treadmill control interface for gait assessment of patients with Parkinson's disease. *IEEE International Conference on Rehabilitation Robotics*, 1–5.

Parkinson's disease - Treatment [online]. 2016. *NHS.* Available from: https://www.nhs.uk/Conditions/Parkinsons-disease/Pages/Treatment.aspx [Accessed 23 Oct 2017].

Pickering, R.M., Grimbergen, Y.A.M., Rigney, U., Ashburn, A., Mazibrada, G., Wood, B., Gray, P., Kerr, G., and Bloem, B.R., 2007. A meta-analysis of six prospective studies of falling in Parkinson's disease. *Movement Disorders*, 22 (13), 1892–1900.

Rochester, L., Hetherington, V., Jones, D., Nieuwboer, A., Willems, A.M., Kwakkel, G., and Van Wegen, E., 2005. The effect of external rhythmic cues (auditory and visual) on walking during a functional task in homes of people with Parkinson's disease. *Archives of Physical Medicine and Rehabilitation*, 86 (5), 999–1006.

Rubinstein, T.C., Giladi, N., and Hausdorff, J.M., 2002. The power of cueing to circumvent dopamine deficits: A review of physical therapy treatment of gait disturbances in Parkinson's disease. *Movement Disorders*, 17 (6), 1148–1160.

Severiano, M.I.R., Zeigelboim, B.S., Teive, H.A.G., Santos, G.J.B., and Fonseca, V.R., 2018. Effect of virtual reality in Parkinson's disease: A prospective observational study. *Arquivos de Neuro-Psiquiatria*, 76 (2), 78–84.

Suteerawattananon, M., Morris, G.S., Etnyre, B.R., Jankovic, J., and Protas, E.J., 2004. Effects of visual and auditory cues on gait in individuals with Parkinson's disease. *Journal of the Neurological Sciences*, 219 (1–2), 63–69.

Takač, B., Chen, W., and Rauterberg, M., 2013. Toward a domestic system to assist people with Parkinson's. *SPIE Newsroom*, 19 (8), 871–884.

University of Rochester. 1999. *Laser Pointer Helps Parkinson's Patients Take Next Step [online]*. Available from: https://www.rochester.edu/news/show.php?id=1175 [Accessed 24 May 2017].

Velik, R., 2012. Effect of on-demand cueing on freezing of gait in Parkinson's patients. *International Journal of Medical, Pharmaceutical Science and Engineering*, 6 (6), 10–15.

Velik, R., Hoffmann, U., Zabaleta, H., Félix, J., Massó, M., and Keller, T., 2012. The effect of visual cues on the number and duration of freezing episodes in Parkinson's patients. *2012 Annual International Conference of the IEEE Engineering in Medicine and Biology Society*, 4656–4659.

Wang, C.Y., Hwang, W.J., Fang, J.J., Sheu, C.F., Leong, I.F., and Ma, H.I., 2011. Comparison of virtual reality versus physical reality on movement characteristics of persons with Parkinson's disease: Effects of moving targets. *Archives of Physical Medicine and Rehabilitation*, 92 (8), 1238–1245.

Yarnall, A., Archibald, N., and Brun, D., 2015. Parkinson's disease. *The Lancet*, 386 (9996), 896–912.

Young, W.R., Shreve, L., Quinn, E.J., Craig, C., and Bronte-Stewart, H., 2016. Auditory cueing in Parkinson's patients with freezing of gait. What matters most: Action-relevance or cue-continuity? *Neuropsychologia*, 87, 54–62.

Zhao, Y., Ramesberger, S., Fietzek, U.M., D'Angelo, L.T., and Luth, T.C., 2013. A novel wearable laser device to regulate stride length in Parkinson's disease. *Proceedings of the Annual International Conference of the IEEE Engineering in Medicine and Biology Society, EMBS*, 5895–5898.

3 Virtual Reality for Stroke Rehabilitation

Kate Laver
Flinders University
Adelaide, Australia

CONTENTS

STROKE REHABILITATION

Stroke is one of the leading causes of death and disability in adults worldwide (World Health Organization, 2012). Rates of survival following stroke are increasing, and the number of stroke survivors is expected to increase due to population growth and ageing of the population (Johnson et al., 2019). Worldwide it is predicted that there are currently over 104 million people who have survived a stroke (Avan et al., 2019). Following stroke, there is a certain amount of spontaneous recovery which is described as the improvement in function which occurs solely due to time (Kwakkel et al., 2003). However, spontaneous recovery can occur slowly and is often incomplete, leaving the person with stroke with remaining impairment and activity restrictions (Cassidy and Cramer, 2017). Rehabilitation services, comprising a multidisciplinary team and restorative therapies, are commonly offered to assist the person with stroke reach their optimal level of functioning.

Much research has been conducted in order to determine the efficacy of different restorative therapies which are available to people after stroke. Guidelines for stroke rehabilitation generally recommend tailored approaches and task-specific practice as a means to improve sensorimotor function (French et al., 2016). Task-specific therapy involves repetitive practice of simple functional tasks (for example, sit to stand or tipping a cup to pour water). Critically, the amount of active task practice is important with more practice related to better outcomes (Schneider et al., 2016). Rehabilitation professionals have reported difficulty in structuring rehabilitation services to ensure adequate doses of therapy are provided. Current studies suggest that rehabilitation ward activity levels are lower than recommended. Common approaches used to increase dose include increasing staffing levels and altering methods of supervision (Stewart et al., 2017). However, these approaches are often dependent on more resources or the re-distribution of resources and may not be sustained over time. A more recent approach which can be used by therapists to increase therapy dose is the use of equipment such as virtual reality applications, computer-based cognitive training programmes and robotics.

POTENTIAL ADVANTAGES OF VIRTUAL REALITY WHEN USED IN STROKE REHABILITATION

There are a number of reasons why virtual reality applications may be desirable in stroke rehabilitation. Research suggests that key principles for effective motor learning include repetitive practice, high dose, task-specific activities, variable practice, increasing level of challenge, multisensory stimulation and feedback (Maier et al., 2019). Virtual environments offer the opportunity to provide therapy tasks which meet many of these criteria.

High dose and 'just right' challenge: People admitted to stroke rehabilitation services spend large periods of time inactive and alone despite evidence showing that greater doses of therapy are beneficial. Rehabilitation is often structured as direct (1:1) therapy which is resource intensive. The therapist is responsible for prescribing the task (to match the person's abilities and goals), structuring the task and supervising and providing feedback. Therapists could prescribe tasks to be performed within a tailored virtual environment and then use the application to provide the instruction and provide feedback reducing the amount of time they need to spend overseeing the person's therapy. Ideally, therapists will have control over the parameters of the virtual task so that they can match the person to the ideal environment and task.

Feedback: Feedback can be provided immediately (through observing the outcome of interaction with the environment) or over time (through timing or scoring performance). Knowledge of performance and knowledge of results may be available to the therapist in real time or conveyed electronically. Feedback is predominantly visual, auditory and haptic.

Engagement in a multisensory environment: Virtual reality applications are designed to be interactive and cognitively stimulating which may enhance motor learning. Traditional therapy approaches are repetitive and conducted in therapy gyms (or the home) and may not be considered enjoyable or motivating by stroke survivors. Furthermore, virtual reality programmes often incorporate gaming features and competition which are likely to be more interesting for people participating in rehabilitation.

Task-specific practice: The environment in which rehabilitation is provided (usually a ward-like environment) offers a limited number of different activity opportunities. Therapists may have difficulty practising activities which are important to the person with stroke. For example, activities like shopping, crossing the road and walking through busy and crowded spaces are often unable to be practiced. Virtual reality applications may offer the opportunity to practice some of these activities in a virtual environment.

Safety: Rehabilitation often involves taking risks. For example, the person with stroke wishing to return to drive may have a complex combination of visual, motor, sensory and cognitive impairments which pose difficulties. However, virtual reality driving simulators offer the opportunity to gain practice and confidence in a safe and supportive environment. Reviews have shown that within stroke rehabilitation settings, adverse events associated with virtual reality are relatively rare and those that do occur tend not to be serious (Laver et al., 2017).

EVOLUTION OF VIRTUAL REALITY IN STROKE REHABILITATION OVER TIME

The first studies in virtual reality were published in the early 2000s and typically involved case series or pilot studies evaluating virtual reality applications for motor rehabilitation (upper limb and gait). The first randomised controlled trials in stroke rehabilitation were published from 2004. Again, these studies mostly tested applications designed to improve motor outcomes however early studies involved evaluation of a driving simulator (Akinwuntan et al., 2005) and practice of using a local public transport system

(Lam et al., 2006). Most early studies involved development of a programme specifically for rehabilitation although there was some research involving the Playstation EyeToy early in the field. Over time, a mix of exergaming and customised programmes have been tested in randomised trials. Commercial applications that have been most popular to date include the Nintendo Wii (released in 2006) and the Microsoft Kinect (released in 2010) programmes. The number of randomised trials has been steadily increasing with 19 studies included in a Cochrane Review published in 2011 (Laver et al., 2012b) and 72 studies included in the updated version of the review published in 2017 (Laver et al., 2017).

Most recently, researchers have tested combinations of virtual reality and other therapeutic interventions. Lee and colleagues compared functional electrical stimulation with virtual reality with cyclic electrical stimulation in a small randomised trial ($N = 48$) which aimed to improve upper limb function (Lee et al., 2018). Those that received electrical stimulation plus virtual reality performed better on the Fugl Meyer Assessment distal component. Therapeutic device developers have combined virtual reality with robotics in applications such as the Armeo® device produced by Hocoma. There is some research suggesting that the addition of virtual reality may result in robotic training that is more acceptable and motivating when compared to standard approaches. Several researchers have examined whether non-invasive brain stimulation may be used in combination with virtual reality to enhance outcomes though there is currently insufficient evidence to determine whether this approach is superior (Lee and Chun, 2014).

To date, there has been limited capacity for stroke survivors using virtual reality programmes to interact with other stroke survivors within the same virtual environment. However, networked multi-user environments are emerging and may facilitate peer support and encouragement (Triandafilou et al., 2018).

WHAT TYPES OF INTERVENTIONS HAVE BEEN CREATED AND TESTED?

Within the field of stroke rehabilitation, virtual reality applications have been used to facilitate activity retraining, improve upper limb function, improve gait and balance, improve global motor function and to enhance cognitive functioning. Most studies have focussed on motor retraining, and there are fewer example of programmes designed to enhance performance of activities of daily living. The most common activity of daily living practiced in virtual environments is driving and while some studies suggest benefits there is currently insufficient evidence available to guide practice (Akinwuntan et al., 2005, Mazer et al., 2005). Virtual reality programmes have also been designed to practice grocery shopping (Laver et al., 2012c) and simple drink and meal preparation (Edmans et al., 2006); however, these virtual environments were not sophisticated enough to closely mimic the task as performed in the real world. Katz and colleagues created a clever application which was designed to assist people with unilateral spatial neglect train to safely cross the street (Katz et al., 2005). Though the application was only tested in a small group of participants' applications such as this have great potential to increase the relevance and complexity of therapy tasks offered in stroke rehabilitation settings.

A novel application recently reported by Threapleton and colleagues was the use of a virtual home to prompt thinking about discharge home for hospitalised people after stroke in the United Kingdom (Threapleton et al., 2018). An occupational therapist led the tour of the virtual home (which could be customised to reflect the person's own home environment) to discuss safety and access issues that the person may encounter on discharge and identify the potential need for assistive equipment. The researchers found that the virtual tour took less than 30 minutes and the use of a tablet device meant that set up time was minimal.

WHAT IS THE EVIDENCE? WHAT LEVEL AND FOR WHAT OUTCOME?

There are now many reviews which examine the efficacy of virtual reality for people after stroke. The first Cochrane Review published in 2011 included 19 studies (Laver et al., 2012b). The second published in 2015 included 37 studies. The most recent updated of the review was published in 2017 and included 72 studies (Laver et al., 2017).

Over time the quality of the evidence of studies testing virtual reality for stroke rehabilitation has only moderately improved. Many studies remain poorly reported, and method of sequence generation and allocation concealment are unclear. A number of studies are at risk of bias due to attrition and incomplete outcome data. Consequently, the evidence for most outcomes is considered very low, low or moderate according the GRADE (Guyatt et al., 2011). Furthermore, sample sizes remain generally small and at risk of being underpowered. The Cochrane Review showed that only 10 of 72 studies involved more than 50 participants (Table 3.1; Laver et al., 2017).

TABLE 3.1
An Overview of the Evidence for Virtual Reality in Stroke Rehabilitation on Key Outcomes

Outcome	Evidence
Upper limb	*Virtual reality vs conventional therapy* Meta-analysis comprising 22 studies with 1038 participants found that offering virtual reality did not result in significantly better outcomes than offering the same dose of other (traditional) therapy approaches (standardized mean difference [SMD] 0.07, 95% CI −0.05 to 0.20). *Virtual reality plus usual care vs usual care* When virtual reality was used to augment existing care and/or rehabilitation, it was shown to have significant benefits on upper limb function (10 studies with 210 participants (SMD 0.49, 95% CI 0.21 to 0.77) ***Clinical implications*** Virtual reality does not appear to be superior (or inferior) to traditional upper limb therapy approaches; however, it can be used to augment existing rehabilitation and lead to better outcomes. As it is not inferior, a virtual reality approach may be well suited to people after stroke who have limited engagement with a traditional therapy approach.
Gait and balance	*Virtual reality vs conventional therapy* Gait speed – Pooling of six studies with 139 participants showed that virtual reality was not superior to traditional therapy (MD 0.09, 95% CI −0.04 to 0.22). Balance – Pooling of three studies with 72 participants showed that virtual reality was not superior to traditional therapy approaches (SMD 0.39, 95% CI −0.09 to 0.86). *Virtual reality plus usual care vs usual care* Gait speed – Three studies comprising only 57 participants did not identify statistically significant gains in function with the addition of virtual reality intervention (SMD 0.08, 95% CI −0.05 to 0.21) Balance – Pooling of seven studies with 173 participants showed that the addition of virtual reality resulted in better balance outcomes (SMD 0.59, 95% CI 0.28 to 0.90) ***Clinical implications*** Fewer studies have been conducted in this area of rehabilitation, and sample sizes of included studies are small therefore evidence is limited. Analyses conducted to date suggest that using virtual reality is a reasonable approach with equivalent outcomes to those achieved via traditional therapy approaches, but use of VR to augment therapy may result in better outcomes.
Activities of daily living	*Virtual reality vs conventional therapy* Pooling of 10 studies with 461 participants revealed a small significant effect in people who received virtual reality intervention (SMD 0.25, 95% CI 0.06 0.43). *Virtual reality plus usual care vs usual care* A small to moderate effect was found when 8 studies with 153 participants were pooled (SMD 0.44, 95% CI 0.11 to 0.76) ***Clinical implications*** For reasons that are not currently clear, use of virtual reality seems to improve activities of daily living function (even when the intervention does not specifically aim to improve performance of activities of daily living). It is possible that the dual task nature of the intervention and combination of cognitive and motor challenges at the same time (dual task) are beneficial for task performance in other activities.

Note: Table content based on Cochrane review, 2017 (Laver et al., 2017).

WHEN AND HOW SHOULD VIRTUAL REALITY BE IMPLEMENTED AFTER STROKE?

Timing: Studies have been conducted with stroke survivors at all stages ranging from those in the subacute phase after stroke to those in the chronic phase. For example, Saposnik and colleagues recruited participants from rehabilitation units in four countries and participants were an average of 27 days after stroke (Saposnik et al., 2016). The intervention was then offered within the rehabilitation facility. In contrast, Housman and colleagues recruited people who were an average of approximately seven years following stroke and invited participants to use the virtual reality programme in a research clinic (Housman et al., 2009).

Dose: The Cochrane Review presented the results of a subgroup analysis amongst the upper limb studies comparing less than 15 hours of therapy with more than 15 hours of therapy. There were no significant differences in the pooled groups although there was a trend for better outcomes in the higher dose. A synthesis of a larger number of studies in stroke rehabilitation showed that more therapy is better (Schneider et al., 2016), and guidelines promote as much practice as possible therefore dose of virtual reality should be as high as feasible.

Type of virtual reality application: A subgroup analysis within the Cochrane review on virtual reality compared outcomes for people who used a virtual reality application designed for rehabilitation settings with those who used an off-the-shelf device found no significant difference between groups. This suggests that the engagement and repetition of the task in the virtual environment are important and the sophistication of the system may be less important.

Applicability: Many of the existing studies testing the efficacy of virtual reality exclude the participation of people with visual, cognitive and communication impairments. These exclusions mean that there are limitations in the generalisability of the current body of evidence. Studies suggest that at least 20% of stroke survivors have aphasia (Kyrozis et al., 2009) and approximately 65% of people have some form of visual impairment (Hepworth et al., 2016). Furthermore, a review published in 2011 showed that virtual reality trialists were only able to recruit 34% of stroke survivors who were screened for eligibility (Laver et al., 2012b).

In an environment where health care resources are scarce, health professionals must be cautious to invest in virtual reality applications that are likely to be widely beneficial. Some of the available interventions may be relevant for a relatively small group of stroke survivors. For example, virtual reality applications for electric mobility scooter driving (Jannink et al., 2008). For these types of interventions (which are rarer) there need to be considerations about benefits for the larger population.

Participants in studies of virtual reality applications in stroke tend be younger than average. Most participants are in their 50s, 60s or maybe 70s whereas the average age of stroke onset was reported to be 69 years old in one study (Kissela et al., 2012). This over-representation of younger stroke survivors may be partially due to eligibility criteria which restricts the participation of older stroke survivors. This may also be due to cultural perceptions that older people have difficulty using technologies. Beliefs about older people and technology use may change over time and current research suggests that older people are increasingly using smartphones, tablets and social media (Pew Research Centre, 2019). Indeed, a study involving exergames on a Geriatric Evaluation and Management unit successfully recruited people with an average age of 85 years and found that adherence to the exergaming intervention (led and supervised by a physiotherapist) was high (Laver et al., 2012a).

A recent study led by Sherrington and colleagues examined a more tailored approach to using virtual reality and interactive video gaming as part of a rehabilitation programme (Hassett et al., 2016). Instead of prescribing and testing one particular technology application, the intervention involved a tailored technology prescription which may have involved devices such as a pedometer, commercial video gaming programme or virtual reality application specifically designed for rehabilitation settings. This tailored and pragmatic approach is more likely to be the approach utilised by health professionals in clinical settings. This is supported by work conducted by Nguyen and colleagues investigating the translation of exergaming into practice found that therapists valued having a range of different applications available for use so that they could match client goals and abilities with applications which would best suit these (Nguyen et al., 2019).

WHAT DO CLINICAL PRACTICE GUIDELINES RECOMMEND IN REGARDS TO VIRTUAL REALITY FOR STROKE REHABILITATION?

Clinical practice guidelines aim to bridge the gap between evidence and practice. They do this by synthesising the findings of existing evidence and providing recommendations for health professionals for clinical practice. Clinical practice guideline committees are charged with the task of considering the evidence related to a particular health technology as well as potential harms, cost, and equity (Moberg et al., 2018). In the case of virtual reality for stroke, evidence suggesting favourable outcomes must be considered alongside the potential for adverse events (which tend to be mild and infrequent) and cost (which varies greatly depending on the programme) and equity. Virtual reality applications may indeed increase equity of rehabilitation by offering rehabilitation options for people who are in rural and remote areas or have limited access to specialist rehabilitation systems. However, there is currently restricted access to both rehabilitation and virtual reality programmes so equity must be considered.

Guidelines from different countries recommend that virtual reality is considered, along with other therapy interventions, as part of a stroke rehabilitation service. Clinical practice guidelines for stroke in Australia recommend that virtual reality may be used in the restoration of balance, upper limb function, walking and activities of daily living (Stroke Foundation, 2019). Guidelines developed in Canada report that virtual reality may be used as an adjunct to other conventional therapy approaches in stroke rehabilitation (Hebert et al., 2016). The American Heart Association and American Stroke Association similarly recommend that virtual reality may be used (in conjunction with other approaches) for gait and upper limb rehabilitation (Winstein et al., 2016). In summary, clinical practice guidelines worldwide suggest that virtual reality may be offered as part of a high-quality stroke rehabilitation service depending on the person's goals and preferences.

ILLUSTRATION OF THE APPLICATION FROM THE USERS' PERSPECTIVE

Although one of the purported benefits of virtual reality is the increased appeal for users of the device, a recent review conducted by Rohrbach and colleagues suggested that this area has been neglected in research to date (Rohrbach et al., 2019). The authors of the review examined studies which referred to motivation, enjoyment, engagement, immersion and presence in stroke rehabilitation virtual reality research studies. They found that only a small number of studies measured these constructs. Most commonly, trialists referred to these constructs as being a factor in why virtual reality might be advantageous. The authors of the review reported that some studies used measures such as the Intrinsic Motivation Inventory or the Presence Questionnaire to measure these constructs although most used self-designed surveys with unknown psychometric properties. They were unable to identify any studies which linked levels of motivation, engagement or enjoyment to the motor learning outcomes of participants.

STROKE SURVIVORS

Interviews and focus groups conducted with stroke survivors who have used virtual reality applications describe some of the benefits as well as some of the challenges. Overall, stroke survivors seem to regard the intervention positively, although this should be interpreted with caution due to selection bias (thoughts capture both those that have agreed to try virtual reality and those that agree to a subsequent interview). Other studies have demonstrated that use is highly variable and depends on usability of the device and concurrent health problems which may act as barriers. On the other hand, flexibility of the intervention, the motivating nature of some games and family support may increase use of virtual reality programmes (Standen et al., 2015).

Many participants have described participation as an engaging and exciting way to exercise (Törnbom and Danielsson, 2018). For example, Törnbom and colleagues found that people reported that walking on an ordinary treadmill was somewhat boring; however, walking on the treadmill linked to a virtual environment was not (Törnbom and Danielsson, 2018). Participants in this study also felt that they appreciated the feeling of being outside in nature in comparison to being in a therapy gym. Memories of previous outdoor walks were elicited. People who have been given access to home based virtual reality training have spoken of the convenience of home-based therapy and experiencing a sense of accomplishment and independence in therapy (Valdes et al., 2018). Virtual environments in which people connect with either health professionals, support workers or their peers are also highly valued as they may help with encouragement and confidence building and increase social contact for people after stroke who commonly report feeling isolated (Amaya et al., 2018).

Much of the negative feedback reported in studies has been specifically related to the characteristics of the technology. For example, difficulty navigating through programmes, limited sensitivity of controls and difficulty reading text are often reported. Other concerns voiced by stroke survivors have been around the need for more training and support as well as assistance in troubleshooting (Valdes et al., 2018). Participants also report that the amount of stimulation and the complexity of the task could be overwhelming and result in more tiredness than after conventional rehabilitation tasks (Törnbom and Danielsson, 2018). Some preferred a quieter and more controlled environment.

HEALTH PROFESSIONALS

Successful implementation of virtual reality in clinical settings is also highly dependent on factors at the institutional and individual therapist levels. Physiotherapists and occupational therapists at a rehabilitation hospital in Canada reported a number of barriers and enablers towards the successful use of an exergaming room linked to the rehabilitation service (Nguyen et al., 2019). At the institutional level, they found that efficient referral processes, discussions about the intervention at multidisciplinary meetings, training in using the equipment, availability of the room and reminders enhanced the utility of the exergaming room and resulted in greater use. However, staff were concerned that their referrals to the room were constrained by the limited amount of staffing and supervision available. At the clinician level, they found that successful use of the room and referrals occurred when staff were familiar with the equipment and the evidence for virtual reality as well as previous positive personal experiences. The availability of an expert clinician within the exergaming room was considered highly beneficial. Clinicians were more likely to refer clients who were seen as highly motivated and less likely to refer people with transportation difficulties and those who were not technologically aware.

The acceptability of virtual reality to therapists is complex with competing thoughts and beliefs. For example, therapists interviewed in a study by Schmid and colleagues made statements such as 'I do not want to play or just toy, I want the patient to benefit when he or she leaves therapy' suggesting that a lack of belief in the therapeutic benefits of the approach. Others spoke of the element of competition, 'I have just realised I am still motivated too, my pulse rises up.... I noticed once when

I switched on the device that I am effectively in the game although I am not playing along'. For most though, decisions about use came down to clinical reasoning, 'Well for me it is the question of what is the therapy goal for the patient and will I reach this goal better with the VR device than with other options' (Schmid et al., 2016).

Drawing on a review of the literature and experience within a large trial, Hamilton and colleagues suggested critical elements of the two phases involved in technology use within rehabilitation (Hamilton et al., 2018). These included initial patient engagement (through a process of introducing the idea, preparing for initial use, trialling the application and reviewing after first use) and maintaining patient engagement (through ongoing learning and skill practice, instructing, practicing, adjusting and managing technical problems).

DISCUSSION

After the emergence of virtual reality in the early 2000s, there was a rapid increase in clinical trials in stroke rehabilitation. In parallel, many rehabilitation settings acquired virtual reality programmes; most being off the shelf commercially available applications.

The evidence suggests that virtual reality is most beneficial when used to augment conventional therapy. In this way, it can be used to increase therapy dose for motor relearning in a way that is interesting and engaging for people after stroke. Applications that facilitate practice of activities of daily living are under developed and not yet ready for wide scale spread. It is also widely accepted that there are some challenges associated with use and that the intervention will not be acceptable or useful for all clients.

As technology becomes more sophisticated, it is expected that virtual reality will become more appealing in stroke rehabilitation settings. Ideally, future systems will be inexpensive, require minimal set up or space, be able to be adjusted by therapists to appropriate challenge level, feature user friendly hardware and appealing graphics and offer a wide range of activities so that fit with the person's interests is likely. In particular, within stroke rehabilitation there are a number of gaps and weaknesses where virtual reality can play a role. This includes returning to past activities and roles (e.g. work and driving) and receiving ongoing therapy beyond traditional rehabilitation services.

REFERENCES

Akinwuntan, A. E., De Weerdt, W., Feys, H., Pauwels, J., Baten, G., Arno, P. & Kiekens, C. 2005. Effect of simulator training on driving after stroke: A randomized controlled trial. *Neurology*, 65, 843–850.

Amaya, A., Woolf, C., Devane, N., Galliers, J., Talbot, R., Wilson, S. & Marshall, J. 2018. Receiving aphasia intervention in a virtual environment: The participants' perspective. *Aphasiology*, 32, 538–558.

Avan, A., Digaleh, H., Di Napoli, M., Stranges, S., Behrouz, R., Shojaeianbabaei, G., Amiri, A., Tabrizi, R., Mokhber, N., Spence, J. D. & Azarpazhooh, M. R. 2019. Socioeconomic status and stroke incidence, prevalence, mortality, and worldwide burden: An ecological analysis from the global burden of disease study 2017. *BMC Medicine*, 17, 191.

Cassidy, J. M. & Cramer, S. C. 2017. Spontaneous and therapeutic-induced mechanisms of functional recovery after stroke. *Translational Stroke Research*, 8, 33–46.

Edmans, J. A., Gladman, J. R., Cobb, S., Sunderland, A., Pridmore, T., Hilton, D. & Walker, M. F. 2006. Validity of a virtual environment for stroke rehabilitation. *Stroke*, 37, 2770–2775.

French, B., Thomas, L. H., Coupe, J., Mcmahon, N. E., Connell, L., Harrison, J., Sutton, C. J., Tishkovskaya, S. & Watkins, C. L. 2016. Repetitive task training for improving functional ability after stroke. *Cochrane Database of Systematic Reviews*, 11, CD006073.

Guyatt, G. H., Oxman, A. D., Schünemann, H. J., Tugwell, P. & Knottnerus, A. 2011. Grade guidelines: A new series of articles in the journal of clinical epidemiology. *Journal of Clinical Epidemiology*, 64, 380–382.

Hamilton, C., Mccluskey, A., Hassett, L., Killington, M. & Lovarini, M. 2018. Patient and therapist experiences of using affordable feedback-based technology in rehabilitation: A qualitative study nested in a randomized controlled trial. *Clinical Rehabilitation*, 32, 1258–1270.

Hassett, L., Van Den Berg, M., Lindley, R. I., Crotty, M., Mccluskey, A., Van Der Ploeg, H. P., Smith, S. T., Schurr, K., Killington, M. & Bongers, B. 2016. Effect of affordable technology on physical activity levels and mobility outcomes in rehabilitation: A protocol for the Activity and Mobility UsiNg Technology (AMOUNT) rehabilitation trial. *BMJ Open*, 6, e012074.

Hebert, D., Lindsay, M. P., Mcintyre, A., Kirton, A., Rumney, P. G., Bagg, S., Bayley, M., Dowlatshahi, D., Dukelow, S. & Garnhum, M. 2016. Canadian stroke best practice recommendations: Stroke rehabilitation practice guidelines, update 2015. *International Journal of Stroke*, 11, 459–484.

Hepworth, L. R., Rowe, F. J., Walker, M. F., Rockliffe, J., Noonan, C., Howard, C. & Currie, J. 2016. Post-stroke visual impairment: A systematic literature review of types and recovery of visual conditions. *Ophthalmology Research: An International Journal*, 5, 1–43.

Housman, S. J., Scott, K. M. & Reinkensmeyer, D. J. 2009. A randomized controlled trial of gravity-supported, computer-enhanced arm exercise for individuals with severe hemiparesis. *Neurorehabilitation and Neural Repair*, 23, 505–514.

Jannink, M. J., Erren-Wolters, C. V., De Kort, A. C. & Van Der Kooij, H. 2008. An electric scooter simulation program for training the driving skills of stroke patients with mobility problems: A pilot study. *Cyber Psychology & Behavior*, 11, 751–754.

Johnson, C. O., Nguyen, M., Roth, G. A., Nichols, E., Alam, T., Abate, D., Abd-Allah, F., Abdelalim, A., Abraha, H. N., Abu-Rmeileh, N. M. E., Adebayo, O. M., Adeoye, A. M., Agarwal, G., Agrawal, S., et al. 2019. Global, regional, and national burden of stroke, 1990-2016: A systematic analysis for the global burden of disease study 2016. *The Lancet Neurology*, 18, 439–458.

Katz, N., Ring, H., Naveh, Y., Kizony, R., Feintuch, U. & Weiss, P. 2005. Interactive virtual environment training for safe street crossing of right hemisphere stroke patients with unilateral spatial neglect. *Disability and rehabilitation*, 27, 1235–1244.

Kissela, B. M., Khoury, J. C., Alwell, K., Moomaw, C. J., Woo, D., Adeoye, O., Flaherty, M. L., Khatri, P., Ferioli, S., De Los Rios La Rosa, F., Broderick, J. P. & Kleindorfer, D. O. 2012. Age at stroke: temporal trends in stroke incidence in a large, biracial population. *Neurology*, 79, 1781–1787.

Kwakkel, G., Kollen, B. J., Van Der Grond, J. & Prevo, A. J. 2003. Probability of regaining dexterity in the flaccid upper limb: Impact of severity of paresis and time since onset in acute stroke. *Stroke*, 34, 2181–2186.

Kyrozis, A., Potagas, C., Ghika, A., Tsimpouris, P., Virvidaki, E. & Vemmos, K. 2009. Incidence and predictors of post-stroke aphasia: The arcadia stroke registry. *European Journal of Neurology*, 16, 733–739.

Lam, Y. S., Man, D. W., Tam, S. F. & Weiss, P. L. 2006. Virtual reality training for stroke rehabilitation. *NeuroRehabilitation*, 21, 245–253.

Laver, K., George, S., Ratcliffe, J., Quinn, S., Whitehead, C., Davies, O. & Crotty, M. 2012a. Use of an interactive video gaming program compared with conventional physiotherapy for hospitalised older adults: A feasibility trial. *Disability and Rehabilitation*, 34, 1802–1808.

Laver, K., George, S., Thomas, S., Deutsch, J. E. & Crotty, M. 2012b. Virtual reality for stroke rehabilitation. *Stroke*, 43, e20–e21.

Laver, K., Lim, F., Reynolds, K., George, S., Ratcliffe, J., Sim, S. & Crotty, M. 2012c. Virtual reality grocery shopping simulator: Development and usability in neurological rehabilitation. *Presence*, 21, 183–191.

Laver, K. E., Lange, B., George, S., Deutsch, J. E., Saposnik, G. & Crotty, M. 2017. Virtual reality for stroke rehabilitation. *Cochrane Database of Systematic Reviews, 11,CD008349*, 10.1002/14651858.CD008349.pub4.

Lee, S. H., Lee, J.-Y., Kim, M.-Y., Jeon, Y.-J., Kim, S. & Shin, J.-H. 2018. Virtual reality rehabilitation with functional electrical stimulation improves upper extremity function in patients with chronic stroke: A pilot randomized controlled study. *Archives of Physical Medicine and Rehabilitation*, 99, 1447–1453. e1.

Lee, S. J. & Chun, M. H. 2014. Combination transcranial direct current stimulation and virtual reality therapy for upper extremity training in patients with subacute stroke. *Archives of Physical Medicine and Rehabilitation*, 95, 431–438.

Maier, M., Ballester, B. R. & Verschure, P. F. M. J. 2019. Principles of neurorehabilitation after stroke based on motor learning and brain plasticity mechanisms. *Frontiers in Systems Neuroscience, 13*, 74.

Mazer, B., Gelinas, I., Vanier, M., Duquette, J., Rainville, C. & Hanley, J. 2005. Poster 57: Effectiveness of retraining using a driving simulator on the driving performance of clients with a neurologic impairment. *Archives of Physical Medicine and Rehabilitation*, 86, e20.

Moberg, J., Oxman, A. D., Rosenbaum, S., Schünemann, H. J., Guyatt, G., Flottorp, S., Glenton, C., Lewin, S., Morelli, A. & Rada, G. 2018. The grade evidence to decision (EtD) framework for health system and public health decisions. *Health Research Policy and Systems*, 16, 45.

Nguyen, A.-V., Ong, Y.-L. A., Luo, C. X., Thuraisingam, T., Rubino, M., Levin, M. F., Kaizer, F. & Archambault, P. S. 2019. Virtual reality exergaming as adjunctive therapy in a sub-acute stroke rehabilitation setting: Facilitators and barriers. *Disability and Rehabilitation: Assistive Technology*, 14, 317–324.

Pew Research Centre. 2019. Millenials stand out for their technology use, but older generations also embrace digital life. https://www.pewresearch.org/fact-tank/2019/09/09/us-generations-technology-use/

Rohrbach, N., Chicklis, E. & Levac, D. E. 2019. What is the impact of user affect on motor learning in virtual environments after stroke? A scoping review. *Journal of NeuroEngineering and Rehabilitation*, 16, 79.

Saposnik, G., Cohen, L. G., Mamdani, M., Pooyania, S., Ploughman, M., Cheung, D., Shaw, J., Hall, J., Nord, P. & Dukelow, S. 2016. Efficacy and safety of non-immersive virtual reality exercising in stroke rehabilitation (EVREST): A randomised, multicentre, single-blind, controlled trial. *The Lancet Neurology*, 15, 1019–1027.

Schmid, L., Glässel, A. & Schuster-Amft, C. 2016. Therapists' perspective on virtual reality training in patients after stroke: A qualitative study reporting focus group results from three hospitals. *Stroke Research and Treatment*, 2016, 1–12.

Schneider, E. J., Lannin, N. A., Ada, L. & Schmidt, J. 2016. Increasing the amount of usual rehabilitation improves activity after stroke: A systematic review. *Journal of Physiotherapy*, 62, 182–187.

Standen, P. J., Threapleton, K., Connell, L., Richardson, A., Brown, D. J., Battersby, S., Sutton, C. J. & Platts, F. 2015. Patients' use of a home-based virtual reality system to provide rehabilitation of the upper limb following stroke. *Physical Therapy*, 95, 350–359.

Stewart, C., Mccluskey, A., Ada, L. & Kuys, S. 2017. Structure and feasibility of extra practice during stroke rehabilitation: A systematic scoping review. *Australian Occupational Therapy Journal*, 64, 204–217.

Stroke Foundation. 2019. *Clinical Guidelines for Stroke Management*. Melbourne, Australia.

Threapleton, K., Newberry, K., Sutton, G., Worthington, E. & Drummond, A. 2018. Virtually home: Feasibility study and pilot randomised controlled trial of a virtual reality intervention to support patient discharge after stroke. *British Journal of Occupational Therapy*, 81, 196–206.

Törnbom, K. & Danielsson, A. 2018. Experiences of treadmill walking with non-immersive virtual reality after stroke or acquired brain injury - A qualitative study. *PLoS One*, 13, e0209214.

Triandafilou, K. M., Tsoupikova, D., Barry, A. J., Thielbar, K. N., Stoykov, N. & Kamper, D. G. 2018. Development of a 3D, networked multi-user virtual reality environment for home therapy after stroke. *Journal of Neuroengineering and Rehabilitation*, 15, 88.

Valdes, B. A., Glegg, S. M. N., Lambert-Shirzad, N., Schneider, A. N., Marr, J., Bernard, R., Lohse, K., Hoens, A. M. & Van Der Loos, H. F. M. 2018. Application of commercial games for home-based rehabilitation for people with hemiparesis: Challenges and lessons learned. *Games Health J*, 7, 197–207.

Winstein, C. J., Stein, J., Arena, R., Bates, B., Cherney, L. R., Cramer, S. C., Deruyter, F., Eng, J. J., Fisher, B. & Harvey, R. L. 2016. Guidelines for adult stroke rehabilitation and recovery: A guideline for healthcare professionals from the American Heart Association/American Stroke Association. *Stroke*, 47, e98–e169.

World Health Organization. 2012. *Global Health Estimates*. Geneva: World Health Organization.

4 VR-Based Assessment and Intervention of Cognitive Functioning after Stroke

Pedro Gamito, Ágata Salvador, Jorge Oliveira,
Teresa Souto and Ana Rita Conde
Universidade Lusófona
HEI-Lab: Digital Human-Environment Interaction Lab
Lisbon, Portugal

João Galhordas
Centro de Medicina de Reabilitação de Alcoitão
Alcoitão, Portugal

CONTENTS

COGNITIVE FUNCTION AFTER STROKE

Stroke is an epidemic disease: one in six people will experience stroke during their lifetime (WSO - World Stroke Organization 2020). This health condition is the second leading cause of death for individuals over 60, the fifth's for individuals aged 15–59 years (Caprio & Sorond 2018; Johnson et al. 2016; WSO 2020), and the third reason of disability worldwide (Caprio & Sorond 2018; Johnson et al. 2016). According to Johnson and colleagues (2016), 70% of stroke and 87% of deaths related to stroke or disability occur in low- and medium-income countries, afflicting individuals at the peak of their productive life.

Stroke leads to sudden death of brain cells due to a lack of oxygen resulting from an interruption of blood flow to the brain. This disruption is caused by obstruction or rupture of a brain artery (Johnson et al. 2016). The updated definition of stroke proposed by the Stroke Council of the American Heart Association/American Stroke Association defines stroke as a result of an acute focal injury to the central nervous system of vascular cause, namely, cerebral infarction, intracerebral hemorrhage and subarachnoid hemorrhage. This event, even in asymptomatic patients, is potentially associated with serious consequences, such as cognitive and functional decline (Sacco et al. 2013).

Deficits in cognitive abilities such as attention, memory, and executive functioning are among the most frequent problems (Tang et al. 2018), conflicting with quality of life and treatment adherence (Cumming et al. 2013). Executive functions compose of a set of integrated skills that include problem solving, planning, shifting, updating, inhibition, self-monitoring, cognitive flexibility, and working memory (De Luca et al. 2016), being crucial for the everyday functioning through the performance of instrumental activities of daily living (IADL) (Sadek et al. 2011). For example, finances, shopping, choosing clothes to wear, or preparing a meal, etc., are aspects that require proper executive functioning due to its cognitive complexity (Lezak et al. 2012). This means that stroke patients may show compromised ability in participating in meaningful activities and social relationships, affecting daily life activities. Therefore, assessing stroke's impact on cognitive functioning, but also on daily life activities, is essential to define a rehabilitation plan that seeks to improve the functional domains affected and to promote the returning to routines and the autonomy in everyday life activities (Bogdanova et al. 2016).

Compromised performance of IADL (Reppermund et al. 2013) can strongly affect the quality of life of both patients (Stites et al. 2018) and family caregivers (Karg et al. 2018), who are the cornerstone of patients' support, representing increased costs for the healthcare delivery system (Prince et al. 2015). Hence, solutions designed to assess cognitive functioning and the performance of IADL may have beneficial effects not only for stroke patients but also for families and society.

VR-BASED NEUROPSYCHOLOGICAL ASSESSMENT

Neuropsychological assessment, while sampling multiple cognitive domains (e.g., attention, memory, and executive functioning), provides countless insight into the neurocognitive, behavioral, and emotional functioning (Parsons 2011).

Traditionally, neuropsychological assessment uses paper-and-pencil tests to examine cognitive functions. For example, cancellation tests are typically used to assess attention abilities, while the Wechsler Memory Scale is among the most widely used tests for memory assessment (Camara et al. 2000). Executive functioning, in turn, is typically assessed using the Frontal Assessment Battery (Lima et al. 2008). However, research has provided evidence on the lack of ecological validity of these standard neuropsychological measures (Spooner and Pachana 2006). In other words, these tests may fail to adequately predict how a person perform IADL, once they do not accurately replicate the complexity and diversity of tasks individuals experience in their real environments (Parsons 2011).

Advanced technology offers a potential solution to the traditional simple and static stimuli presentation that does not capture the features of the real-world context (Parsons 2015). Virtual reality (VR) is a special case of a technology that can be adopted for computer-based neuropsychological assessment and has been increasingly considered as providing more ecologically valid assessments (Parsons 2011). Virtual environments are easily designed to reproduce aspects of the real world, namely, IADL (i.e., shopping, dressing, or preparing meals; Gamito et al. 2016a, 2016b). As such, when compared to traditional tests, VR-based assessment may closely resemble real-life performance, while maintaining rigorous experimental control (Rizzo et al. 2004). Furthermore, computer-based approaches offer a unique set of advantages, including increased standardization and ease of test administration; increased precision of timing presentation and response latency measurement (Parsons 2011; Cernich et al. 2007); greater ease in scoring and accessing data (Cernich et al. 2007); minimization of floor and ceiling effects; and quantification of a wider range of abilities (Tarnanas et al. 2014).

VR applications that track and monitor behavioral responses while individuals perform IADL have been developed in recent years (Parsons 2015). Research has generally suggested that these VR tools allow for the assessment of cognitive functioning, particularly IADL, in people with different conditions. For example, previous studies have found that the performance in VR tasks predicts the scores in traditional neuropsychological tests from patients with epilepsy (Grewe et al. 2014), acquired brain injuries (Jovanovski et al. 2012), and even from normative samples (Oliveira et al. 2017). Furthermore, VR IADL tasks seem to be able to discriminate cognitive deficits. One example is the Virtual Kitchen Test, which is a VR IADL test that reflects a real-life activity of food preparation. The results from a sample of patients with cognitive deficits have shown that scores (errors and task execution times) were positively correlated with the performance on traditional neuropsychological tests, being able to differentiate patients from healthy controls (Gamito et al. 2015).

Regarding stroke patients, two studies were found assessing difficulties related to neglect, a common disorder after stroke. Broeren and colleagues (2007) report that using a virtual environment depicting a cancelation task may be useful to assess visual search deficits associated with neglect of stroke patients. These results were aligned with a previous study for neglect (Kim et al. 2004). Another study with stroke patients explored the usability and performance in a supermarket test, suggesting differences in patients' performance comparing to healthy controls (Kang et al. 2008). In addition, performance on a virtual shopping task was also poor in post-stroke patients when compared to different healthy groups (Rand et al. 2007). Taken together, these results highlight the potential usability of VR IADL tasks for cognitive assessment in stroke patients. Neuropsychological assessment not only plays a central role in assisting diagnosis but also in providing the guidelines for appropriate therapeutic and personalized rehabilitation plans patients with cognitive deficits (Parsons 2011).

STROKE REHABILITATION

Addressing stroke's sequelae requires an interdisciplinary approach with key implications to patients' success considering the medical, surgical, and psychological effects of this diagnosis (Capizzi et al. 2020). Furthermore, the focus should be placed on the individual's functioning level (i.e., ability to perform IADL) rather than on the disease (diagnosis).

World Health Organization (WHO) defines rehabilitation as a set of interventions designed to optimize functioning and reduce disability in individuals with health conditions in interaction with their environment (WHO 2019). After conducting a scoping review on rehabilitation services, WHO authored "The rehabilitation in health systems: Guide for action." This guide set the emphasis of rehabilitation on the impact of health condition on the person's life by primarily focusing on the improving of functioning and on reducing of the experience of disability (WHO 2019).

Aiming at offering options for rehabilitation that replace traditional paper-and-pencil approach, new interventions, which some are still on an embryony stage, are commonly framed within the following categories (Bogdanova et al. 2016; Dang et al. 2017; Storm & Utesch 2019):

- **Hyperbaric Oxygen Therapy:** This type of treatment describes the inhalation of 100% oxygen under the pressure greater than atmosphere absolute, currently used for neurological diseases due to its therapeutic benefits (inhibition of apoptosis, suppression of inflammation, protection of the integrity of blood–brain barrier, and promotion of angiogenesis and neurogenesis).
- **Non-invasive Brain Stimulation:** Repetitive transcranial magnetic stimulation characterizes a painless, noninvasive, easily operated treatment with few adverse reactions and relevant positive effects in rehabilitation of stroke patients. Modification of neuronal excitability, by generating excitatory or inhibitory activity according to the frequency used, providing potential benefits in memory deficits, attention impairments, and depression symptoms.
- **Limb or Organ Function Reconstruction:** This approach is based on low-frequency pulse current to stimulate limb or organ dysfunction that favors replacement or correction of lost function in limbs and organs, promoting functional reconstruction by adjusting the advanced nerve center.
- **Behavioral and Emotional Therapies:** Behavioral changes post stroke often include not only mood disorders (e.g., anger, depression, anxiety) but also physical or verbal aggression.

Emotional instability of stroke patients needs to be addressed; otherwise, it may hamper rehabilitation outcomes. Psychological intervention (individual or group modalities) may promote coping skills training with a benefit in emotional regulation, namely, in anger/ rage management.

*   **Mental Practice Interventions:** Frequently used with stroke survivors to improve motor function. Mental practice is supported by the imaginary representation of a motor action or skill – motor imagery by mentally performing an action instead of physically – sparing repetition/training as well as supervision or the resort to equipment.

Although, some of these alternatives to paper-and-pencil assessment are still controversial and under debate (Ding et al. 2013, Cramer 2018), the use of computers and VR are being increasingly used in stroke rehabilitation. Virtual reality cognitive training (VRCT) allows an intuitive training, supported by combined auditive and visual stimulations, which can engage different components of the cognitive apparatus, namely, memory, attention, and executive functioning. And because VR provides a three-dimensional experience, training IADL in virtual contexts that resembles real day-to-day activities is possible. These properties of VR caught the attention of therapists and researchers alike and are going to be further discussed in the next section.

VRCT REHABILITATION AFTER STROKE

All the above mentioned interventions rest on the underlying assumption of neuroplasticity, meaning that cognitive stimulation can potentially modify brain structure and function in reaction to environmental diversity, an adaptive potential that may promote the ability to learn new skills and behaviors. In fact, studies have provided evidence of neuroplasticity following cognitive interventions with mild cognitive impairment (Belleville et al. 2011). Systematic reviews suggested significant gains in cognitive function through cognitive training, an empirical evidence that has supported the growing interest and development of cognitive-based interventions (Martin et al. 2011).

Within this scope, VRCT has emerged as an alternative to paper-and-pencil exercises. VRCT seems to be a valuable approach offering the following advantages: the adaptation of exercises to specific needs and cognitive status offering personalization of treatment (Coyle et al. 2014); the use of serious games, which can promote individuals' motivation (Kueider et al. 2012) and positively impact mental and physical health (Gamito et al. 2011); the provision of behavioral feedback to the therapist (Gamito et al. 2017); cost-effective, less labor-intensive, and easily disseminated solution (Kueider et al. 2012); and ecological validity, once the exercises in virtual environment can be similar to activities of daily living, promoting transfer of learned skills to everyday living (Garcia-Betances et al. 2015; Doniger et al. 2018; Gamito et al. 2019).

Accordingly, VRCT seems to be the only available option when it comes to simulate IADL. The adoption of IADL-based VRCT for cognitive rehabilitation appears to promote the inclusion of functional aspects of cognition while improving generalization of the desired effects via training in these activities. Also, the ability to create standardized treatment plans where repetition and task-specific learning may be easily implemented. Another advantage is related to personalization of exercises to meet the patients' needs that are specific of each phase of stroke recovery (Ferreira-Brito et al. 2019).

There are diverse VR environments targeting cognitive improvement, but few have adopted a multidomain approach. One example of the former is the Systemic Lisbon Battery (SLB; Gamito et al. 2016a, 2016b). SLB has been used by these authors for cognitive assessment and rehabilitation in different populations with cognitive deficits (Gamito et al. 2014, 2015; Oliveira et al. 2017). This battery comprises IADL exercises, such as (a) morning hygiene task (basic cognitive functioning), using the brush teeth and shower; (b) wardrobe task (executive functions), to choose clothes; (c) shoe rack test (attention), for pairing shoes; (d) virtual kitchen task (attention and executive functions), to prepare a cake; (e) supermarket and pharmacy tasks (memory and executive functions), for purchases of articles; (f) virtual gallery task (attention and executive functions), for observation and identification of details in virtual paintings (for further insights on SLB, please see: https://www.youtube.com/watch?v=hW8cqrp3CMs). Studies with SLB have been suggesting that VR may be an efficient tool to improve cognitive functions in the context of different conditions, namely, substance use disorders (Gamito et al. 2014), aging (Gamito et al. 2019, 2020), and following acquired brain injuries as traumatic brain injury (Gamito et al. 2011) or stroke (Gamito et al. 2015; Oliveira et al. 2020).

In stroke patients, SLB was the cradle for two studies. In a controlled study on VR vs treatment as usual, the results showed significant improvements in memory and attention abilities (Gamito et al. 2015). However, in a recent study, consisting in a brief intervention (approximately of 240 h), improvements on global cognitive functioning were found but not in executive functions or specific cognitive abilities (Oliveira et al. 2020). These results suggest that the use of functional behaviors on activities of daily living, involving different cognitive abilities and demands, is useful to promote cognitive functioning of stroke patients. However, longer rehabilitation plans may be needed in order to provide more consistent enhancements in cognitive and executive functioning.

Complementary contributions of the SLB are discussed in the next section, where the reader can find the testimony of a therapist that has been using SLB for several years in the largest rehabilitation center in Portugal.

The effectiveness of VRCT among stroke patients was also supported in a recent meta-analysis. Aminov and colleagues (2018) reveal significant small to moderate effect sizes of virtual rehabilitation on patients' cognition. Another recent study (Rogers et al. 2018), comparing computer-based rehabilitation with classical interventions, suggested a small overall effect on cognition but did not reveal specific effects related to cognitive rehabilitation approach (i.e., computer-based rehabilitation did not produce higher effects when compared to classical interventions).

A randomized controlled trial comparing VR with classical intervention in acute stroke patients suggested similar effects both at the level of attention and memory, but superior effects of VR for improving daily living performance were found, supporting far transfer effects of this kind of training to overall functionality (Cho & Lee 2019). Overall, these findings suggest that VR may be more effective at producing global improvements in cognition.

However, a further challenge besides demonstrating the efficacy of cognitive rehabilitation approaches is to determine the optimal timing to implement such interventions (Bernhardt et al. 2017). Also, no standardized treatment protocols or clinical guidelines are yet established (Bognadova et al. 2016), which is a significant barrier to the implementation of this approach directed at improving patients' cognition and psychological functioning. Despite the promising results, consistent evidence based on multidomain interventions (Ballesteros et al. 2015) with larger samples (Kueider et al. 2012; Ballesteros et al. 2015) are required to provide more robust empirical support for the efficacy and the mechanisms underlying VRCT interventions.

VRCT: A TESTIMONY FROM A THERAPIST

VRCT exercises started at the Alcoitão Rehabilitation Medicine Center (CMRA, http://cmra.pt/) in 2009 through the "Casa" program, in the adult stroke and traumatic brain injury of the inpatient population, to provide cognitive stimulation, namely, memory and attention. In 2016, SLB was introduced, and the exercises extended to the adult population on an outpatient basis and to adolescents attending the Center's Pediatric Rehabilitation and Development Service. The consultation is carried out frequently by psychologists and, in the outpatient clinic, by physicians. There has been a higher frequency of the adult population, especially those who are hospitalized.

As inclusion criteria, patients need to have a good verbal comprehension and good vision ability. Patients with a significant state of cognitive deterioration that prevents the normal prosecution of sessions and continues with the traditional rehabilitation approach.

TABLE 4.1

Mr. A., 58 Years Old, Hospitalized, Suffered an Ischemic Stroke, 10 Sessions

Cognitive Status	SLB Performance	Therapeutic Intervention
• Slight Anosognosia.	• In the description of the activities, he showed lack of inhibitory control, while being careless with the activities, along with memory difficulties.	• Awareness of the difficulties and alternatives to deal with them (revealed little critical judgment).
• Impairment of executive functions. Lack of cognitive flexibility. Perseveration.	• The patient forgot about shopping at the supermarket and pharmacy but remembered with the use of cues.	• The team talked about the need for supervision in the personal management of Mr. A's treatments.
• Lack of attention, attention dispersion, and tendency to error without correction.	• Revealed difficulties in learning less familiar and more complex information.	• The team spoke with the wife about the need to help managing Mr. A's personal life, reinforcing the good initiative and availability shown by Mr. A.
• Difficulties in spontaneous information retrieval.	• In replying to specific questions related to the tasks (for instance, what is the door number of the apartment)	• The promotion of tasks related to memory difficulties, while using cues related to the target information for helping memory retrieval.

In the inpatient adults, there is a previously established protocol of ten sessions with an increasing level of difficulty. Intervention with the adult population of the outpatient clinic and with adolescents is longer and flexible, with the possibility of having about 20 sessions, while respecting the "base standard" of the first ten. In the case of adolescents, sometimes activities are adapted to the age group, for example, going to the pharmacy can be replaced by another activity where the patients try to remember the brands and colors of the cars and, sometimes, by the execution of memory and attention exercises on the virtual computer that is in the bedroom of the SLB scenario.

To illustrate the sessions, we briefly exemplify the activities of two patients that have used SLB, during their hospitalization following ischemic stroke (see Tables 4.1 and 4.2).

In these sessions, it was observed positive aspects associated with SLB and some negative that may give rise to some reflection in order to improve the program and the hardware. These aspects are common to many other sessions that we have had with other patients.

A first positive aspect is the fact that this scenario is closer to reality, allowing one to try to relate the tasks performed in SLB with the day to day. For example, we found that Mr. A. is an impulsive person and little care for managing his money, while Mr. B. is more careful and has a good sense of prices, being also methodical with regard to the ingredients needed for cooking and careful when choosing the drugs at the pharmacy.

Another positive aspect of the program relates to its non-verbal characteristics, allowing the patient to put into practice non-verbal thinking, more concrete and effective for carrying out daily activities. This is an important aspect of SLB in the sense that it allowed creating, throughout the sessions, a parallelism with real life, by talking to the participants, watching them during hospitalization, and talking to family members and the rehabilitation team.

When comparing SLB with the most traditional assessment and cognitive stimulation tests, namely, Montreal Cognitive Assessment and Weschler Memory Scale, we found that this battery, being closer to reality and less verbal, is quite suitable for people with aphasia, making it possible to show more clearly the real abilities and difficulties of these people. This was particularly evident with Mr. B., who showed difficulties in participating in these more verbal and abstract tests.

TABLE 4.2

Mr. B., 58 Years Old, Hospitalized, Suffered an Ischemic Stroke, Ten Sessions

Cognitive Status	SLB Performance	Therapeutic Intervention
• Expressive aphasia.	• Difficulty in remembering the door number, street name, brand of grocery, and pharmacy products (he remembered better with visual stimuli).	• The new tasks were introduced gradually.
• Generalized cognitive impairment.	• Slowness in carrying out activities, especially those that involved alternating attention.	• The use of non-verbal communication was reinforced.
• Slight difficulties in spontaneous information retrieval, especially verbal information.	• Anxious and worried about not doing the activities properly.	• These difficulties and Mr. B.'s awareness about these difficulties were discussed with the team.
• Low self-esteem and self-confidence.		
• Lack of cognitive flexibility.	• Difficulties in set shifting between tasks, for instance by having difficulties in choosing clothes from the wardrobe according to different criteria.	• Providing exercises based on set shifting ability of increasing difficulty.

SLB also proved to be both a playful and therapeutic tool, reinforcing the motivation and the ability to fantasize (they were often enthusiastic about good cars) and allowing, unlike other tests, to have an immediate answer about performance.

The difficulties were mainly related to hardware interaction, because due to neurological injuries, hemiparesis in Mr. A. and Mr. B., they revealed motor difficulties in the use of the mouse, especially Mr. B. who was not familiar to technologies. It should be noted that these motor limitations and respective difficulties in the practical use of the program have been a constant in the population of CMRA, which were overcome with motivation and with the help from therapists in using the mouse.

Finally, it is important to emphasize that SLB has been an important instrument of mediation of the therapeutic relationship, which makes it possible to play and be more comfortable with people and at the same time to access their psychological characteristics.

VRCT with SLB allows the therapist to have an immediate feedback about the abilities and difficulties encountered, permitting to find compensatory strategies, together with the patient, with the people close to him/her and with the rehabilitation team.

As previously mentioned, VR has a good ecological validity, as it is close to the reality of everyday life and uses many non-verbal tasks, being an important aspect in the cognitive rehabilitation of people with aphasia. Some authors (Gamito et al. 2015; Larson et al. 2014; Oliveira et al. 2014, 2017) had made reference to these important characteristics, an aspect that has been frequently demonstrated in clinical practice at CMRA.

FINAL REMARKS

Overall, the drive behind this chapter was to highlight the need for a leap forward into a new paradigm of assessment and rehabilitation of cognitive functions after stroke. Traditional procedures, although may be able to assess and rehabilitate cognitive dimensions, have little impact on people's day to day lives. They do not tackle the functionality; they just assess the construct. The cognitive status, as measured by old-style instruments, reflects the performance of completing a paper-and-pencil test, no indications on how the individual is going to perform when a meal needs to be cooked or a trip to the supermarket is required is provided. Hence, we looked into some of the efforts taken in the last decade or so to design, develop, and validate ecological options to assess and stimulate cognitive functionality through VR. The bulk of the studies point toward the capacity of VR-based protocols in reaching out for veridicality: the ability to predict daily functioning and veri-similitude: the similitude between measures and daily activities (Parsons 2011), laying the path for the democratization of evaluation and interventions strategies that will promote the performance of IADL. We close the chapter with a different outlook. We asked one therapist that work with VR for about ten years to give his testimony. What is reported is difficult to assess through the traditional instruments that academics use. The relational component of the intervention and the adaptability of the VR platform to patients and other participants' needs were brought to light, revealing other characteristics that, again, make VR-based assessment and intervention of no match.

ACKNOWLEDGMENTS

Some of the studies reported in this chapter were funded by the Foundation for Science and Technology – FCT (Portuguese Ministry of Science, Technology and Higher Education), under the grant UIDB/05380/2020.

REFERENCES

Aminov, A., Rogers, J. M., Middleton, S., Caeyenberghs, K., & Wilson, P. H. (2018). What do randomized controlled trials say about virtual rehabilitation in stroke? A systematic literature review and meta-analysis of upper-limb and cognitive outcomes. *Journal of Neuroengineering and Rehabilitation*. 15(1), 29.

Ballesteros, S., Kraft, E., Santana, S., & Tziraki, C. (2015). Maintaining older brain functionality: A targeted review. *Neuroscience and Biobehavioral Reviews*. 55, 453–477.

Belleville, S., Clément, F., Mellah, S., Gilbert, B., Fontaine, F., & Gauthier, S. (2011). Training-related brain plasticity in subjects at risk of developing Alzheimer's disease. *Brain*. 134, 1623–1634.

Bernhardt, J., Hayward, K. S., Kwakkel, G., Ward, N. S., Wolf, S. L., Borschmann, K., Krakauer, J. W., Boyd, L. A., Carmichael, S. T., Corbett, D., & Cramer, S. C. (2017). Agreed definitions and a shared vision for new standards in stroke recovery research: The stroke recovery and rehabilitation roundtable taskforce. *International Journal of Stroke*. 12(5), 444–450.

Bogdanova, Y., Yee, M. K., Ho, V. T., & Cicerone, K. D. (2016). Computerized cognitive rehabilitation of attention and executive function in acquired brain injury: A systematic review. *The Journal of Head Trauma Rehabilitation*. 31, 419–433.

Broeren, J., Samuelsson, H., Stibrant-Sunnerhagen, K., Blomstrand, C., & Rydmark, M. (2007). Neglect assessment as an application of virtual reality. *Acta Neurologica Scandinavica*. 116, 157–163. doi:10.1111/j.1600-0404.2007.00821.x

Camara, W. J., Nathan, J. S., & Puente, A. E. (2000). Psychological test usage: Implications in professional psychology. *Professional Psychology: Research and Practice*. 31(2), 141–154. https://doi.org/10.1037/0735-7028.31.2.141

Capizzi, A., Woo, J., & Verduzco-Gutierrez, M. (2020). Traumatic brain injury: An overview of epidemiology, pathophysiology, and medical management. *Medical Clinics*. 104(2), 213–238.

Caprio, Z. F., & Sorond, F. A. (2018). Cerebrovascular disease: Primary and secondary stroke prevention. *The Medical Clinics of North America*. 103(2), 295–308.

Cernich, A. N., Brennana, D. M., Barker, L. M., & Bleiberg, J. (2007). Sources of error in computerized neuropsychological assessment. *Archives of Clinical Neuropsychology*. 22:S39–S48. https://doi.org/10.1016/j.acn.2006.10.004

Cho, D. R., & Lee, S. H. (2019). Effects of virtual reality immersive training with computerized cognitive training on cognitive function and activities of daily living performance in patients with acute stage stroke: A preliminary randomized controlled trial. *Medicine (Baltimore)*. 98(11), e14752.

Coyle, H., Traynor, V., & Solowij, N. (2014). Computerised and virtual reality cognitive training for individuals at high risk of cognitive decline: Systematic review of the literature. *American Journal of Geriatric Psychiatry*. 23, 335–359. cpb.2004.7.742

Cramer, S. (2018). Treatments to promote neural repair after stroke. *Journal Stroke.* 20(1), 57–70. doi:10.5853/jos.2017.02796

Cumming, T. B., Marshall, R. S., & Lazar, R. M. (2013). Stroke, cognitive deficits, and rehabilitation: Still an incomplete picture. *International Journal of Stroke.* 8(1), 38–45.

Dang, B., Chen, W., He, W., & Chen, G. (2017). Rehabilitation treatment and progress of traumatic brain injury dysfunction. *Neural Plasticity,* 2017, 1582182.

Ding, Q., Stevenson, I. H., & Wang, N. (2013). Motion games improve balance control in stroke survivors: A preliminary study based on the principle of constraint induced movement therapy. *Displays* 34:125–131.

De Luca, R., Calabrò, R. S., & Bramanti, P. (2016). Cognitive rehabilitation after severe acquired brain injury: Current evidence and future directions. *Neuropsychological Rehabilitation.* 25, 1–20.

Doniger, G. M., Beeri, M. S., Bahar-Fuchs, A., Gottlieb, A., Tkachov, A., Kenan, H., Livny, A., Bahat, Y., Sharon, H., Ben-Gal, O., Cohen, M., Zeilig, G., & Plotnik, M. (2018). Virtual reality-based cognitive-motor training for middle-aged adults at high Alzheimer's disease risk: A randomized controlled trial. *Alzheimer's and Dementia.* 4, 118–129.

Ferreira-Brito, F., Fialho, M., Virgolino, A., Neves, I., Miranda, A. C., Sousa-Santos, N., Caneiras, C., Carriço, L., Verdelho, A., & Santos, O. (2019). Game-based interventions for neuropsychological assessment, training and rehabilitation: Which game-elements to use? A systematic review. *Journal of Biomedical Informatics.* 98, 103287.

Gamito, P., Morais, D., Oliveira, J., Lopes, P., Picarelli, F., Correia, S., & Matias, M. (2016a). Systemic Lisbon battery: Normative data for memory and attention assessments. *Journal of Medical Internet Research: Rehabilitation and Assistive Technologies.* 55(1), 93–97. doi:10.2196/rehab.4155

Gamito, P., Oliveira, J., Pinto, L., Rodelo, L., Lopes, P., Brito, R., & Morais, D. (2014). Normative data for a cognitive VR rehab serious games-based approach. *Pervasive Health 14 Proceeding of 8th International Conferences on Pervasive Computing Technologies for Healthcare,* May 2014, Oldenburg. pp. 443–446.

Gamito, P., Oliveira, J., Alghazzawi, D., Fardoun, H., Rosa, P., & Sousa, T. (2017). The art gallery test: A preliminary comparison between traditional neuropsychological and ecological VR-based tests. *Frontiers in Psychology.* 8, 191.

Gamito, P., Oliveira, J., Alves, C., Santos, N., Coelho, C., & Brito, R. (2020). Virtual reality-based cognitive stimulation to improve cognitive functioning in community elderly: A controlled study. *Cyberpsychology, Behavior and Social Networking.* 23(3), 150–156. doi: 10.1089/cyber.2019.0271.

Gamito, P., Oliveira, J., Brito, R., Lopes, P., Rodelo, L., Pinto, L., & Morais, D. (2016b). Evaluation of cognitive functions through the systemic Lisbon battery: Normative data. *Methods of Information in Medicine.* 55(1), 93–97. https://doi.org/10.3414/ME14-02-0021

Gamito, P., Oliveira, J., Caires, C., Morais, D., Brito, R., Lopes, P., Saraiva, T., Soares, F., Sottomayor, C., Barata, F., Picareli, F., Prates, M., & Santos, C. (2015). Virtual kitchen test. Assessing frontal lobe functions in patients with alcohol dependence syndrome. *Methods of Information in Medicine.* 54(2), 122–126. https://doi.org/10.3414/ME14-01-0003

Gamito, P., Oliveira, J., Morais, D., Coelho, C., Santos, N., Alves, C., Galamba, A., Soeiro, M., Yerra, M., French, H., Talmers, L., Gomes, T., & Brito, R. (2019). Cognitive stimulation of elderly individuals with instrumental virtual reality-based activities of daily life: Pre-post treatment study. *Cyberpsychology, Behavior and Social Networking.* 22, 69–75.

Gamito, P., Oliveira, J., Morais, D., Rosa, P., & Saraiva, T. (2011). Serious games for serious problems: From Ludicus to Therapeuticus. In Kim J. J., ed. *Virtual Reality.* London, UK: IntechOpen, pp. 527–548.

Garcia-Betances, R. I., Arredondo, W. M. T., Fico, G., & Cabrera-Umpiérrez, M. F. (2015). A succinct overview of virtual reality technology use in Alzheimer's disease. *Frontiers in Aging Neuroscience.* 7, 80.

Grewe, P., Lahr, D., Kohsik, A., Dyck, E., Markowitsch, H. J., Bien, C. G., Botsch, M., & Piefke, M. (2014). Real-life memory and spatial navigation in patients with focal epilepsy: Ecological validity of a virtual reality supermarket task. *Epilepsy and Behavior.* 231, 57–66. https://doi.org/10.1016/j.yebeh.2013.11.014

Johnson, W., Onuma, O., Owolabi, M., & Sachdev, S. (2016). Stroke: A global response is needed. *Bulletin of the World Health Organization.* 94, 634–634A.

Jovanovski, D., Zakzanis, K., Campbell, Z., Erb, S., & Nussbaum, D. (2012). Development of a novel, ecologically oriented virtual reality measure of executive function: The multitasking in the city test. *Applied Neuropsychology: Adult.* 19(3), 171–182. https://doi.org/10.1080/09084282.2011.643955

Kang, Y. J., Ku, J., Han, K., Kim, S. I., Yu, T. W., Lee, J. H., & Park, C. I. (2008). Development and clinical trial of virtual reality-based cognitive assessment in people with stroke: Preliminary study. *Cyberpsychology & Behavior,* 11, 329–339. doi:10.1089/cpb.2007.0116

Karg, N., Graessel, E., Randzio, O., & Pendergrass, A. (2018). Dementia as a predictor of care-related quality of life in informal caregivers: A cross-sectional study to investigate differences in health-related outcomes between dementia and non-dementia caregivers. *BMC Geriatrics.* 18(1), 189. https://doi.org/10.1186/s12877-018-0885-1

Kim, K., Kim, J., Ku, J., Kim, D. Y., Chang, W. H., Shin, D. I., Lee, J. H., Kim, I. H., & Kim, S. I. (2004). A virtual reality assessment and training system for unilateral neglect. *CyberPsychology & Behavior*, 7, 742–749. doi: 10.1089/cpb.2004.7.742

Kueider, A. M., Parisi, J. M., Gross, A. L., & Rebok, G. W. (2012). Computerized cognitive training with older adults: A systematic review. *PLoS One.* 7, e40588.

Larson, E., Feignon, M., Gagliardo, P., & Dvorkin, A. (2014). Virtual reality and cognitive rehabilitation: A review of current outcome research. *NeuroRehabilitation.* 34(4), 759–72. doi:10.3233/NRE-141078

Lezak, M. D., Howieson, D .B., Bigler, E. D., & Tranel, D. (2012). *Neuropsychological Assessment* (5th ed.). New York: Oxford University Press.

Lima, C. F., Meireles, L. P., Fonseca, R., Castro, S. L., & Garrett, C. (2008). The Frontal Assessment Battery (FAB) in Parkinson's disease and correlations with formal measures of executive functioning. *Journal of Neurology.* 255(11), 1756–1761. https://doi.org/10.1007/s00415-008-0024-6

Martin, M., Clare, L., Altgassen, A. M., Cameron, M. H., & Zenhder, F. et al., (2011). Cognition- based interventions for healthy older people and people with mild cognitive impairment. *The Cochrane Database of Systematic Reviews.* 1, CD006220.

Oliveira, J., Gamito, P., Alghazzawi, D. M., Fardoun, H. M., Rosa, P. J., Sousa, T., Picareli, L. F., Morais, D., & Lopes, P. (2017). Performance on naturalistic virtual reality tasks depends on global cognitive functioning as assessed via traditional neurocognitive tests. *Applied Neuropsychology: Adult.* 25(6), 555–561.

Oliveira, J., Gamito, P., Lopes, B., Silva, A. R., Galhordas, J., Pereira, E., Ramos, E., Silva, A. P., Jorge, A., & Fantasia, A. (2020). Computerized cognitive training using virtual reality on everyday life activities for patients recovering from stroke. *Disability and Rehabilitation: Assistive Technology.* Apr 7, 1-6. doi: 10.1080/17483107.2020.1749891

Oliveira, J., Gamito, P., Morais, D., Brito, R., Lopes, P., & Norberto, L. (2014). Cognitive assessment of stroke patients with mobile apps: A controlled study. *Annual Review of Cybertherapy and Telemedecine.* 103–107. doi:10.3233/978-1-1499-401-5-103.

Parsons, T. (2011). Neuropsychological assessment using virtual environments: Enhanced assessment technology for improved ecological validity. In. S. Brahnam, & L. C. Jain (Eds.), *Advanced Computational Intelligence Paradigms in Healthcare 6.* Virtual Reality in Psychotherapy, Rehabilitation, and Assessment (271–289). Springer. https://doi.org/10.1007/978-3-642-17824-5_13

Parsons, T. D. (2015). Virtual reality for enhanced ecological validity and experimental control in the clinical, affective and social neurosciences. *Frontiers in Human Neuroscience.* 9, 660. https://doi.org/10.3389/fnhum.2015.00660

Prince, M., Wimo, A. G. M., Ali, G. C., Wu, Y. T., & Prina, M (2015). World Alzheimer Report 2015 - The global impact of dementia: An analysis of prevalence, incidence, cost and trends. London, *Alzheimer's Disease International retrieved on 17/05/2020 from* https://www.alz.co.uk/research/WorldAlzheimerReport2015.pdf.

Rand, D., Katz, N., & Weiss, P. L. (2007). Evaluation of virtual shopping in the VMall: Comparison of post-stroke participants to healthy control groups. *Disability and Rehabilitation.* 29, 1710–1719. doi:10.1080/09638280601107450

Reppermund, S., Brodaty, H., Crawford, J. D., Kochan, N. A., Draper, B., Slavin, M. J., Trollor, J. N., & Sachdev, P. S. (2013). Impairment in instrumental activities of daily living with high cognitive demand is an early marker of mild cognitive impairment: The Sydney memory and ageing study. *Psychological Medicine.* 43(11), 2437–2445. https://doi.org/10.1017/S003329171200308X

Rizzo, A., Schulteis, M., Kerns, K., & Mateer, C. (2004). Analysis of assets for virtual reality applications in neuropsychology. *Neuropsychological Rehabilitation*, 14(1–2), 207–239. https://doi.org/10.1080/09602010343000183

Rogers, J. M., Foord, R., Stolwyk, R. J., Wong, D., & Wilson, P. (2018). General and domain-specific effectiveness of cognitive remediation after stroke: Systematic literature review and meta-analysis. *Neuropsychology Review.* 28(3), 285–309.

Sacco, R. L., Kasner, S., Broderick, J. P., Caplan, L. R., Connors, J. J. B., Culebras, A., Elkind, M.S., George, M.G., Hamdan, A.D., Higashida, R.T., Hoh, B.L., Janis, L.S, Kase, C.S., Kleindorfer, D.O., Lee, J.M., Moseley, M.E., Peterson, E.D., Turan, T.N., Valderrama, A.L., Vinters, H.V., American Heart Association Stroke Council, Council on Cardiovascular Surgery and Anesthesia, Council on Cardiovascular Radiology and Intervention, Council on Cardiovascular and Stroke Nursing, Council on Epidemiology and Prevention, Council on Peripheral Vascular Disease, & Council on Nutrition, Physical Activity and Metabolism (2013). An updated definition of stroke for the 21st Century: A statement for healthcare professionals from the American Heart Association/American Stroke Association. AHA, ASA expert consensus document. *Stroke.* 44(7), 2064–2089.

Sadek, J. R., Stricker, N., Adair, J. C., & Haaland, K. Y. (2011). Performance-based everyday functioning after stroke: Relationship with IADL questionnaire and neurocognitive performance. *Journal of the International Neuropsychological Society.* 17(5), 832–840.

Stites, S. D., Harkins, K., Rubright, J. D., & Karlawish, J. (2018). Relationships between cognitive complaints and quality of life in older adults with mild cognitive impairment, mild Alzheimer disease dementia, and normal cognition. *Alzheimer Disease and Associated Disorders.* 32(4), 276–283. https://doi.org/10.1097/WAD.0000000000000262

Storm, V., & Utesch, T. (2019). Mental practice ability among stroke survivors: Investigation of gender and age. *Frontiers in Psychology,* 10, 1568.

Tang, E. Y., Amiesimaca O, Harrison S, et al. (2018). Longitudinal effect of stroke on cognition: A systematic review. *Journal of the American Heart Association.* 7(2), 006443.

Tarnanas, I., Tsolaki, M., Nef, T., Müri, R., & Mosimann, U. P. (2014). Can a novel computerized cognitive screening test provide additional information for early detection of Alzheimer's disease? *Alzheimer's & Dementia.* 10(6), 790–798. https://doi.org/10.1016/j.jalz.2014.01.002

Traumatic Brain Injury and Spinal Cord Injury Collaborators. (2018). Global, regional, and national burden of traumatic brain injury and spinal cord injury, 1990–2016: A systematic analysis for the Global Burden of Disease Study 2016. *The Lancet, Neurology,* 18, 56–87.

World Health Organization. (2019). Rehabilitation in health systems: guide for action. Geneva retrieved on 17/05/2020 from https://www.who.int/rehabilitation/rehabilitation-guide-for-action/en/

World Stroke Organization. (2020). Learn about stroke. Retrieved on 17/05/2020 from https://www.world-stroke.org/world-stroke-day-campaign/why-stroke-matters/learn-about-stroke

5 Virtual Reality in Robotic Neurorehabilitation

Nicolas Wenk and Karin A. Buetler
University of Bern
Bern, Switzerland

Laura Marchal-Crespo
Delft University of Technology
Delft, The Netherlands
University of Bern
Bern, Switzerland

CONTENTS

INTRODUCTION

Patients with neurological dysfunction (e.g., stroke or Parkinson's disease) suffer from cognitive and motor impairments, including muscle weakness, reduced movement workspace, and loss of movement quality. Stroke remains a major source of permanent disability worldwide, because of its frequency (~17 million people worldwide) and consequences on the capability of stroke survivors to perform activities of daily living ("WHO | International Classification of Functioning, Disability and Health (ICF)," 2016). This number is expected to rise with increasing life expectancy, as the risk of stroke doubles every decade after age 55 ("WHO | Life expectancy"). Injuries and diseases of the nervous system have an enormous negative effect on the quality of life of patients. On average, one-third of stroke survivors require assistance to perform activities of daily living (Mercier et al., 2001). This dependency leads to a tremendous financial burden on society, throughout the patients' lifetime.

Evidence suggests that the brain has the ability to reorganize (neuroplasticity) and that this reorganization allows the lasting recovery of lost functions (Cramer et al., 2011). To regain part of their former motor function, neurologic patients should engage in an intensive neurorehabilitation process. Some conventional therapy treatments are especially promising, particularly those that focus on high-intensity (Kwakkel et al., 2004) and repetitive task-specific practice (French et al., 2016). However, high-intensity therapy is labor-intensive, often requiring more than one therapist to support the patient's movements. As a result, the training duration might be limited by the cost and endurance of the therapists, reducing the potential of motor recovery. There has been an increasing interest in using robotic devices to provide more cost-effective and intensive rehabilitation therapy following brain injury (Marchal-Crespo and Reinkensmeyer, 2009). Robots can support neurorehabilitation by delivering a high dose of intensive training. Besides, they allow therapists to train several patients simultaneously, therefore, reducing personnel costs.

There has been a progression in the development of robotic training strategies that specify how robotic devices interact with patients – see (Marchal-Crespo and Reinkensmeyer, 2009) for a review. The most accepted paradigm is to physically guide patients' limbs during movement training (i.e., haptic guidance). It is thought that moving the limb in ways that patients would otherwise not be able to move, would provide novel somatosensory stimulation that drives brain plasticity (Poon, 2004; Rossini and Dal Forno, 2004). However, robot-guided movements might, in some cases, decrease patients' physical and mental effort during training, and therefore, limit its rehabilitation potential (Israel et al., 2006; Marchal-Crespo et al., 2014, 2017).

Research in neurorehabilitation has demonstrated that patients' active engagement and motivation during rehabilitation interventions are crucial to recover arm function post-stroke (Maclean et al., 2000; Maclean and Pound, 2000; Putrino et al., 2017). Importantly, robot-assisted training can incorporate other technologies to increase patients' engagement and motivation to participate in the intensive and (generally) long rehabilitation programs (Brütsch et al., 2010; Freivogel et al., 2009; Gobron et al., 2015). Virtual reality (VR) offers the possibility to simulate many different imaginary or real activities of daily living that can be adapted to the patients' special needs while providing a motivating (Lee et al., 2003), safe (Marchal-Crespo and Reinkensmeyer, 2008), and consistent environment (Rose et al., 2005). Robotic and VR-based rehabilitation interventions – alongside with mental practice, mirror therapy, and sensory stimulation – have been shown to improve upper limb function after stroke (Pollock et al., 2014). VR-based interventions have a positive impact on regaining upper limb function (Domínguez-Téllez et al., 2020; Laver et al., 2017), especially when the VR-based interventions are designed for rehabilitation purposes (specific-VR), compared to commercial VR (non-specific-VR) (Aminov et al., 2018; Laver et al., 2017; Maier et al., 2019; Thomson et al., 2014).

Within the therapy context, Laver et al. defined VR as: *"An advanced form of human–computer interface that allows the user to 'interact' with and become 'immersed' in a computer-generated environment in a naturalistic fashion"* (Laver et al., 2017). The visual representation of the users' movement, i.e., **movement visualization** (MV), plays an important role in the interaction with the virtual environment (VE) in a naturalistic fashion. Although the more naturalistic MV is to map the patients' movements on a virtual humanoid (i.e., avatar) in real-time, to date, only 10% of employed therapy-based VR incorporates an avatar, while the majority of VRs employ indirect or abstract MVs (Ferreira dos Santos et al., 2016). Another important aspect stated in Lavers et al.'s definition of VR is **immersion**, defined as *"The extent to which the user perceives that they are in the virtual environment rather than the real world and is related to the design of the software and hardware."* The most immersive hardware nowadays is head-mounted displays (HMDs). However, from over 72 reviewed studies (Laver et al., 2017), only four used an HMD, while 56 employed 2D screens (e.g., television, computer screen, or projection). The level of immersion can also be modulated through software, e.g., with the interactivity level of virtual elements, the MV type, or the user's perspective (Debarba et al., 2017). However, it is still an open question how the level of immersion and MV type might affect neurorehabilitation outcomes (Rose et al., 2018).

Besides neurorehabilitation, VR therapy has been evidenced to be a valid tool for non-pharmacological pain and anxiety reduction by providing distraction from chronic and acute pain (e.g., Martini, 2016; Pourmand et al., 2018) associated with rheumatism (e.g., Tack, 2019), phantom limb pain after amputation (Matamala-Gomez et al., 2020; Murray et al., 2006; Osumi et al., 2020; Perry et al., 2018), and regional pain from (past) injuries (Feyzioğlu et al., 2020; Hoffman et al., 2011). Further, VR interventions have been applied in the field of psychotherapy, especially in the treatment of anxiety-related (e.g., specific phobias, Boeldt et al., 2019; Freeman et al., 2017) and post-traumatic stress disorders (Gonçalves et al., 2012; Oing and Prescott, 2018). For these patients, immersion in various safe, confidential, and controllable computer-generated interactive scenarios is a promising alternative to standard therapy sessions in public space. Finally, enriched VEs have become a unique opportunity to transport patients to both, stimulating and calming places, e.g., during intensive care after critical illness (Gerber et al., 2019, 2017; Hirota, 2020), or at different stages of neurodegenerative diseases (Dockx et al., 2016; Ferguson et al., 2020; Gates et al., 2019; Kim et al., 2019; Moreno et al., 2019; Sokolov et al., 2020).

Together, VR therapy proved to be a safe, inexpensive, non-pharmacological means of retraining motor function, managing pain and anxiety, mental health problems, and neurologic and geriatric syndromes by providing cognitive/motor engagement and enjoyment/motivation to a rapidly growing aging demographic.

STATE-OF-THE-ART VR TECHNOLOGY

CONVENTIONAL VR TECHNOLOGY EMPLOYED IN NEUROREHABILITATION

VR has been proposed as a promising tool to improve neurorehabilitation because it allows the provision of meaningful, versatile, and individualized motor tasks that can enhance patients' motivation, enjoyment, and engagement during training (Rohrbach et al., 2019). Several commercial computerized rehabilitation systems (e.g., "Augmented Performance Feedback & Challenge Package," Hocoma, China; "MindMotion," Mindmaze, Switzerland, etc.) already incorporate a library of VR games, based on audiovisual elements inspired by games.

Gamification is an approach to make non-game tasks more engaging and fun by applying motivational techniques derived from games. It primarily targets patients' extrinsic motivation (i.e., originated by external elements) by including game scores, achievements, etc. (Holmes et al., 2015; Perez-Marcos et al., 2018). **Serious games** are games designed for a primary purpose (e.g., rehabilitation) other than entertainment (Djaouti et al., 2011). Serious games modulate patients' intrinsic motivation (i.e., driven by internal rewards, such as satisfaction and joy) by having a gameplay based on, for example, the *Flow Theory* (Csikszentmihalyi, 1990) and/or the *Self-Determination Theory* (Crittenden and Deci, 1982). The *Flow Theory* states that optimal motivation is achieved when the patient's experience is in a balance between feeling that the task is "too easy" or "too hard." The *Self-Determination Theory* specifies that motivation relies on three psychological needs: autonomy, competence, and relatedness. These motivation techniques are crucial to empower patients to positively influence their recovery (Perez-Marcos et al., 2018).

The use of VR during robotic intervention allows patients to perform motivating and purposely goal/task-oriented exercises (i.e., exercises that are driven by a purpose and/or focused on the completion of a particular task). Furthermore, numerous studies have demonstrated the efficacy of VR to provide visual feedback (i.e., visual information on the user's performance as a basis for improvement) to promote neurorehabilitation (Adamovich et al., 2004; Deutsch et al., 2004; Jang et al., 2005) and to (re)learn motor tasks (Liebermann et al., 2002; Shea and Wulf, 1999; Todorov et al., 1997). Visual feedback during motor training (i.e., concurrent feedback) is especially effective in the early phase of learning (Sigrist et al., 2013) and when learning complex movements (Liebermann et al., 2002; Marchal-Crespo et al., 2013). Adding robotic (haptic) assistance to the visual feedback (i.e., multimodal feedback) allows to enrich the sensory information in the training environment and to further enhance motor learning (Marchal-Crespo et al., 2019; Sigrist et al., 2013; Wei and Patton, 2004) and function recovery in brain-injured patients (Krishnan et al., 2012, 2013).

Finally, the versatility of the VR environments allows the adaptation of the training interventions to the patient's individual physical and cognitive capabilities (Baur et al., 2018; Fu et al., 2015; Laver et al., 2017; Maier et al., 2020). Adapting the task difficulty to the patient's specific needs ensures not only an optimal motivational level throughout the intensive neurorehabilitation process

(Csikszentmihalyi, 2014) but also enhances motor learning (Marchal-Crespo et al., 2017; Milot et al., 2010). In line with the *Flow Theory*, the *Challenge Point Theory* states that optimal learning is achieved when the difficulty of the task is appropriate for the individual participant's level of expertise (Guadagnoli and Lee, 2004). Therefore, severely impaired patients may benefit from VR strategies that facilitate task completion (e.g., error reduction) (Ballester et al., 2015), while less severely impaired/more skilled patients may benefit from more challenging training strategies that make movement tasks more difficult (Patton et al., 2013).

Limitations of Conventional VR Technology in Robotic Neurorehabilitation

Conventional VR-based interventions involve movements and interactions with virtual objects that are far from those required in the real world. There are clear differences in the sensory-motor integration during object interaction in 2D-displayed VRs and the physical real environment that might limit the transfer of acquired skills into real activities of daily living (Bezerra et al., 2018; de Mello Monteiro et al., 2014; Levac and Jovanovic, 2017). Within the 2D-displayed VR, the patients interact via a symbolic virtual representation of their limb (e.g., a cursor) (Ferreira dos Santos et al., 2016). While this provides useful visual feedback, it has several drawbacks. First, 2D screens lack depth cues to help patients to perform 3D movements. Furthermore, 2D screens draw patients' attention away from their real limbs and limit the possibility to fully embody the virtual arm, negatively affecting motor performance. Importantly, the spatial and sensory mismatch between the virtual 2D and real 3D environments might increase cognitive demands from neurological patients, limiting their adherence to VR-based motor training.

Novel Immersive VR Technologies

With the recent advances in technology, VR now allows for more immersive experiences. Stereoscopic displays (also 3D displays) offer the possibility to have a three-dimensional visualization of the patients' movement. Within the last years, commercial HMDs started to be available at a relatively low cost. With sensors and an integrated stereoscopic display, HMDs allow a natural visual exploration of the environment by tracking head movements and providing 3D visualization with the eyes' vergence.

Augmented (AR) or mixed reality HMDs allow patients to interact with virtual elements and simultaneously visualize their limbs and interact with the therapists in the real world. Video see-through AR HMDs display the surrounding real-world using cameras and a stereoscopic screen and allow the rendering of virtual elements in the patient's field of view. However, current off-the-shelves HMDs (e.g., Oculus Quest, Facebook, USA) or compatible stereo cameras (e.g., ZED Mini, SereoLabs, USA) offer a distorted representation of the real world, e.g., the misalignment between the cameras viewpoint and the patients' perspective translates into a distorted perception of space (Cutolo et al., 2018). Besides, all the displayed (real-world and virtual) elements have the same focal distance, losing a depth cue and leading to a vergence-accommodation conflict (Condino et al., 2019). Optical see-through AR HMDs offer a more natural visualization of the real environment through a transparent glass, onto which virtual elements can be projected. However, they also suffer from several pitfalls: the virtual elements, being just a reflection of the light on a transparent surface, cannot be opaque and, therefore, rendering dark elements is not possible. Besides, with current off-the-shelf HMDs (e.g., Meta, Meta View, USA; HoloLens, Microsoft, USA), only the virtual elements have a shared and fixed focal distance (Condino et al., 2019). This leads to focal rivalry, i.e., patients should adjust the focus between real and virtual elements, even if they are at the same distance based on the eye vergence. These limitations may create discomfort in healthy users (Condino et al., 2019) and hamper the adoption of AR in clinical settings.

Immersive virtual reality HMDs immerse participants in a 3D computer-generated VE, providing the illusion of being inside the VE with the ability to interact with its elements – a sensation commonly referred to as "presence" (Sanchez-Vives and Slater, 2005). The current off-the-shelf immersive VR HMDs (e.g., Oculus, Facebook, USA; HTC Vive, HTC, Taiwan & Valve, USA) are well accepted by healthy young and older users, and no association with cybersickness, i.e., motion sickness symptoms developed when being in VR (LaViola, 2000), has been reported (Huygelier et al., 2019). Although immersive VR HMDs also suffer from the vergence-accommodation conflict, there is neither perspective shift nor focal rivalry as they show a purely virtual environment. Commercial immersive VR HMDs offer an (almost) total audiovisual immersion in the VE. Although this immersion might have the drawback of "disconnecting" patients from their body and/or the therapists, it gathers all of the previously identified potential benefits of VR (impact on patients' motivation and engagement, adaptation of the task/environment to patients' needs, and provision of visual feedback) and reduce possible incongruences between the perception of real and virtual objects, facilitating the transfer of acquired skills (Levac et al., 2019a).

Potential Advantages of Immersive VR in Robotic Neurorehabilitation

Although VEs experienced through HMDs offer a high visual immersion and a natural depth visualization, they are seldom used in the robotic rehabilitation context (Laver et al., 2017). Nevertheless, the use of immersive VR together with robotic-assisted neurorehabilitation offers several potential advantages to enhance neuroplasticity compared to conventional 2D screens.

In a highly immersive 3D VR, the symbolic virtual representation of the patients' limb viewed from a first-person perspective may become a *self*-representation of their own body. Thus, in contrast to conventional rehabilitation setups using 2D screens, immersive visualization technologies may promote body ownership over the virtual limb. **Body ownership** is the cognition that a body and its parts belong to oneself (Blanke, 2012; Tsakiris, 2010). The feeling of body ownership results from the integration and interpretation of multimodal sensory information in the brain, namely, visual, tactile and proprioceptive signals (Botvinick and Cohen, 1998; Ehrsson et al., 2004). Body ownership is regarded as one out of three components of embodiment, together with self-location and agency (Kilteni et al., 2012; Longo et al., 2008). **Self-location** describes the experience of where we are in space (Blanke, 2012). **Agency** refers to the sensation of being the cause and in control of actions and movements of a body (Braun et al., 2018). In VR, the sense of **embodiment** describes the phenomenon that the avatar's limbs are processed like the users' own limbs (Kilteni et al., 2012).

To date, less is known about the neural and behavioral correlates of embodying an avatar during robotic motor training. Yet, motor training in VR may crucially benefit from enhancing body ownership over the avatar. Importantly, brain areas involved in body ownership are shared with brain areas linked to motor learning, namely, multimodal areas associated with motor control and movement planning, i.e., frontal premotor cortices, temporal parietal junction, and insula; (Ehrsson et al., 2004, 2005; Tsakiris et al., 2007; Wise, 1985; Zeller et al., 2016). Previous studies suggest that this anatomical coupling might be functional. For example, an increase in body ownership over a virtual limb was found to be associated with faster reaction times (Grechuta et al., 2017) and better motor performance (Shibuya et al., 2018).

Immersive VR also provides a means to lower the **cognitive demands** required from brain-injured patients during motor training. Over half of stroke survivors suffer from cognitive impairments (Mellon et al., 2015; Patel and Birns, 2015). Therefore, every virtual element adding cognitive effort during movement training and/or decreasing the patients' attention on the key elements of the exercise (distractors) must be treated carefully. More cognitively disabled patients or patients in their early stages of training could better focus on the training task if they are not overwhelmed/distracted by too many/complex visual stimuli (Holden, 2005). Yet, enriched VEs, i.e., environments that facilitate enhanced sensory, cognitive, and motor stimulations (Nithianantharajah and Hannan, 2006), have been shown to improve engagement in stroke patients (Janssen et al., 2014). Therefore, except for the social stimulations that might be hindered by the immersion in a VE, rich VEs might be more suitable for cognitively less impaired patients. The VE richness could be adapted by adjusting the quantity, complexity, or diversity of the virtual elements. Using immersive VR with HMD offers the possibility of having almost complete control over all the audiovisual stimuli presented to the patient. Thus, HMDs might broaden the possibilities of using VR in the clinic, as they could be used in patients with a wide range of disability levels and at different stages of recovery.

Immersive VR supports motor recovery because it allows training impaired motor functions in highly realistic virtual scenarios (Levac et al., 2019a). However, the interaction with virtual elements lacks **haptic sensations**, which may create incongruences between the perception of real and virtual objects, potentially limiting the transfer of acquired skills (Levac et al., 2019a). Furthermore, proprioceptive and tactile information from the somatosensory system generated at the muscle and skin mechanoreceptors is fundamental to induce brain plasticity (Poon, 2005; Siegel and Sapru, 2011). Therefore, the addition of haptic rendering which allows users to "feel" tangible virtual objects could enhance motor recovery by stimulating the somatosensory system and reinforcing sensorimotor integration, i.e., the brain's ability to process and weight the information provided by all senses to increase its reliability (Ernst and Bülthoff, 2004).

Taking together, the use of immersive VR in the rehabilitation of neurologic patients may be a powerful tool to promote brain plasticity when recovering and compensating for affected sensorimotor abilities. The use of HMD offers a more direct visuospatial transformation between the real and visualized movement that might enhance movement quality and reduce the patients' cognitive load during training. A first-person immersive virtual reality that induces embodiment might promote brain plasticity in relevant motor areas. Furthermore, immersive VR allows full control of the delivered audiovisual stimuli and can be employed to individualize the VR to enhance patients' motivation and engagement. Finally, the use of robotic devices that finely render the interactions with virtual objects could further enhance sensorimotor recovery and skill transfer through multisensory integration.

ILLUSTRATION OF THE APPLICATION FROM THE USERS PERSPECTIVE: IMMERSIVE VR IN NEUROREHABILITATION SETTINGS

IMMERSIVE VR FOR A MORE DIRECT VISUOSPATIAL TRANSFORMATION

The use of VR during robotic neurorehabilitation intervention allows patients to perform functional task-oriented exercises and obtain concurrent visual feedback about their task performance. Goal-directed reaching movements in real life are enabled through visual depth cues. However, in conventional virtual environments, the available visual depth cues may be limited (e.g., when employing 2D screens), potentially increasing patients' cognitive effort and hampering their movement quality.

In a recent study, Wenk et al. (2019) compared the impact of novel, i.e., immersive VR HMD (HTC Vive, HTC, Taiwan and Valve, USA) and optical see-through AR HMD (Meta 2, Meta View, USA), vs. conventional visualization, i.e., computer screen, technologies on movement quality and users' cognitive load (Figure 5.1). Twenty healthy participants performed a reaching task in 3D space and a simultaneous cognitive counting task. The authors did not find significant differences across visualization technologies in the cognitive task performance and concluded that participants may have prioritized performance in the cognitive over the motor task. Indeed, better movement quality was observed when visualizing the movements in immersive VR compared to the computer screen, especially in movements that convey more than one dimension. Similarly, Gerig et al. (2018) reported that healthy participants performed significantly better reaching movements in immersive VR compared to conventional computer screens. Interestingly, Gerig et al. (2018) found that virtually recreated visual depth cues on the computer screen (e.g., aerial and linear perspective, occlusion, and shadows) had a minor impact on movement quality.

FIGURE 5.1 Experimental setup employed in Wenk et al. (2019) to compare the effect of different task visualization on movement quality and participants' cognitive load. Left: computer screen; Middle: immersive VR HMD (HTC Vive, HTC, Taiwan and Valve, USA); Right: optical see-through AR HMD (Meta 2, Meta View, USA). The HMDs allowed participants to visualize movements with a first-person perspective for a more direct visuospatial transformation.

Similar beneficial effects on movement quality were observed when the movement was visualized with AR. In the study by Wenk et al. (2019), a statistical trend suggested better movement quality in AR compared to computer screens. Similarly, in a clinical study with chronic stroke survivors (Mousavi Hondori et al., 2016), the movement quality and game performance were superior when a goal-directed reaching task was visualized in an AR projection on a surface compared to a conventional computer screen and a cursor. The authors concluded that the lower motor performance with the computer screen is in part due to the greater cognitive demands imposed by the computer screen visualization. However, no direct measurements of cognitive load were reported.

The relative benefits of AR visualization on neurorehabilitation, compared to immersive VR, are still inconclusive. The front-facing cameras in video see-through AR and transparent displays in optical see-through AR HMDs enable the visualization of the real environment (including the therapist), which might minimize collisions with the environment while still providing interactions with virtual objects. However, they also come with their drawbacks (e.g., viewpoint's misalignment, focal rivalry, etc.). In the study by Wenk et al. (2019), a statistical trend suggested better movement quality when performing a reaching task in immersive VR compared to AR. Further, Chicchi Giglioli et al. (2019) compared motor performance and psychophysiological measures (heart rate and skin conductance response) in healthy participants during a cooking task visualized through immersive VR HMD (HTC Vive, HTC, Taiwan and Valve, USA) or AR HMD (HoloLens, Microsoft, USA). Participants completed the task faster with immersive VR and physiological responses indicated that visualizing the task in AR produced more individual excitement (i.e., increased heart rate) and activation (i.e., increased electrodermal activity) than immersive VR, while immersive VR produced higher levels of sense of presence than AR. The authors concluded that current immersive VR represents VEs with greater realism, usability, and feasibility than AR, likely due to differences in the human–computer interaction, i.e., the wider field of view in the immersive VE and the use of controllers, which users were more familiar to interact with. In contrast, the natural interaction using the own hands could increase the user's interaction fidelity. This is in line with a recent study that reported better performance in a motor task that involved object selection and manipulation when using AR compared to immersive VR HMDs (Krichenbauer et al., 2018).

Together, recent research suggests that HMDs should be preferred over standard 2D visualization technologies to train multidimensional movements in VR. Task performance when using 2D screens is likely hampered by the limited depth perception and the increased cognitive demand and not just by participants' motor deficits/skills. However, more studies are needed to better evaluate differences in cognitive load across technologies.

First-Person Immersive Virtual Reality to Enhance Embodiment

Virtual reality can immerse participants in a computer-generated VE that they can explore and interact with. Advances in computer graphics and HMD-based VR systems, together with decreasing costs of associated equipment, have led scientists to consider immersive VR as a useful tool for conducting experimental studies with high ecological validity and generalizability in fields such as cognitive neuroscience and experimental psychology (Bohil et al., 2011). In particular, VR has been employed to systematically modulate the sense of "virtual body ownership", where the feeling of ownership over a virtual body is elicited. The use of immersive VR HMDs in the therapy of neurologic patients may be a powerful tool to promote motor learning and neurorehabilitation since body ownership and the motor system share neural correlates (Ehrsson et al., 2004, 2005; Wise, 1985; Zeller et al., 2016). Yet, in order to develop novel immersive VR-based training tasks to improve neurorehabilitation, it is important to understand how body ownership in the virtual training environment interacts with motor performance.

Using paradigms such as the rubber hand illusion (RHI; Botvinick and Cohen, 1998) or the virtual hand illusion (VHI; Slater et al., 2008), where body ownership over a fake/virtual hand is induced by applying visuo-tactile congruent information to the real and the fake/virtual hand, neuroimaging studies have found enhanced neural activation in the motor system when participants were observing an embodied fake hand that was moving spontaneously (Shibuya et al., 2018). Moreover, this neural activation correlated with the reported feeling of body ownership.

Few studies investigated if a higher level of body ownership enhances performance in motor tasks. Grechuta et al. (2017) modulated the level of virtual body ownership through visuo-tactile stimulation, i.e., a brush stroked the participants' fingers congruently or incongruently with a virtual brush stroking the avatar hand in the VE. Whenever the brush stroked the index finger of the real hand, participants were instructed to press a button with the other hand. The results showed faster response times with increasing level of reported body ownership. In a further study, Grechuta et al. (2019) asked participants to perform an air-hockey task in VR while receiving either congruent or incongruent auditory feedback. In line with their previous results, incongruent auditive feedback decreased the level of body ownership and worsened motor performance. Similarly, Shibuya et al. (2018) modulated the level of body ownership through visuo-motor feedback delays while participants draw circles at a given speed. They observed that, as the visual delay increased, body ownership decreased and motor performance worsened.

The integration of the various technologies necessary to induce virtual body ownership illusions is, with the currently more advanced technology, easily accessible. To achieve a full immersion, visual, auditory, and haptic displays, together with a head-tracking system, are required to induce the illusion of being in a place and that what occurs in this place is considered plausible by the participant (Slater, 2009). In their comprehensive paper, Spanlang et al. (2014) presented the technical infrastructure needed to induce virtual body ownership (Spanlang et al., 2014): a core VR system that represents the minimum infrastructure needed to induce the illusion and a list of extensions for more advanced experiences. The core system consists of a VR module (i.e., a computer with a graphics card that can handle the VE rendering together with the avatar animation and that integrates all other modules), a head-tracking module (to render the VE from the first-person perspective), and a display module (the HMD). These minimal requirements can be achieved with a computer with VR rendering capabilities (using game engines) and coupled with an off-the-shelf HMD (as they all incorporate a head-tracking system). Some HMDs (e.g., HTC Vive Focus, HTC, Taiwan and Valve, USA; Oculus Quest, Facebook, USA) integrate the three modules, significantly simplifying the set-up. However, special caution is needed before using smartphone-based HMD as many use inertial systems that only track the head's orientation but not its position.

The proposed extensions to the core system by Spanlang et al. (2014) include, among others, a full-body tracking system, haptics module for sensory stimulation, quantitative measurement tools to record physiological signals, and brain–computer interfaces (e.g., electroencephalography).

These extensions generally need dedicated computers to run at high frequency and exchange data with the VR module. However, current off-the-shelf HMD solutions already integrate some of these extra modules. For example, the HTC Vive and HTC Vive Pro (HTC, Taiwan and Valve, USA) allow the tracking of body parts through trackers that can be attached to the users' limbs. Other HMDs, e.g., Oculus Quest (Facebook, USA); HTC Vive Cosmo (HTC, Taiwan and Valve, USA), employ simultaneous localization and mapping (SLAM) to track the user's hands. Several haptic modules are now commercially available, e.g., hand exoskeletons to render haptic interactions with virtual elements (e.g., Dexmo, Dexta Robotics, China; SenseGlove, SenseGlove, Netherlands; VRgluv, VRgluv, USA; HaptX glove, HaptX Inc., USA) or gloves that provide tactile stimulation through vibrations (Senso, Senso Device Inc., USA; Hi5, Noitom International Inc., USA). Although some off-the-shelf VR solutions incorporate eye movement tracking systems, e.g., HTC Vive Pro Eye (HTC, Taiwan and Valve, USA), to date, no HMD devices incorporate physiological measurements and/or brain–computer interfaces.

Together, the few studies investigating the potential benefits of virtual body ownership on motor performance support the notion of a functional link between body ownership and motor brain areas. Enforcing body ownership by using immersive virtual training environments may, therefore, be a powerful tool to boost (re)learning of motor tasks in robotic neurorehabilitation of neurologic patients. The most advanced off-the-shelf HMDs might facilitate the integration of the required (core) technology to induce virtual body ownership illusions in neurorehabilitation interventions.

Individualized Immersive VR to Enhance Motivation and Engagement

Motivation and engagement are two closely connected constructs with a positive indirect (i.e., through an increased amount of practice) but also a direct effect on motor learning (by enhancing the consolidation process) (Lohse et al., 2016). Based on the *Flow Theory* (Csikszentmihalyi, 1990), VR-based therapeutic games have been developed to achieve an optimal motivation level to maximize recovery by incorporating algorithms that dynamically adapt the game difficulty to meet the patients' specific **motor disabilities** (Cameirão et al., 2010; Hocine et al., 2015; Wittmann et al., 2016). Adaptive difficulty adjustment also allows users with different skill levels to play together, with their therapist, or with family members, potentially also improving home-based rehabilitation (Baur et al., 2018). Multiplayer games (Figure 5.2) integrate the social aspect of motivation into the rehabilitation therapy and are a promising approach to increase patients' recovery (Baur et al., 2018; Goršič et al., 2017; Novak et al., 2014; Octavia and Coninx, 2014).

However, only few studies aimed at understanding how to best adapt the VE to the patients' specific **cognitive deficits**, besides the high prevalence of cognitive impairments among stroke survivors (Mellon et al., 2015; Patel and Birns, 2015). In commercial games for 2D screens (e.g., "Augmented Performance Feedback & Challenge Package", Hocoma, China) and immersive VR with HMD (Peruzzi et al., 2013), the level of the VE richness is left to the therapists' choice. These games offer a selection of poor VEs with fewer distractors (i.e., a small number of visual details) for severely impaired patients and richer VEs for less impaired patients who could engage in more complex cognitive tasks. The VE richness can be adapted by changing the number of distractors, i.e., using audiovisual environmental elements (e.g., bird's chirping, moving animals, weather), varying the size and amount of virtual objects, and/or the number of choices/interactions from the user (Peruzzi et al., 2013). In some studies, the VE richness was employed as a cognitive challenge to train cognition at the same time as (re)learning to walk (Mirelman et al., 2011). Using VR to train simultaneously various cognitive abilities was shown to not only improve some of the trained cognitive aspects but also to improve the mental well-being of patients (Maier et al., 2020). To the best of our knowledge, only one study (Levac et al., 2019b) investigated the effect of VE richness on motor learning, namely in children with cerebral palsy. No differences in motor performance, engagement, nor motivation were reported across different VE richness. However, further analyses regarding this study are still to be published. Finally, this study only changed the amount of

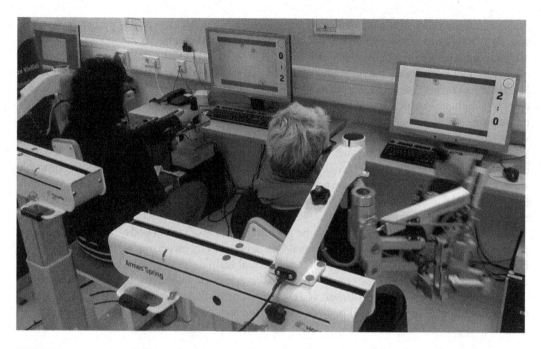

FIGURE 5.2 Multiplayer adaptive VR system. A mechanical device supports patients and allows them to interact with the VR. Adaptive difficulty adjustments in the VR allow patients with different degrees of impairment to play together a virtual air-hockey game targeting the social aspect of motivation. (Courtesy of the SMS lab at ETH Zurich.)

audiovisual stimuli and did not use a highly immersive display but a computer screen. Therefore, the experimental setup did not allow for full control of distractors (i.e., potential distractors in the real world were still visible) and it is difficult to conclude on the overall effect of VE richness on motor learning.

Together, a powerful and not yet exploited aspect of VR for neurorehabilitation relies on its possible adaptability to the patient's cognitive needs. Individually adapting task difficulty or VE richness to the patient's motor and/or cognitive skill level may be a promising tool to optimize task engagement and motivation, positively impacting therapy outcomes. HMDs provide almost full immersion in the VE and could facilitate the control of virtual and real distractors.

Enhancing Skill Transfer Through Multisensory Integration

A major advantage of VR is the possibility to realistically mimic real scenarios, supporting the recovery of trained function and transfer of skills (Levac et al., 2019a). Realism and, therefore, recovery and skill transfer could be further supported by complementing the visual information from the VR with other sensory information, namely, audio and haptics.

Most of the activities of daily living involve interaction with the environment, such as grasping and releasing objects. Research on neurorehabilitation has emphasized that somatosensory information, i.e., the information coming from our moving body parts about the interaction with the environment, plays a fundamental role in motor learning and error correction during movement execution (Schmidt and Lee, 2010) and is essential to induce brain plasticity (Poon, 2005). However, to date, the lack of somatosensory information during robot-assisted VR interventions compel participants to rely on visual information, leading to unrealistic and conflicting sensorimotor integrations compared to the real physical task, hampering motor performance and function recovery (Carey et al., 1997) and limiting skill transfer (de Mello Monteiro et al., 2014).

Considering its great potential, haptic rendering, together with immersive VR, could enhance the neurorehabilitation outcomes if implemented in robotic training setups as a standard. This would allow the training of functional tasks, such as grasping and releasing tangible objects with rendered virtual dynamics, stimulating the somatosensory system and reinforcing sensorimotor integration. However, only few studies have examined the role of haptic rendering in robotic-assisted neuro-rehabilitation (Gassert and Dietz, 2018; Metzger et al., 2012). Several robotic devices have been designed, either to assist severely affected patients to perform hand and arm movements (Hofmann et al., 2018; Klein et al., 2010; Nef et al., 2009; Randazzo et al., 2018; Rowe et al., 2017), or render haptic interactions ("Dexta Robotics - Touch the Untouchable"; "SenseGlove | Make the Digital Feel Real") in healthy users. However, there is only a limited number of robotic devices that can apply both robotic assistance and haptic rendering in an immersive VR. This can only be achieved with low inherent impedance/high transparent robotic devices (Gassert and Dietz, 2018; Marchal-Crespo et al., 2012; Özen et al., 2020).

An example is the Reactive Rope Robot (r^3) developed at ETH Zurich (Figure 5.3) (Zitzewitz et al., 2013). This versatile tendon-based robot system can provide robotic assistance and/or render realistic haptic interactions with virtual elements (e.g., water). The system is surrounded by three high-resolution screens that display visual information by rendering the same scene from three different perspectives. The robot is actuated through ropes, driven from motors located outside of the robot's workspace. Thus, it is especially suitable for VR integration, since the robot can render a large bandwidth of haptic interactions while ropes can hardly occlude the surrounding screens. The r^3 system has been employed in several studies to evaluate the effect of haptic rendering/guidance on motor learning (Basalp et al., 2019; Marchal-Crespo et al., 2013; Sigrist et al., 2015). In Basalp et al. (2019), authors trained healthy naïve participants in a sweep rowing task by providing

FIGURE 5.3 The tendon-based robotic system from ETH Zurich integrates immersive visual feedback and a tendon-based robotic device to train a sweep rowing task by varying the haptic rendering of the water–oar interaction. (Courtesy of the SMS lab at ETH Zurich.)

concurrent visual feedback about the desired and actual oar movement on the surrounding screens (Figure 5.3). An experimental group trained with different heaviness of the haptically rendered water–oar interaction (i.e., with different rendered water densities) that increased the task variability, while a control group performed the task under constant water density. The authors concluded that training with different water densities could enhance learning, highlighting the potential benefit of haptic rendering in motor learning and neurorehabilitation.

Together, the new generation of tailored robotic neurorehabilitation technologies should allow brain-injured patients to train functional activities of daily living in a realistic immersive VR with fine rendered haptic sensations to boost sensorimotor recovery and transfer of trained function.

DISCUSSION AND FUTURE TRENDS/DEVELOPMENTS

Rehabilitation robots can deliver high-intensity and consistent training, while VR offers the possibility to simulate different real or imaginary activities of daily living. However, to date, the use of robots together with VR in rehabilitation is not the standard of care and benefits are limited to a narrow group of patients with severe motor impairment and high sensory and cognitive abilities. Many VRs currently employed in clinics are on the one hand simplistic and unengaging but on the other hand cognitively too demanding for brain-injured patients, making them both tedious and clinically less effective.

Conventional robot-assisted exercises commonly involve goal-oriented tasks that are displayed on a 2D screen and trained via a symbolic virtual representation of the patient's limb. Although the virtual training environments and symbolic self-representation provide useful visual guidance through concurrent visual and/or auditory feedback, the lack of haptic sensation and depth cues results in the training of movements and object interactions that are far from those required in real conditions, limiting the transfer of acquired skills and function recovery.

Recent research suggests that the use of immersive VR systems experienced through HMDs may be especially beneficial to boost function recovery in brain-injured patients. In this chapter, we have discussed the major advantages of immersive VR for neurorehabilitation, highlighting open points and promising research directions and future trends.

First, VR HMDs offer immersion in a 3D virtual training environment. In contrast to the abstract or indirect MV in conventional 2D screens, immersive VRs offer depth cues and may reduce the cognitive load associated with visuospatial transformations. Several HMDs are currently available at a relatively low cost to provide a more immersive, realistic, and direct MV. However, few studies have investigated the specific benefits associated with different HMD types (e.g., immersive VR vs. AR) on neurorehabilitation. Yet, the optimal visualization technology may depend on several factors. The use of AR may be preferred if the interaction with the real environment (e.g., with the therapist) is important. In contrast, immersive VR may be the first choice to treat patients with severe cognitive deficits, since they allow to render exercises more naturally (fewer inconsistencies between the real and virtual world) and without real world's distractors, lowering the cognitive demand. Thus, the selection of a visualization technology has to be carefully matched with the therapy aim and the cognitive abilities of the patient. Nevertheless, more studies are needed to evaluate the benefits of different technologies on neurorehabilitation.

Second, immersive VR realistically mimics real and fictive scenarios. Crucially, training with a highly realistic first-person perspective avatar in an immersive VE may enhance body ownership over the virtual self-representation. Whereas few studies have shown that increased body ownership is associated with better motor performance, other psychological effects of embodying an avatar during robotic training have not yet been studied. For example, body ownership over the virtual arm may increase the patient's trust in the virtual representation. Similarly, embodying a virtual representation of the limb that is not visually attached to the robot may reduce biases toward robotic training and increase patients' compliance and movement quality.

Third, immersive VEs are highly controllable and versatile training environments. A major benefit of immersive VR relies on enforcing patients' engagement by adapting VEs to the patient's individual motor and cognitive capabilities. Adaptive difficulty adjustment also allows users with different skill levels to enroll in engaging multiplayer games. Different elements of the VR (e.g., difficulty level or VE richness) can be adapted to patients' abilities. However, up to date, it is still an open question which VE elements should be adapted (e.g., textures, sounds, light effects, animations) and how this adaptation should be performed. For example, severely cognitively impaired patients might benefit from a poor environment with few distractors, while less impaired patients could benefit from more stimuli in a richer environment. However, some impoverishment might also negatively impact therapy if, by removing too many stimuli, the environment becomes less coherent. HMDs provide almost complete control over all the delivered real and virtual audiovisual stimuli, boosting its potential to enhance neurorehabilitation in a broader neurologic population.

Finally, a promising advantage of robotic VR neurorehabilitation – and currently almost entirely neglected – is the combination of VR with somatosensory feedback, for example, provided by the robotic device. Somatosensory information plays a fundamental role in motor learning and is essential to induce brain plasticity. The addition of somatic sensation during the training of virtual tasks may be used to support sensorimotor integration, i.e., linking the visual information with the motor task and, therefore, more realistically mimicking real-world scenarios that might support skill transfer and recovery.

Together, VR has an enormous potential to boost the effectiveness of robotic neurorehabilitation. In this chapter, we have reviewed the state of the art of VR technology and its current use in conventional robotic-assisted rehabilitation. We highlighted the limitations of conventional VR therapy setups that rely on 2D screens and discussed in detail the rationale behind the use of immersive VR to enhance robotic-assisted interventions. The most effective robot-assisted VR therapies seem to be the ones that support active task engagement/motivation, multisensory signal integration, realistic scenarios, and full immersion. Therefore, immersive and adaptive virtual training environments using HMDs that enforce multisensory processing have the greatest potential to induce brain plasticity and motor recovery in brain-injured patients. With the recent fast development of HMD, we should expect an increase in their use in clinics within the next years.

REFERENCES

Adamovich, S.V., Merians, A.S., Boian, R., Tremaine, M., Burdea, G.S., Recce, M., Poizner, H., 2004. A virtual reality based exercise system for hand rehabilitation post-stroke: transfer to function. Conf. Proc. Annu. Int. Conf. IEEE Eng. Med. Biol. Soc. IEEE Eng. Med. Biol. Soc. Annu. Conf. 7, 4936–4939. https://doi.org/10.1109/IEMBS.2004.1404364

Aminov, A., Rogers, J.M., Middleton, S., Caeyenberghs, K., Wilson, P.H., 2018. What do randomized controlled trials say about virtual rehabilitation in stroke? A systematic literature review and meta-analysis of upper-limb and cognitive outcomes. J. Neuroeng. Rehabil. 15, 29. https://doi.org/10.1186/s12984-018-0370-2

Ballester, B.R., Nirme, J., Duarte, E., Cuxart, A., Rodriguez, S., Verschure, P., Duff, A., 2015. The visual amplification of goal-oriented movements counteracts acquired non-use in hemiparetic stroke patients. J. Neuroeng. Rehabil. 12, 50. https://doi.org/10.1186/s12984-015-0039-z

Basalp, E., Marchal-Crespo, L., Rauter, G., Riener, R., Wolf, P., 2019. Rowing simulator modulates water density to foster motor learning. Front. Robot. AI. 6. https://doi.org/10.3389/frobt.2019.00074

Baur, K., Schättin, A., de Bruin, E.D., Riener, R., Duarte, J.E., Wolf, P., 2018. Trends in robot-assisted and virtual reality-assisted neuromuscular therapy: a systematic review of health-related multiplayer games. J. Neuroeng. Rehabil. 15, 107. https://doi.org/10.1186/s12984-018-0449-9

Bezerra, Í.M.P., Crocetta, T.B., Massetti, T., da Silva, T.D., Guarnieri, R., Meira, C.M., Arab, C., de Abreu, L.C., de Araujo, L.V., de Mello Monteiro, C.B., 2018. Functional performance comparison between real and virtual tasks in older adults. Medicine (Baltimore). 97. https://doi.org/10.1097/MD.0000000000009612

Blanke, O., 2012. Multisensory brain mechanisms of bodily self-consciousness. Nat. Rev. Neurosci. 13, 556–571. https://doi.org/10.1038/nrn3292

Boeldt, D., McMahon, E., McFaul, M., Greenleaf, W., 2019. Using virtual reality exposure therapy to enhance treatment of anxiety disorders: identifying areas of clinical adoption and potential obstacles. *Front. Psychiatry.* 10. https://doi.org/10.3389/fpsyt.2019.00773

Bohil, C.J., Alicea, B., Biocca, F.A., 2011. Virtual reality in neuroscience research and therapy. *Nat. Rev. Neurosci.* 12, 752–762. https://doi.org/10.1038/nrn3122

Botvinick, M., Cohen, J., 1998. Rubber hands 'feel' touch that eyes see. *Nature.* 391, 756–756. https://doi.org/10.1038/35784

Braun, N., Debener, S., Spychala, N., Bongartz, E., Sörös, P., Müller, H.H.O., Philipsen, A., 2018. The senses of agency and ownership: a review. *Front. Psychol.* 9. https://doi.org/10.3389/fpsyg.2018.00535

Brütsch, K., Schuler, T., Koenig, A., Zimmerli, L., Koeneke, S.M., Lünenburger, L., Riener, R., Jäncke, L., Meyer-Heim, A., 2010. Influence of virtual reality soccer game on walking performance in robotic assisted gait training for children. *J. Neuroeng. Rehabil.* 7, 15. https://doi.org/10.1186/1743-0003-7-15

Cameirão, M.S., Badia, S.B.i., Oller, E.D., Verschure, P.F., 2010. Neurorehabilitation using the virtual reality based rehabilitation gaming system: methodology, design, psychometrics, usability and validation. *J. Neuroeng. Rehabil.* 7, 48. https://doi.org/10.1186/1743-0003-7-48

Carey, L.M., Oke, L.E., Matyas, T.A., 1997. Impaired touch discrimination after stroke: a quantitative test. *J. Neurol. Rehabil.* 11, 219–232. https://doi.org/10.1177/154596839701100404

Chicchi Giglioli, I., Bermejo Vidal, C., Alcañiz Raya, M., 2019. A virtual versus an augmented reality cooking task based-tools: a behavioral and physiological study on the assessment of executive functions. *Front Psychol.* 10, 1–12. https://doi.org/10.3389/fpsyg.2019.02529

Condino, S., Carbone, M., Piazza, R., Ferrari, M., Ferrari, V., 2019. Perceptual limits of optical see-through visors for augmented reality guidance of manual tasks. *IEEE Trans. Biomed. Eng.* 9294, 1. https://doi.org/10.1109/TBME.2019.2914517

Cramer, S.C., Sur, M., Dobkin, B.H., O'Brien, C., Sanger, T.D., Trojanowski, J.Q., Rumsey, J.M., Hicks, R., Cameron, J., Chen, D., Chen, W.G., Cohen, L.G., deCharms, C., Duffy, C.J., Eden, G.F., Fetz, E.E., Filart, R., Freund, M., Grant, S.J., Haber, S., Kalivas, P.W., Kolb, B., Kramer, A.F., Lynch, M., Mayberg, H.S., McQuillen, P.S., Nitkin, R., Pascual-Leone, A., Reuter-Lorenz, P., Schiff, N., Sharma, A., Shekim, L., Stryker, M., Sullivan, E.V., Vinogradov, S., 2011. Harnessing neuroplasticity for clinical applications. *Brain J. Neurol.* 134, 1591–1609. https://doi.org/10.1093/brain/awr039

Crittenden, K.S., Deci, E.L., 1982. The psychology of self-determination. *Contemp. Sociol.* 11, 343. https://doi.org/10.2307/2067164

Csikszentmihalyi, M., 2014. Toward a psychology of optimal experience, in: *Flow and the Foundations of Positive Psychology.* Dordrecht: Springer, pp. 209–226. https://doi.org/10.1007/978-94-017-9088-8_14

Csikszentmihalyi, M., 1990. *The Psychology of Optimal Experience.* New York: Harper and Row.

Cutolo, F., Fontana, U., Ferrari, V., 2018. Perspective preserving solution for quasi-orthoscopic video see-through HMDs. *Technologies.* 6, 9. https://doi.org/10.3390/technologies6010009

de Mello Monteiro, C.B., Massetti, T., da Silva, T.D., van der Kamp, J., de Abreu, L.C., Leone, C., Savelsbergh, G.J.P., 2014. Transfer of motor learning from virtual to natural environments in individuals with cerebral palsy. *Res. Dev. Disabil.* 35, 2430–2437. https://doi.org/10.1016/j.ridd.2014.06.006

Debarba, H.G., Bovet, S., Salomon, R., Blanke, O., Herbelin, B., Boulic, R., 2017. Characterizing first and third person viewpoints and their alternation for embodied interaction in virtual reality. *PLOS One.* 12, e0190109. https://doi.org/10.1371/journal.pone.0190109

Deutsch, J.E., Merians, A.S., Adamovich, S., Poizner, H., Burdea, G.C., 2004. Development and application of virtual reality technology to improve hand use and gait of individuals post-stroke. *Restor. Neurol. Neurosci.* 22, 371–386.

Dexta Robotics - Touch the Untouchable [WWW Document], n.d. URL https://origin.dextarobotics.com/en-us/#product (accessed 10.13.19).

Djaouti, D., Alvarez, J., Jessel, J.-P., 2011. Classifying serious games: the G/P/S model, in: Felicia, P. (Ed.), *Handbook of Research on Improving Learning and Motivation through Educational Games: Multidisciplinary Approaches.* Hershey: IGI Global, pp. 118–136. https://doi.org/10.4018/978-1-60960-495-0.ch006

Dockx, K., Bekkers, E.M., Van den Bergh, V., Ginis, P., Rochester, L., Hausdorff, J.M., Mirelman, A., Nieuwboer, A., 2016. Virtual reality for rehabilitation in Parkinson's disease. *Cochrane Database Syst. Rev.* 2016. https://doi.org/10.1002/14651858.CD010760.pub2

Domínguez-Téllez, P., Moral-Muñoz, J.A., Salazar, A., Casado-Fernández, E., Lucena-Antón, D., 2020. Game-based virtual reality interventions to improve upper limb motor function and quality of life after stroke: systematic review and meta-analysis. *Games Health J.* 9, 1–10. https://doi.org/10.1089/g4h.2019.0043

Ehrsson, H.H., Holmes, N.P., Passingham, R.E., 2005. Touching a rubber hand: feeling of body ownership is associated with activity in multisensory brain areas. *J. Neurosci. Off. J. Soc. Neurosci.* 25, 10564–10573. https://doi.org/10.1523/JNEUROSCI.0800-05.2005

Ehrsson, H.H., Spence, C., Passingham, R.E., 2004. That's my hand! Activity in premotor cortex reflects feeling of ownership of a limb. *Science.* 305, 875–877. https://doi.org/10.1126/science.1097011

Ernst, M.O., Bülthoff, H.H., 2004. Merging the senses into a robust percept. *Trends Cogn. Sci.* 8, 162–169. https://doi.org/10.1016/j.tics.2004.02.002

Ferguson, C., Shade, M.Y., Blaskewicz Boron, J., Lyden, E., Manley, N.A., 2020. Virtual reality for therapeutic recreation in dementia hospice care: a feasibility study. *Am. J. Hosp. Palliat. Med.* 1049909120901525. https://doi.org/10.1177/1049909120901525

Ferreira dos Santos, L., Christ, O., Mate, K., Schmidt, H., Krüger, J., Dohle, C., 2016. Movement visualisation in virtual reality rehabilitation of the lower limb: a systematic review. *Biomed. Eng. Online.* 15, 144. https://doi.org/10.1186/s12938-016-0289-4

Feyzioğlu, Ö., Dinçer, S., Akan, A., Algun, Z.C., 2020. Is Xbox 360 Kinect-based virtual reality training as effective as standard physiotherapy in patients undergoing breast cancer surgery? Support. *Care Cancer.* https://doi.org/10.1007/s00520-019-05287-x

Freeman, D., Reeve, S., Robinson, A., Ehlers, A., Clark, D., Spanlang, B., Slater, M., 2017. Virtual reality in the assessment, understanding, and treatment of mental health disorders. *Psychol. Med.* 47, 2393–2400. https://doi.org/10.1017/S003329171700040X

Freivogel, S., Schmalohr, D., Mehrholz, J., 2009. Improved walking ability and reduced therapeutic stress with an electromechanical gait device. *J. Rehabil. Med.* 41, 734–739. https://doi.org/10.2340/16501977-0422

French, B., Thomas, L.H., Coupe, J., McMahon, N.E., Connell, L., Harrison, J., Sutton, C.J., Tishkovskaya, S., Watkins, C.L., 2016. Repetitive task training for improving functional ability after stroke. *Cochrane Database Syst. Rev.* 11, CD006073. https://doi.org/10.1002/14651858.CD006073.pub3

Fu, M.J., Knutson, J.S., Chae, J., 2015. Stroke rehabilitation using virtual environments. *Phys. Med. Rehabil. Clin. N. Am.* 26, 747–757. https://doi.org/10.1016/j.pmr.2015.06.001

Gassert, R., Dietz, V., 2018. Rehabilitation robots for the treatment of sensorimotor deficits: a neurophysiological perspective. *J. Neuroeng. Rehabil.* 15, 46. https://doi.org/10.1186/s12984-018-0383-x

Gates, N.J., Vernooij, R.W., Di Nisio, M., Karim, S., March, E., Martínez, G., Rutjes, A.W., 2019. Computerised cognitive training for preventing dementia in people with mild cognitive impairment. *Cochrane Database Syst. Rev.* 3, CD012279. https://doi.org/10.1002/14651858.CD012279.pub2

Gerber, S.M., Jeitziner, M.-M., Knobel, S.E.J., Mosimann, U.P., Müri, R.M., Jakob, S.M., Nef, T., 2019. Perception and performance on a virtual reality cognitive stimulation for use in the intensive care unit: a non-randomized trial in critically ill patients. *Front. Med.* 6. https://doi.org/10.3389/fmed.2019.00287

Gerber, S.M., Jeitziner, M.-M., Wyss, P., Chesham, A., Urwyler, P., Müri, R.M., Jakob, S.M., Nef, T., 2017. Visuo-acoustic stimulation that helps you to relax: a virtual reality setup for patients in the intensive care unit. *Sci. Rep.* 7, 13228. https://doi.org/10.1038/s41598-017-13153-1

Gerig, N., Mayo, J., Baur, K., Wittmann, F., Riener, R., Wolf, P., 2018. Missing depth cues in virtual reality limit performance and quality of three dimensional reaching movements. *PLoS ONE.* 19, 1–18. https://doi.org/10.1371/journal.pone.0189275

Gobron, S.C., Zannini, N., Wenk, N., Schmitt, C., Charrotton, Y., Fauquex, A., Lauria, M., Degache, F., Frischknecht, R., 2015. Serious games for rehabilitation using head-mounted display and haptic devices, in: De Paolis, L.T., Mongelli, A. (Eds.), *Augmented and Virtual Reality.* Cham: Springer International Publishing, pp. 199–219. https://doi.org/10.1007/978-3-319-22888-4_15

Gonçalves, R., Pedrozo, A.L., Coutinho, E.S.F., Figueira, I., Ventura, P., 2012. Efficacy of virtual reality exposure therapy in the treatment of PTSD: a systematic review. *PLoS One.* 7. https://doi.org/10.1371/journal.pone.0048469

Goršič, M., Cikajlo, I., Goljar, N., Novak, D., 2017. A multisession evaluation of an adaptive competitive arm rehabilitation game. *J. Neuroeng. Rehabil.* 14, 128. https://doi.org/10.1186/s12984-017-0336-9

Grechuta, K., Guga, J., Maffei, G., Ballester, B.R., Verschure, P.F.M.J., 2017. Visuotactile integration modulates motor performance in a perceptual decision-making task. *Sci. Rep.* 7, 3333. https://doi.org/10.1038/s41598-017-03488-0

Grechuta, K., Ulysse, L., Rubio Ballester, B., Verschure, P.F.M.J., 2019. Self Beyond the Body: Action-Driven and Task-Relevant Purely Distal Cues Modulate Performance and Body Ownership. *Front. Hum Neurosci.* 13, 1–12. https://doi.org/10.3389/fnhum.2019.00091

Guadagnoli, M.A., Lee, T.D., 2004. Challenge point: a framework for conceptualizing the effects of various practice conditions in motor learning. *J. Motor Behav.* 36, 212–224. https://doi.org/10.3200/JMBR.36.2.212-224

Hirota, K., 2020. Preoperative management and postoperative delirium. *J. Anesth.* 34, 1–4. https://doi. org/10.1007/s00540-019-02660-2

Hocine, N., Gouaïch, A., Cerri, S.A., Mottet, D., Froger, J., Laffont, I., 2015. Adaptation in serious games for upper-limb rehabilitation: an approach to improve training outcomes. *User Model. User Adapt. Interact.* 25, 65–98. https://doi.org/10.1007/s11257-015-9154-6

Hoffman, H.G., Chambers, G.T., Meyer, W.J., Arceneaux, L.L., Russell, W.J., Seibel, E.J., Richards, T.L., Sharar, S.R., Patterson, D.R., 2011. Virtual reality as an adjunctive non-pharmacologic analgesic for acute burn pain during medical procedures. *Ann. Behav. Med. Publ. Soc. Behav. Med.* 41, 183–191. https://doi.org/10.1007/s12160-010-9248-7

Hofmann, U.A.T., Bützer, T., Lambercy, O., Gassert, R., 2018. Design and evaluation of a Bowden-cable-based remote actuation system for wearable robotics. *IEEE Robot. Autom. Lett.* 3, 2101–2108. https:// doi.org/10.1109/LRA.2018.2809625

Holden, M.K., 2005. Virtual environments for motor rehabilitation: review. *Cyberpsychol. Behav.* 8, 187–211. https://doi.org/10.1089/cpb.2005.8.187

Holmes, D., Charles, D., Morrow, P., McClean, S., McDonough, S., 2015. Rehabilitation Game Model for Personalised Exercise, in: 2015 International Conference on Interactive Technologies and Games. Presented at the 2015 International Conference on Interactive Technologies and Games (iTAG), IEEE, Nottingham, United Kingdom, pp. 41–48. https://doi.org/10.1109/iTAG.2015.11

Huygelier, H., Schraepen, B., van Ee, R., Vanden Abeele, V., Gillebert, C.R., 2019. Acceptance of immersive head-mounted virtual reality in older adults. *Sci. Rep.* 9, 4519. https://doi.org/10.1038/s41598-019-41200-6

Israel, J.F., Campbell, D.D., Kahn, J.H., Hornby, T.G., 2006. Metabolic costs and muscle activity patterns during robotic- and therapist-assisted treadmill walking in individuals with incomplete spinal cord injury. *Phys. Ther.* 86, 1466–1478. https://doi.org/10.2522/ptj.20050266

Jang, S.H., You, S.H., Hallett, M., Cho, Y.W., Park, C.-M., Cho, S.-H., Lee, H.-Y., Kim, T.-H., 2005. Cortical reorganization and associated functional motor recovery after virtual reality in patients with chronic stroke: an experimenter-blind preliminary study. *Arch. Phys. Med. Rehabil.* 86, 2218–2223. https://doi. org/10.1016/j.apmr.2005.04.015

Janssen, H., Ada, L., Bernhardt, J., McElduff, P., Pollack, M., Nilsson, M., Spratt, N.J., 2014. An enriched environment increases activity in stroke patients undergoing rehabilitation in a mixed rehabilitation unit: a pilot non-randomized controlled trial. *Disabil. Rehabil.* 36, 255–262. https://doi.org/10.3109/ 09638288.2013.788218

Kilteni, K., Groten, R., Slater, M., 2012. The sense of embodiment in virtual reality. *Presence Teleoper. Virt. Environ.* 21, 373–387. https://doi.org/10.1162/PRES_a_00124

Kim, O., Pang, Y., Kim, J.-H., 2019. The effectiveness of virtual reality for people with mild cognitive impairment or dementia: a meta-analysis. *BMC Psychiatry.* 19. https://doi.org/10.1186/s12888-019-2180-x

Klein, J., Spencer, S., Allington, J., Bobrow, J.E., Reinkensmeyer, D., 2010. Optimization of a parallel shoulder mechanism to achieve a high-force, low-mass, robotic-arm exoskeleton. *IEEE Trans. Robot.* 26, 710–715. https://doi.org/10.1109/TRO.2010.2052170

Krichenbauer, M., Yamamoto, G., Taketom, T., Sandor, C., Kato, H., 2018. Augmented reality versus virtual reality for 3D object manipulation. *IEEE Trans. Vis. Comput. Graph.* 24, 1038–1048. https://doi. org/10.1109/TVCG.2017.2658570

Krishnan, C., Ranganathan, R., Dhaher, Y.Y., Rymer, W.Z., 2013. A pilot study on the feasibility of robot-aided leg motor training to facilitate active participation. *PLoS One.* 8, e77370. https://doi.org/10.1371/ journal.pone.0077370

Krishnan, C., Ranganathan, R., Kantak, S.S., Dhaher, Y.Y., Rymer, W.Z., 2012. Active robotic training improves locomotor function in a stroke survivor. *J. Neuroeng. Rehabil.* 9, 57. https://doi. org/10.1186/1743-0003-9-57

Kwakkel, G., van Peppen, R., Wagenaar, R.C., Wood Dauphinee, S., Richards, C., Ashburn, A., Miller, K., Lincoln, N., Partridge, C., Wellwood, I., Langhorne, P., 2004. Effects of augmented exercise therapy time after stroke: a meta-analysis. *Stroke.* 35, 2529–2539. https://doi.org/10.1161/01.STR.0000143153.76460.7d

Laver, K.E., Lange, B., George, S., Deutsch, J.E., Saposnik, G., Crotty, M., 2017. Virtual reality for stroke rehabilitation. *Cochrane Database Syst. Rev.* https://doi.org/10.1002/14651858.CD008349.pub4

LaViola, J.J., 2000. A discussion of cybersickness in virtual environments. *ACM SIGCHI Bull.* 32, 47–56. https://doi.org/10.1145/333329.333344

Lee, J.H., Ku, J., Cho, W., Hahn, W.Y., Kim, I.Y., Lee, S.-M., Kang, Y., Kim, D.Y., Yu, T., Wiederhold, B.K., Wiederhold, M.D., Kim, S.I., 2003. A virtual reality system for the assessment and rehabilitation of the activities of daily living. *Cyberpsychol. Behav. Impact Internet Multimed. Virt. Real. Behav. Soc.* 6, 383–388. https://doi.org/10.1089/109493103322278763

Levac, D.E., Huber, M.E., Sternad, D., 2019a. Learning and transfer of complex motor skills in virtual reality: a perspective review. *J. Neuroeng. Rehabil.* 16, 121. https://doi.org/10.1186/s12984-019-0587-8

Levac, D.E., Jovanovic, B.B., 2017. Is children's motor learning of a postural reaching task enhanced by practice in a virtual environment?, in: 2017 International Conference on Virtual Rehabilitation (ICVR). Presented at the 2017 International Conference on Virtual Rehabilitation (ICVR), pp. 1–7. https://doi.org/10.1109/ICVR.2017.8007489

Levac, D.E., Taylor, M.M., Payne, B., Ward, N., 2019b. Influence of virtual environment complexity on motor learning in typically developing children and children with cerebral palsy, in: 2019 International Conference on Virtual Rehabilitation (ICVR). Presented at the 2019 International Conference on Virtual Rehabilitation (ICVR), IEEE, Tel Aviv, Israel, pp. 1–7. https://doi.org/10.1109/ICVR46560.2019.8994487

Liebermann, D.G., Katz, L., Hughes, M.D., Bartlett, R.M., McClements, J., Franks, I.M., 2002. Advances in the application of information technology to sport performance. *J. Sports Sci.* 20, 755–769. https://doi.org/10.1080/026404102320675611

Lohse, K.R., Boyd, L.A., Hodges, N.J., 2016, Engaging environments enhance motor skill learning in a computer gaming task. *J. Motor Behav.* 48, 172–182. https://doi.org/10.1080/00222895.2015.1068158

Longo, M.R., Schüür, F., Kammers, M.P.M., Tsakiris, M., Haggard, P., 2008. What is embodiment? A psychometric approach. *Cognition.* 107, 978–998. https://doi.org/10.1016/j.cognition.2007.12.004

Maclean, N., Pound, P., 2000. A critical review of the concept of patient motivation in the literature on physical rehabilitation. *Soc. Sci. Med. 1982* 50, 495–506. https://doi.org/10.1016/s0277-9536(99)00334-2

Maclean, N., Pound, P., Wolfe, C., Rudd, A., 2000. Qualitative analysis of stroke patients' motivation for rehabilitation. *BMJ.* 321, 1051–1054.

Maier, M., Ballester, B.R., Leiva Bañuelos, N., Duarte Oller, E., Verschure, P.F.M.J., 2020. Adaptive conjunctive cognitive training (ACCT) in virtual reality for chronic stroke patients: a randomized controlled pilot trial. *J. Neuroeng. Rehabil.* 17, 42. https://doi.org/10.1186/s12984-020-0652-3

Maier, M., Rubio Ballester, B., Duff, A., Duarte Oller, E., Verschure, P.F.M.J., 2019. Effect of specific over nonspecific VR-based rehabilitation on poststroke motor recovery: a systematic meta-analysis. *Neurorehabil. Neural Repair.* 33, 112–129. https://doi.org/10.1177/1545968318820169

Marchal-Crespo, L., Michels, L., Jaeger, L., Lopez-Oloriz, J., Riener, R., 2017. Effect of error augmentation on brain activation and motor learning of a complex locomotor task. *Front. Neurosci.* 11. https://doi.org/10.3389/fnins.2017.00526

Marchal-Crespo, L., Rauter, G., Wyss, D., Zitzewitz, J. von, Riener, R., 2012. Synthesis and control of an assistive robotic tennis trainer, in: 2012 4th IEEE RAS EMBS International Conference on Biomedical Robotics and Biomechatronics (BioRob). Presented at the 2012 4th IEEE RAS EMBS International Conference on Biomedical Robotics and Biomechatronics (BioRob), pp. 355–360. https://doi.org/10.1109/BioRob.2012.6290262

Marchal-Crespo, L., Reinkensmeyer, D.J., 2008. Haptic guidance can enhance motor learning of a steering task. *J. Motor Behav.* 40, 545–556. https://doi.org/10.3200/JMBR.40.6.545-557

Marchal-Crespo, L., Reinkensmeyer, D.J., 2009. Review of control strategies for robotic movement training after neurologic injury. *J. Neuroeng. Rehabil.* 6, 20. https://doi.org/10.1186/1743-0003-6-20

Marchal-Crespo, L., Schneider, J., Jaeger, L., Riener, R., 2014. Learning a locomotor task: with or without errors? *J. Neuroeng. Rehabil.* 11, 25. https://doi.org/10.1186/1743-0003-11-25

Marchal-Crespo, L., Tsangaridis, P., Obwegeser, D., Maggioni, S., Riener, R., 2019. Haptic error modulation outperforms visual error amplification when learning a modified gait pattern. *Front. Neurosci.* 13. https://doi.org/10.3389/fnins.2019.00061

Marchal-Crespo, L., van Raai, M., Rauter, G., Wolf, P., Riener, R., 2013. The effect of haptic guidance and visual feedback on learning a complex tennis task. *Exp. Brain Res.* 231, 277–291. https://doi.org/10.1007/s00221-013-3690-2

Martini, M., 2016. Real, rubber or virtual: the vision of "one's own" body as a means for pain modulation. a narrative review. *Conscious. Cogn.* 43, 143–151. https://doi.org/10.1016/j.concog.2016.06.005

Matamala-Gomez, M., Nierula, B., Donegan, T., Slater, M., Sanchez-Vives, M.V., 2020. Manipulating the perceived shape and color of a virtual limb can modulate pain responses. *J. Clin. Med.* 9, 291. https://doi.org/10.3390/jcm9020291

Mellon, L., Brewer, L., Hall, P., Horgan, F., Williams, D., Hickey, A., ASPIRE-S Study Group, 2015. Cognitive impairment six months after ischaemic stroke: a profile from the ASPIRE-S study. *BMC Neurol.* 15, 31. https://doi.org/10.1186/s12883-015-0288-2

Mercier, L., Audet, T., Hébert, R., Rochette, A., Dubois, M.F., 2001. Impact of motor, cognitive, and perceptual disorders on ability to perform activities of daily living after stroke. *Stroke.* 32, 2602–2608.

Metzger, J.C., Lambercy, O., Gassert, R., 2012. High-fidelity rendering of virtual objects with the ReHapticKnob - novel avenues in robot-assisted rehabilitation of hand function, in: 2012 IEEE Haptics Symposium (HAPTICS). Presented at the 2012 IEEE Haptics Symposium (HAPTICS), pp. 51–56. https://doi.org/10.1109/HAPTIC.2012.6183769

Milot, M.-H., Marchal-Crespo, L., Green, C.S., Cramer, S.C., Reinkensmeyer, D.J., 2010. Comparison of error-amplification and haptic-guidance training techniques for learning of a timing-based motor task by healthy individuals. *Exp. Brain Res.* 201, 119–131. https://doi.org/10.1007/s00221-009-2014-z

Mirelman, A., Maidan, I., Herman, T., Deutsch, J.E., Giladi, N., Hausdorff, J.M., 2011. Virtual reality for gait training: can it induce motor learning to enhance complex walking and reduce fall risk in patients with Parkinson's disease? *J. Gerontol. A. Biol. Sci. Med. Sci.* 66A, 234–240. https://doi.org/10.1093/gerona/glq201

Moreno, A., Wall, K.J., Thangavelu, K., Craven, L., Ward, E., Dissanayaka, N.N., 2019. A systematic review of the use of virtual reality and its effects on cognition in individuals with neurocognitive disorders. *Alzheimers Dement. Transl. Res. Clin. Interv.* 5, 834–850. https://doi.org/10.1016/j.trci.2019.09.016

Mousavi Hondori, H., Khademi, M., Dodakian, L., McKenzie, A., Lopes, C.V., Cramer, S.C., 2016. Choice of human–computer interaction mode in stroke rehabilitation. *Neurorehabil. Neural Repair.* 30, 258–265. https://doi.org/10.1177/1545968315593805

Murray, C.D., Patchick, E.L., Caillette, F., Howard, T., Pettifer, S., 2006. Can immersive virtual reality reduce phantom limb pain? *Stud. Health Technol. Inform.* 119, 407–412.

Nef, T., Guidali, M., Riener, R., 2009. ARMin III: arm therapy exoskeleton with an ergonomic shoulder actuation. *Appl. Bionics Biomech.* 6, 127–142. https://doi.org/10.1080/11762320902840179

Nithiananatharajah, J., Hannan, A.J., 2006. Enriched environments, experience-dependent plasticity and disorders of the nervous system. *Nat. Rev. Neurosci.* 7, 697–709. https://doi.org/10.1038/nrn1970

Novak, D., Nagle, A., Keller, U., Riener, R., 2014. Increasing motivation in robot-aided arm rehabilitation with competitive and cooperative gameplay. *J. Neuroeng. Rehabil.* 11, 64. https://doi.org/10.1186/1743-0003-11-64

Octavia, J.R., Coninx, K., 2014. Adaptive personalized training games for individual and collaborative rehabilitation of people with multiple sclerosis. *BioMed. Res. Int.* 2014, 345728. https://doi.org/10.1155/2014/345728

Oing, T., Prescott, J., 2018. Implementations of virtual reality for anxiety-related disorders: systematic review. *JMIR Serious Games.* 6. https://doi.org/10.2196/10965

Osumi, M., Sano, Y., Ichinose, A., Wake, N., Yozu, A., Kumagaya, S.-I., Kuniyoshi, Y., Morioka, S., Sumitani, M., 2020. Direct evidence of EEG coherence in alleviating phantom limb pain by virtual referred sensation: case report. *Neurocase.* 26, 55–59. https://doi.org/10.1080/13554794.2019.1696368

Özen, Ö., Penalver-Andres, J., Ortega, E.V., Buetler, K.A., Marchal-Crespo, L., 2020. Haptic Rendering Modulates Task Performance, Physical Effort and Movement Strategy during Robot-Assisted Training. 2020 8th IEEE RAS/EMBS International Conference for Biomedical Robotics and Biomechatronics (BioRob), 1223–1228. https://doi.org/10.1109/BioRob49111.2020.9224317.

Patel, B., Birns, J., 2015. *Post-Stroke Cognitive Impairment, in: Management of Post-Stroke Complications.* Cham: Springer International Publishing, pp. 277–306. https://doi.org/10.1007/978-3-319-17855-4_12

Patton, J.L., Wei, Y.J., Bajaj, P., Scheidt, R.A., 2013. Visuomotor learning enhanced by augmenting instantaneous trajectory error feedback during reaching. *PLoS One.* 8, e46466. https://doi.org/10.1371/journal.pone.0046466

Perez-Marcos, D., Bieler-Aeschlimann, M., Serino, A., 2018. Virtual reality as a vehicle to empower motor-cognitive neurorehabilitation. *Front. Psychol.* 9, 1–8. https://doi.org/10.3389/fpsyg.2018.02120

Perry, B.N., Armiger, R.S., Wolde, M., McFarland, K.A., Alphonso, A.L., Monson, B.T., Pasquina, P.F., Tsao, J.W., 2018. Clinical trial of the virtual integration environment to treat phantom limb pain with upper extremity amputation. *Front. Neurol.* 9. https://doi.org/10.3389/fneur.2018.00770

Peruzzi, A., Cereatti, A., Mirelman, A., Della Croce, U., 2013. Feasibility and acceptance of a virtual reality system for gait training of individuals with multiple sclerosis. *Eur. Int. J. Sci. Technol.* 2, 171–181.

Pollock, A., Farmer, S.E., Brady, M.C., Langhorne, P., Mead, G.E., Mehrholz, J., van Wijck, F., 2014. Interventions for improving upper limb function after stroke. *Cochrane Database Syst. Rev.* https://doi.org/10.1002/14651858.CD010820.pub2

Poon, C.-S., 2005. Sensorimotor learning and information processing by Bayesian internal models 4481–4482. https://doi.org/10.1109/iembs.2004.1404245

Poon, C.-S., 2004. Sensorimotor learning and information processing by Bayesian internal models. Conf. Proc. Annu. Int. Conf. IEEE Eng. Med. Biol. Soc. IEEE Eng. Med. Biol. Soc. Annu. Conf. 6, 4481–4482. https://doi.org/10.1109/IEMBS.2004.1404245

Pourmand, A., Davis, S., Marchak, A., Whiteside, T., Sikka, N., 2018. Virtual reality as a clinical tool for pain management. *Curr. Pain Headache Rep.* 22, 53. https://doi.org/10.1007/s11916-018-0708-2

Putrino, D., Zanders, H., Hamilton, T., Rykman, A., Lee, P., Edwards, D.J., 2017. Patient engagement is related to impairment reduction during digital game-based therapy in stroke. *Games Health J.* 6, 295–302. https://doi.org/10.1089/g4h.2016.0108

Randazzo, L., Iturrate, I., Perdikis, S., Millán, J. d R., 2018. Mano: a wearable hand exoskeleton for activities of daily living and neurorehabilitation. *IEEE Robot. Autom. Lett.* 3, 500–507. https://doi.org/10.1109/LRA.2017.2771329

Rohrbach, N., Chicklis, E., Levac, D.E., 2019. What is the impact of user affect on motor learning in virtual environments after stroke? A scoping review. *J. Neuroeng. Rehabil.* 16, 79. https://doi.org/10.1186/s12984-019-0546-4

Rose, F.D., Brooks, Barbara.M., Rizzo, A.A., 2005. Virtual reality in brain damage rehabilitation: review. *Cyberpsychol. Behav.* 8, 241–262. https://doi.org/10.1089/cpb.2005.8.241

Rose, T., Nam, C.S., Chen, K.B., 2018. Immersion of virtual reality for rehabilitation - review. *Appl. Ergon.* 69, 153–161. https://doi.org/10.1016/j.apergo.2018.01.009

Rossini, P.M., Dal Forno, G., 2004. Integrated technology for evaluation of brain function and neural plasticity. *Phys. Med. Rehabil. Clin. N. Am.* 15, 263–306.

Rowe, J.B., Chan, V., Ingemanson, M.L., Cramer, S.C., Wolbrecht, E.T., Reinkensmeyer, D.J., 2017. Robotic assistance for training finger movement using a Hebbian model: a randomized controlled trial. *Neurorehabil. Neural Repair.* 31, 769–780. https://doi.org/10.1177/1545968317721975

Sanchez-Vives, M.V., Slater, M., 2005. From presence to consciousness through virtual reality. *Nat. Rev. Neurosci.* 6, 332–339. https://doi.org/10.1038/nrn1651

Schmidt, R., Lee, T., 2010. *Motor Control and Learning: A Behavioral Emphasis.* Champaign, IL: Human Kinetics Publishers, 2005.

SenseGlove | Make the Digital Feel Real [WWW Document], n.d. Senseglove. URL https://www.senseglove.com (accessed 10.13.19).

Shea, C.H., Wulf, G., 1999. Enhancing motor learning through external-focus instructions and feedback. *Hum. Mov. Sci.* 18, 553–571. https://doi.org/10.1016/S0167-9457(99)00031-7

Shibuya, S., Unenaka, S., Zama, T., Shimada, S., Ohki, Y., 2018. Spontaneous imitative movements induced by an illusory embodied fake hand. *Neuropsychologia.* https://doi.org/10.1016/j.neuropsychologia.2018.01.023

Siegel, A., Sapru, H.N., 2011. *Essential Neuroscience.* Philadelphia: Wolters Kluwer Health/Lippincott Williams & Wilkins.

Sigrist, R., Rauter, G., Marchal-Crespo, L., Riener, R., Wolf, P., 2015. Sonification and haptic feedback in addition to visual feedback enhances complex motor task learning. *Exp. Brain Res.* 233, 909–925. https://doi.org/10.1007/s00221-014-4167-7

Sigrist, R., Rauter, G., Riener, R., Wolf, P., 2013. Augmented visual, auditory, haptic, and multimodal feedback in motor learning: a review. *Psychon. Bull. Rev.* 20, 21–53. https://doi.org/10.3758/s13423-012-0333-8

Slater, M., 2009. Place illusion and plausibility can lead to realistic behaviour in immersive virtual environments. *Philos. Trans. R. Soc. B Biol. Sci.* 364, 3549–3557. https://doi.org/10.1098/rstb.2009.0138

Slater, M., Perez-Marcos, D., Ehrsson, H.H., Sanchez-Vives, M.V., 2008. Towards a digital body: the virtual arm illusion. *Front. Hum. Neurosci.* 2. https://doi.org/10.3389/neuro.09.006.2008

Sokolov, A.A., Collignon, A., Bieler-Aeschlimann, M., 2020. Serious video games and virtual reality for prevention and neurorehabilitation of cognitive decline because of aging and neurodegeneration. *Curr. Opin. Neurol.* 33, 239–248. https://doi.org/10.1097/WCO.0000000000000791

Spanlang, B., Normand, J.-M., Borland, D., Kilteni, K., Giannopoulos, E., PomÃ©s, A. s, GonzÃ¡lez-Franco, M., Perez-Marcos, D., Arroyo-Palacios, J., Muncunill, X.N., Slater, M., 2014. How to build an embodiment lab: achieving body representation illusions in virtual reality. *Front. Robot. AI.* 1. https://doi.org/10.3389/frobt.2014.00009

Tack, C., 2019. Virtual reality and chronic low back pain. *Disabil. Rehabil. Assist. Technol.* 1–9. https://doi.org/10.1080/17483107.2019.1688399

Thomson, K., Pollock, A., Bugge, C., Brady, M., 2014. Commercial gaming devices for stroke upper limb rehabilitation: a systematic review. *Int. J. Stroke.* 9, 479–488. https://doi.org/10.1111/ijs.12263

Todorov, E., Shadmehr, R., Bizzi, E., 1997. Augmented feedback presented in a virtual environment accelerates learning of a difficult motor task. *J. Motor Behav.* 29, 147–158. https://doi.org/10.1080/00222899709600829

Tsakiris, M., 2010. My body in the brain: a neurocognitive model of body-ownership. *Neuropsychologia.* https://doi.org/10.1016/j.neuropsychologia.2009.09.034

Tsakiris, M., Hesse, M.D., Boy, C., Haggard, P., Fink, G.R., 2007. Neural signatures of body ownership: a sensory network for bodily self-consciousness. *Cereb. Cortex.* 17, 2235–2244. https://doi.org/10.1093/cercor/bhl131

Wei, Y., Patton, J.L., 2004. Force field training to facilitate learning visual distortions: a "sensory crossover" experiment, in: 12th International Symposium on Haptic Interfaces for Virtual Environment and Teleoperator Systems, 2004. HAPTICS '04. Proceedings. Presented at the 12th International Symposium on Haptic Interfaces for Virtual Environment and Teleoperator Systems, 2004. HAPTICS '04. Proceedings, pp. 194–199. https://doi.org/10.1109/HAPTIC.2004.1287196

Wenk, N., Penalver-Andres, J., Palma, R., Buetler, K.A., Muri, R., Nef, T., Marchal-Crespo, L., 2019. Reaching in several realities: motor and cognitive benefits of different visualization technologies. *IEEE Int. Conf. Rehabil. Robot. Proc.* 2019, 1037–1042. https://doi.org/10.1109/ICORR.2019.8779366

WHO | International Classification of Functioning, Disability and Health (ICF) [WWW Document], 2016. WHO. URL http://www.who.int/classifications/icf/en/ (accessed 10.3.16).

WHO | Life expectancy [WWW Document], n.d. WHO. URL http://www.who.int/gho/mortality_burden_disease/life_tables/situation_trends_text/en/ (accessed 10.11.19).

Wise, S.P., 1985. The primate premotor cortex: past, present, and preparatory. *Annu. Rev. Neurosci.* 8, 1–19. https://doi.org/10.1146/annurev.ne.08.030185.000245

Wittmann, F., Held, J.P., Lambercy, O., Starkey, M.L., Curt, A., Höver, R., Gassert, R., Luft, A.R., Gonzenbach, R.R., 2016. Self-directed arm therapy at home after stroke with a sensor-based virtual reality training system. *J. Neuroeng. Rehabil.* 13. https://doi.org/10.1186/s12984-016-0182-1

Zeller, D., Friston, K.J., Classen, J., 2016. Dynamic causal modeling of touch-evoked potentials in the rubber hand illusion. *NeuroImage.* 138, 266–273. https://doi.org/10.1016/j.neuroimage.2016.05.065

Zitzewitz, J. von, Morger, A., Rauter, G., Marchal-Crespo, L., Crivelli, F., Wyss, D., Bruckmann, T., Riener, R., 2013. A reconfigurable, tendon-based haptic interface for research into human-environment interactions. *Robotica.* 31, 441–453. https://doi.org/10.1017/S026357471200046X

6 Virtual Reality Interventions' Effects on Functional Outcomes for Children with Neurodevelopmental Disorders

Jorge Lopes Cavalcante Neto
State University of Bahia
Bahia, Brazil

CONTENTS

INTRODUCTION

Interventions based on virtual reality (VR) have greatly increased over the last decades, and these interventions employ key elements of this approach to enable gains in people's health and functionality. VR is characterized as a technology resource that allows a dynamic form of human–machine interaction through several types of scenarios designed by a sophisticated computer apparatus, simulating, in most cases, a real-word context (Schultheis and Rizzo 2001).

VR resources are commonly characterized as immersive or non-immersive according to the level of immersion provided by the apparatus and the level of immersion is determined by how effectively the technology creates a perception of reality, supporting sensorimotor demands (O'Regan and Noë 2001). The differences between immersive and non-immersive VR resources can also be determined by characterizing the spatial presence of the user (Ventura *et al.* 2019). Finally, the closer the apparatus gets the user to the perception of real-life scenarios, the more immersive the technology is.

However, equipment based on immersive VR resources is very expensive, which might make it unfeasible for many intervention settings as well as the effectiveness of interventions based on VR is related more to the engagement than to the degree of immersion afforded by the VR apparatus (O'Brien and Toms 2008). Therefore, non-immersive VR seems to be a promising rehabilitation option for some patients, including pediatric ones (Deutsch *et al.* 2008). Rehabilitation using VR technology is built on distinctive advantages not observed in many traditional approaches (Palacios-Navarro *et al.* 2016), particularly with children (Snider *et al.* 2010).

Among the advantages observed with VR technology are the augmentative feedback of the action required (Wang and Reid 2011); the more controlled and safe environment provided; and the increased control of the duration, frequency, and intensity of the movements performed during the rehabilitation session (Jack *et al.* 2001). Taken together, all these factors enable more flexibility within rehabilitation sessions, which might optimize the possibilities of transferring the skills learned during intervention sessions to the real world (Rizzo and Galen Buckwalter 1997).

Another distinctive advantage of VR is the huge capacity to promote fun (Mouatt *et al.* 2019), which might be crucial for helping children maintain their motivation throughout the rehabilitation sessions, consequently children can challenge themselves within an interactive relationship between them and the machine (Lillard 1993), the targeted functional outcomes can then be reached faster and better. Given this previous contextualization, the literature has highlighted the intervention effects of VR or synonymous terms (active video games, VR games, exergames) on health and functional outcomes for children with different conditions, such as cancer (Li *et al.* 2011), pain and anxiety (Eijlers *et al.* 2019), and neurodevelopmental disorders (Hickman *et al.* 2017). Among these conditions, evidence about VR interventions for children with neurodevelopmental disorders has been highlighted in recent systematic reviews (Hickman *et al.* 2017, Ravi *et al.* 2017, Mesa-Gresa *et al.* 2018, Cavalcante Neto, de Oliveira *et al.* 2019), particularly because such conditions and their corresponding treatments are complex, and updates about rehabilitation approaches for this group are provided frequently.

Neurodevelopmental disorders are characterized by delays related to the central or peripheral nervous system (Golden 1987) and compose a group of the most prevalent, complex, and disabling disorders during childhood (Cardoso 2014). Additionally, the impairments associated with these conditions interfere with children's health, self-esteem, and quality of life (Cardoso 2014) and, consequently, in their functionality over adulthood. As examples among several types of neurodevelopmental disorders, cerebral palsy (CP), developmental coordination disorder (DCD), and autism spectrum disorder share the most challenges with respect to both diagnostic processes and rehabilitation strategies, which has aroused intriguing findings in the context of VR from the sources of evidence available. Therefore, the objective of this chapter is to present an overview of the effects of VR interventions/rehabilitation on functional outcomes for children with neurodevelopmental disorders, particularly CP, DCD, and autism spectrum disorder. In order to achieve this, the results of this chapter are divided into three topics, one for each condition.

In total, seven systematic reviews were selected to help and illustrate the use of VR for children with neurodevelopmental disorders. Table 6.1 presents the main characteristics about the systematic reviews included in this study.

Broadly, the outcomes targeted for the majority of studies were related to motor function or motor skills recovery, which emphasizes that VR's use in the rehabilitation process for children with neurodevelopmental disorders is focused primarily on their body function. Studies regarding autism spectrum disorder are the exception, since the one review retrieved about this primarily focused on behavioral outcomes. However, it did not neglect the role of motor function for these children.

Virtual Reality Interventions' Effects for Children with Cerebral Palsy

Four out of the seven systematic reviews included focused on CP, which suggests that VR rehabilitation approaches for this neurodevelopmental disorder have been addressed more frequently in the literature than the others included here. In part, the reasons for this might be that the characteristics of CP are more clearly detectable than other neurodevelopmental disorders during childhood and that clinicians and scientists have accumulated more experience and observed more cases related to CP (Baxter 2015) in their practices. Also, the prevalence of CP worldwide is large, since about three to four children out of 1,000 live births are affected by it (Jonsson *et al.* 2019). Taken together, this information may encourage new forms of rehabilitation, such as those based on VR.

According to the information from the four systematic reviews, commercial VR equipment was used more often than engineer-built systems (Ravi *et al.* 2017, Chen *et al.* 2018, Rathinam *et al.* 2019, Warnier *et al.* 2019). However, engineer-built systems better promoted high levels of evidence than commercials ones did (Chen *et al.* 2018), since commercial systems are not made for the

TABLE 6.1

Characteristics of the Systematic Reviews Included

Study/Year	Objectives	Outcomes	Target Neurodevelopmental Disorder
Warnier et al. (2019)	To investigate the effect of virtual reality therapy (VRT) on balance and walking in children with cerebral palsy (CP)	Balance and walking	Cerebral palsy
Rathinam et al. (2019)	To determine the effectiveness of VR as an intervention to improve hand function in children with CP compared to either conventional physiotherapy or other therapeutic interventions. The secondary purpose was to classify the outcomes evaluated according to the International Classification of Functioning, Disability and Health (ICF) dimensions	Hand function	Cerebral palsy
Chen et al. (2018)	To update the current evidence about VR by systematically examining the research literature	Movement-related outcomes	Cerebral palsy
Ravi et al. (2017)	To provide updated evidence-based guidance for virtual reality rehabilitation in sensory and functional motor skills of children and adolescents with cerebral palsy	Balance and motor skills	Cerebral palsy
Cavalcante Neto, de Oliveira et al. (2019)	To synthesize evidence on the effectiveness of VR interventions for motor performance improvement in children with DCD	Motor performance	Developmental coordination disorder
Mentiplay et al. (2019)	To systematically review the literature on virtual reality or video game interventions that aim to improve motor outcomes in children with DCD	Motor performance	Developmental coordination disorder
Mesa-Gresa et al. (2018)	To carry out an evidence-based systematic review including both clinical and technical databases about the effectiveness of VR-based intervention in ASD	Social and emotional skills	Autism spectrum disorder

specific disability demands present in children with CP. As most of the CP children had impaired hand function and unbalanced postural control, commercial VR accessories were not able to fit these children's body profiles, whereas engineer-built ones were. Nintendo's Wii Fit and other Nintendo's Wii equivalent systems were the most frequently used brands in the studies (Ravi et al. 2017, Chen et al. 2018, Rathinam et al. 2019, Warnier et al. 2019). However, the improvements observed for those studies were achieved precisely due to the Wii Fit accessories, such as balance boards. Along with the fact that VR promotes a motivational environment (Harris and Reid 2005), this accessory appears to be key for improving motor skills outcomes in CP children mainly by helping them gain better balance and postural control, which generally lag behind in these children.

There was no consensus among studies regarding the dosage of VR treatment for children with CP. All studies synthetized various durations, frequency, and number of sessions. The total duration of therapy ranged from one day to 20 weeks with several intensities and per-week frequencies, ranging from 15 to 120 minutes a session and from one single session to seven days per week. All the included studies observed better results with high-intensity sessions than with a lower intensity (Ravi *et al.* 2017, Chen *et al.* 2018, Rathinam *et al.* 2019, Warnier *et al.* 2019). Also, more major improvements occurred in studies under a therapist or researcher's supervision than in those with games performed at home by children alone (Ravi *et al.* 2017, Chen *et al.* 2018, Rathinam *et al.* 2019, Warnier *et al.* 2019).

Regarding CP classification, different types were observed across studies, such as hemiplegia, diplegia, spastic, mixed or still not specified CP, slightly prevailing spastic, and mixed cases. In addition, most studies considered the level of functionality according to the Gross Motor Function Classification System (GMFCS), particularly levels I and II, as inclusion criteria for children be part of the study. GMFCS is a tool used to assess the level of functionality based on children's ability to self-initiate gross motor movement as well as their need to be assisted by technology or other mobility resources (Palisano *et al.* 1997). The reported use of this classification for children with CP across studies not only legitimates this tool as a standard procedure to be adopted with this group but also demonstrates that the focus is mainly on the functionality of these children. In addition, the majority of systematic reviews (Ravi *et al.* 2017, Chen *et al.* 2018, Rathinam *et al.* 2019) employed the International Classification of Functioning, Disability, and Health (ICF) to classify the outcomes measured. This tool has facilitated the understanding of health-related outcomes and functional improvements in children with CP related to the demands of their daily lives (World Health Organization (WHO) 2001), even using a technology resource like VR equipment. Regarding the effects of interventions, all systematic reviews agreed that VR improved the majority of outcomes evaluated, such as balance and walking performance (Warnier *et al.* 2019); hand function (Rathinam *et al.* 2019); arm function, ambulation, and postural control (Chen *et al.* 2018); and balance and motor skills (Ravi *et al.* 2017). However, it is necessary to discuss in more detail the specifics among these studies' results, taking into account their VR intervention effects calculated across studies.

For example, the effect sizes found by Chen *et al.* (2018) ranged from medium to large. For the general analyses considering the intervention effects of VR versus other interventions, the effect size was large ($d = 0.861$), as it was when considering the effects on postural control ($d = 1.003$) and arm function ($d = 0.835$). For ambulation, the effect size was medium ($d = 0.755$) but still clinically important. Warnier *et al.* (2019) performed a meta-analysis of five studies considering balance as an outcome and another meta-analysis with four studies considering walking as an outcome. For the first analysis, the standardized mean difference (SMD) was 0.89 (95% CI, SD 0.14, 1.63), while for the second, the SMD was 3.10 (95% CI, SD 0.78, 5.35). Both analyses attested to a large effect size (using SMD as effect measure) in favor of VR for these outcomes compared to the control group, who performed no VR therapy. In contrast, Rathinam *et al.* (2019) did not perform a meta-analysis since procedures among studies were heterogeneous. The authors concluded the study with conflicting results and limited evidence because 66% of their included studies showed improvements on children's hand function while 34% showed no improvements. Similarly, Ravi *et al.* (2017) highlighted that the evidence about the effects of VR for other motor skills is inconclusive, since the authors only found promising moderate evidence regarding improvements in balance and overall motor development after VR intervention.

Therefore, when comparing traditional rehabilitation approaches like conventional physiotherapy, all studies had only weak evidence to support superior benefits from one rehabilitation type to another (Ravi *et al.* 2017, Chen *et al.* 2018, Rathinam *et al.* 2019, Warnier *et al.* 2019). Additionally, regarding the improvements in the outcomes of VR rehabilitation for children with CP, the best recommendation from all systematic reviews accessed is toward the use of VR approaches as complementary therapeutic modalities. Because of the heterogeneity among the studies retrieved by the four systematic reviews used here, proper evidence about the effectiveness of VR rehabilitation treatment for children with CP remains to be found.

Virtual Reality Interventions' Effects for Children with Developmental Coordination Disorder (DCD)

DCD is a motor impairment condition affecting around 5% of school-age children (Blank *et al.* 2019). The current diagnosis is based on four interrelated criteria according to the Diagnostic and Statistical Manual of Mental Disorders—fifth edition (DSM-5): A—the children have insufficient motor performance for age and practice opportunities; B—the lack in motor performance interferes in daily, school, and leisure activities; C—children present these characteristics from their early development; D—the insufficient motor performance is not explained by disabilities or other known neurological conditions (American Psychiatric Association 2013). Since DCD interferes in several life components, negative consequences in children's quality of life are expected (Zwicker *et al.* 2013). In that context, several researchers have tried to test intervention types in order to find the best option for helping these children improve motor outcomes, reaching a consensus that any type of intervention would be better than none (Hillier 2007). However, taking into account the motivation necessary to maintain children's involvement in the intervention, approaches based on VR have been endorsed in recent investigations with DCD (Bonney, Ferguson *et al.* 2017, Bonney, Jelsma *et al.* 2017, Cavalcante Neto, Steenbergen *et al.* 2019). Despite this, further evidence is still needed regarding the effectiveness of VR for children with DCD.

Only two systematic reviews regarding VR interventions' effects for children with DCD were retrieved. Given that DCD is still a new term used in this field and has a multifactorial and complex diagnosis, this small number of systematic reviews is expected. Also, the evidence about the effectiveness of VR for children with DCD has been explored in combination with other intervention types (Smits-Engelsman *et al.* 2018) or along with other neuromotor conditions (Hickman *et al.* 2017, Page *et al.* 2017), which would not be relevant to the purpose of this chapter. Therefore, the two systematic reviews included here exclusively considered studies that evaluated the effectiveness of VR interventions for children with DCD.

According to the information from the two systematic reviews, Nintendo Wii was the most frequently used VR device (Cavalcante Neto, de Oliveira *et al.* 2019, Mentiplay *et al.* 2019), and motor performance as assessed by the Movement Assessment Battery for Children (MABC) was the most frequent outcome across studies. Particularly, the systematic review conducted by Mentiplay *et al.* (2019) classified outcomes from studies included according to the ICF framework, which attested to more benefits for body structure and function and activity. Although the other systematic review (Cavalcante Neto, de Oliveira *et al.* 2019) did not focus on the ICF, it is possible to notice these outcomes were in fact prioritized in interventions with VR for children with DCD.

Both systematic reviews (Cavalcante Neto, de Oliveira *et al.* 2019, Mentiplay *et al.* 2019) decided not to conduct a meta-analysis due to the heterogeneity among studies. Regarding the dosage of VR interventions for this condition, the total length of therapy ranged from one day to 16 weeks, with intensities ranging from 4 to 60 minutes per session, occurring from one single session to five times per week. In summary, the most common weekly frequency is estimated at around two to three times per week.

Both systematic reviews (Cavalcante Neto, de Oliveira *et al.* 2019, Mentiplay *et al.* 2019) also attested that VR interventions are effective in improving motor performance in children with DCD, but the effectiveness of this intervention type compared to conventional approaches, such as task-oriented training, is still uncertain. Task-oriented training is a therapeutic modality based on a specific movement task, aiming to improve corresponding skills (Hubbard *et al.* 2009, Miyahara *et al.* 2017). Despite finding no significant differences in motor performance between VR and no-VR (task-specific) intervention groups, a recent study (Cavalcante Neto, Steenbergen *et al.* 2019) found greater effect size in favor of no-VR intervention, based on task-oriented training principles, for children with DCD. Similarly, in those studies in which task-oriented training principles were used as the control intervention group, greater improvements were observed and synthesized by one systematic review included here (Mentiplay *et al.* 2019).

In addition, a recent systematic review comparing various intervention types for children with DCD found the greatest benefits for motor performance with task-oriented training compared to other approaches, including VR (Smits-Engelsman *et al.* 2018). This finding does not mean that VR should be dropped, but clinicians and researchers must be cautious in recommending this approach as an exclusive therapeutic modality for the motor rehabilitation of children with DCD. Therefore, both systematic reviews (Cavalcante Neto, de Oliveira *et al.* 2019, Mentiplay *et al.* 2019) also agreed that VR should be used as a complementary therapeutic modality combined with other conventional modalities, particularly those with stronger evidence. So far, those including task-oriented training principles appear to be the best approaches.

Virtual Reality Interventions' Effects for Children with Autism Spectrum Disorder

Autism spectrum disorder (ASD) is another complex neurodevelopmental disorder referred to as a persistent impairment, interfering in communication and social interactions (American Psychiatric Association 2013). One of the most pronounced characteristics of this disorder is related to patterns of behaviors repetitive actions or movements and restricted and specific interests (Campisi *et al.* 2018). The global prevalence of ASD is around 7.6 per 1,000 persons (Baxter *et al.* 2015), and the impact on daily life or academic activities is huge (Campisi *et al.* 2018). Since communication and social skills are necessary to function and have autonomy, children with ASD are at a great disadvantage compared to their peers, because communication and social skills are, in fact, remarkable in those children.

Given these features, the diagnostic process for ASD is complex and difficult. There is scarce evidence about the etiology of ASD, but genetic components are present in more than 20% of the cases (Ivanov *et al.* 2015). Therefore, the earlier the diagnosis is made, the better the prognosis will be. Early intervention should be done as soon as possible in order to avoid severe consequences in all aspects of the children's development (Zwaigenbaum *et al.* 2015). Since interaction and communication are elements involved in therapeutic approaches for children, these deficiencies along with restrictive interests may be a barrier during the rehabilitation process. In contrast, VR approaches may afford a suitable therapeutic environment with ideal dosages of sensorial, social, and motor stimuli.

Although there are other systematic reviews available, only one systematic review (Mesa-Gresa *et al.* 2018) about VR for children with ASD fulfilled the purpose of this chapter, since it focused exclusively on children with ASD and interventions based on VR. Unfortunately, this systematic review (Mesa-Gresa *et al.* 2018) did not synthesize either the VR tools used among studies during interventions or the wide information about the dosage of treatment with those children. The majority of studies were interested in social (45%) and emotional (21%) skills as outcomes. Despite the improvements observed in the majority of studies (30 out 31 studies) in at least one outcome assessed, only in ten studies were these improvements significant. Regarding the level of evidence attested to by this systematic review, the authors (Mesa-Gresa *et al.* 2018) found only a moderate effect of VR intervention for children with ASD. This implies that despite the improvements observed, moderate evidence is not sufficient to recommend rehabilitation treatment based on VR as a substitute for traditional treatments for this population.

Moreover, this systematic review (Mesa-Gresa *et al.* 2018) compiled useful and promising information that widens our possibilities for helping children with ASD achieve greater quality of life and functionality. However, stronger evidence is necessary to consider VR as a good choice to treat ASD symptomatology, particularly related to aspects of social and emotional skills. In addition, following the suggestions given by authors of this systematic review, more studies are necessary, and new rehabilitation techniques using VR should be encouraged and developed for children with ASD.

CONCLUSION

Considering all evidence from the seven systematic reviews retrieved here, VR interventions are promising approaches for the rehabilitation process of children with neurodevelopmental disorders, particularly CP, DCD, and autism spectrum disorder. VR has brought benefits for functional outcomes for these children, such as motor skills, motor performance, balance, walking, social skills, and emotional skills. However, these benefits are not superior to those achieved by traditional interventions, such as conventional physiotherapy. Despite that, better effects of VR interventions appear to be achieved when combined with traditional no-VR interventions, which would increase functional benefits for children with neurodevelopmental disorders while simultaneously maintaining engagement with the therapeutic process. Thus, more investigations regarding VR interventions' effects for children with neurodevelopmental disorders should be carried out, particularly taking into account the advantages of technology and the unique and complex characteristics and needs of these children.

REFERENCES

American Psychiatric Association, 2013. *Diagnostic and Statistical Manual of Mental Disorders, DSM-5*. 5th ed. Washington, DC: American Psychiatric Association.

Baxter, P., 2015. For and against the term cerebral palsy. *Developmental Medicine & Child Neurology*, 57 (7), 592–592.

Baxter, A.J., Brugha, T.S., Erskine, H.E., Scheurer, R.W., Vos, T., and Scott, J.G., 2015. The epidemiology and global burden of autism spectrum disorders. *Psychological Medicine*, 45 (3), 601–613.

Blank, R., Barnett, A.L., Cairney, J., Green, D., Kirby, A., Polatajko, H., Rosenblum, S., Smits-Engelsman, B., Sugden, D., Wilson, P., and Vinçon, S., 2019. International clinical practice recommendations on the definition, diagnosis, assessment, intervention, and psychosocial aspects of developmental coordination disorder. *Developmental Medicine & Child Neurology*, 61 (3), 242–285.

Bonney, E., Ferguson, G., and Smits-Engelsman, B., 2017. The efficacy of two activity-based interventions in adolescents with developmental coordination disorder. *Research in Developmental Disabilities*, 71, 223–236.

Bonney, E., Jelsma, L.D., Ferguson, G.D., and Smits-Engelsman, B.C.M., 2017. Learning better by repetition or variation? Is transfer at odds with task specific training? *PLOS One*, 12 (3), e0174214.

Campisi, L., Imran, N., Nazeer, A., Skokauskas, N., and Azeem, M.W., 2018. Autism spectrum disorder. *British Medical Bulletin*, 127, 91–100.

Cardoso, F., 2014. Movement disorders in childhood. *Parkinsonism and Related Disorders*, 20 (Suppl.1), S13–S16.

Cavalcante Neto, J.L., de Oliveira, C.C., Greco, A.L., Zamunér, A.R., Moreira, R.C., and Tudella, E., 2019. Is virtual reality effective in improving the motor performance of children with developmental coordination disorder? A systematic review. *European Journal of Physical and Rehabilitation Medicine*, 55 (2), 291–300.

Cavalcante Neto, J.L., Steenbergen, B., Wilson, P., Zamunér, A.R., and Tudella, E., 2020. Is Wii-based motor training better than task-specific matched training for children with developmental coordination disorder? A randomized controlled trial. *Disability and Rehabilitation*, 42, 2611–2620.

Chen, Y., Fanchiang, H.D., and Howard, A., 2018. Effectiveness of virtual reality in children with cerebral palsy: A systematic review and meta-analysis of randomized controlled trials. *Physical Therapy*, 98 (1), 63–77.

Deutsch, J.E., Borbely, M., Filler, J., Huhn, K., and Guarrera-Bowlby, P., 2008. Use of a low-cost, commercially available gaming console (Wii) for rehabilitation of an adolescent with cerebral palsy. *Physical Therapy*, 88 (10), 1196–1207.

Eijlers, R., Utens, E.M.W.J., Staals, L.M., De Nijs, P.F.A., Berghmans, J.M., Wijnen, R.M.H., Hillegers, M.H.J., Dierckx, B., and Legerstee, J.S., 2019. Systematic review and meta-analysis of virtual reality in pediatrics: Effects on pain and anxiety. *Anesthesia and Analgesia*, 129 (5), 1344–1353.

Golden, G.S., 1987. Common neuromotor disorders. *In*: Gottlieb, M.I. and Williams, J.E. (eds) *Textbook of Developmental Pediatrics*. Boston, MA: Springer US, 27–40.

Harris, K. and Reid, D., 2005. The influence of virtual reality play on children's motivation. *Education*, 72 (1), 21–29.

Hickman, R., Popescu, L., Manzanares, R., Morris, B., Lee, S.-P., and Dufek, J.S., 2017. Use of active video gaming in children with neuromotor dysfunction: A systematic review. *Developmental Medicine & Child Neurology*, 59 (9), 903–911.

Hillier, S., 2007. Intervention for children with developmental coordination disorder: A systematic review. *The Internet Journal of Allied Health Sciences and Practice*, 5 (3), 1–11.

Hubbard, I.J., Parsons, M.W., Neilson, C., and Carey, L.M., 2009. Task-specific training: Evidence for and translation to clinical practice. *Occupational Therapy International*, 16 (3–4), 175–189.

Ivanov, H.Y., Stoyanova, V.K., Popov, N.T., and Vachev, T.I., 2015. Autism spectrum disorder - A complex genetic disorder. *Folia Medica*, 57 (1), 19–28.

Jack, D., Boian, R., Merians, A.S., Tremaine, M., Burdea, G.C., Adamovich, S. V., Recce, M., and Poizner, H., 2001. Virtual reality-enhanced stroke rehabilitation. *IEEE Transactions on Neural Systems and Rehabilitation Engineering*, 9 (3), 308–318.

Jonsson, U., Eek, M.N., Sunnerhagen, K.S., and Himmelmann, K., 2019. Cerebral palsy prevalence, sub-types, and associated impairments: A population-based comparison study of adults and children. *Developmental Medicine and Child Neurology*, 61 (10), 1162–1167.

Li, W.H., Chung, J.O., and Ho, E.K., 2011. The effectiveness of therapeutic play, using virtual reality computer games, in promoting the psychological well-being of children hospitalised with cancer. *Journal of Clinical Nursing*, 20 (15–16), 2135–2143.

Lillard, A.S., 1993. Pretend play skills and the child's theory of mind. *Child Development*, 64 (2), 348.

Mentiplay, B.F., Fitzgerald, T.L., Clark, R.A., Bower, K.J., Denehy, L., and Spittle, A.J., 2019. Do video game interventions improve motor outcomes in children with developmental coordination disorder? A systematic review using the ICF framework. *BMC Pediatrics*, 19 (1), 22.

Mesa-Gresa, P., Gil-Gómez, H., Lozano-Quilis, J.A., and Gil-Gómez, J.A., 2018. Effectiveness of virtual reality for children and adolescents with autism spectrum disorder: An evidence-based systematic review. *Sensors (Switzerland)*, 18 (8), 2486.

Miyahara, M., Hillier, S.L., Pridham, L., and Nakagawa, S., 2017. Task-oriented interventions for children with developmental co-ordination disorder. *Cochrane Database of Systematic Reviews*, 7 (7), CD010914.

Mouatt, B., Smith, A., Mellow, M., Parfitt, G., Smith, R., and Stanton, T., 2019. The use of virtual reality to influence engagement and enjoyment during exercise: A scoping review. *Journal of Science and Medicine in Sport*, 22, S98.

O'Brien, H.L. and Toms, E.G., 2008. What is user engagement? A conceptual framework for defining user engagement with technology. *Journal of the American Society for Information Science and Technology*, 59 (6), 938–955.

O'Regan, J.K. and Noë, A., 2001. A sensorimotor account of vision and visual consciousness. *Behavioral and Brain Sciences*, 24 (5), 939–973.

Page, Z.E., Barrington, S., Edwards, J., and Barnett, L.M., 2017. Do active video games benefit the motor skill development of non-typically developing children and adolescents: A systematic review. *Journal of Science and Medicine in Sport*, 20 (12), 1087–1100.

Palacios-Navarro, G., Albiol-Pérez, S., and García-Magariño García, I., 2016. Effects of sensory cueing in virtual motor rehabilitation. A review. *Journal of Biomedical Informatics*, 60, 49–57.

Palisano, R., Rosenbaum, P., Bartlett, D., Livingston, M., Walter, S., Russell, D., Wood, E., and Galuppi, B., 1997. Development and reliability of a system to classify gross motor function in children with cerebral palsy. *Developmental Medicine and Child Neurology*, 39 (4), 214–223.

Rathinam, C., Mohan, V., Peirson, J., Skinner, J., Nethaji, K.S., and Kuhn, I., 2019. Effectiveness of virtual reality in the treatment of hand function in children with cerebral palsy: A systematic review. *Journal of Hand Therapy*, 32 (4), 426–434.e1.

Ravi, D.K., Kumar, N., and Singhi, P., 2017. Effectiveness of virtual reality rehabilitation for children and adolescents with cerebral palsy: An updated evidence-based systematic review. *Physiotherapy (United Kingdom)*, 103 (3), 245–258.

Rizzo, A.A. and Galen Buckwalter, J., 1997. Virtual reality and cognitive assessment and rehabilitation: The state of the art. *Studies in Health Technology and Informatics*, 44, 123–145.

Schultheis, M.T. and Rizzo, A.A., 2001. The application of virtual reality technology in rehabilitation. *Rehabilitation Psychology*, 46 (3), 296–311.

Smits-Engelsman, B., Vinçon, S., Blank, R., Quadrado, V.H., Polatajko, H., and Wilson, P.H., 2018. Evaluating the evidence for motor-based interventions in developmental coordination disorder: A systematic review and meta-analysis. *Research in Developmental Disabilities*, 74, 72–102.

Snider, L., Majnemer, A., and Darsaklis, V., 2010. Virtual reality as a therapeutic modality for children with cerebral palsy. *Developmental Neurorehabilitation*, 13 (2), 120–128.

Ventura, S., Brivio, E., Riva, G., and Baños, R.M., 2019. Immersive versus non-immersive experience: Exploring the feasibility of memory assessment through 360° technology. *Frontiers in Psychology*, 10, 2509.

Wang, M. and Reid, D., 2011. Virtual reality in pediatric neurorehabilitation: Attention deficit hyperactivity disorder, autism and cerebral palsy. *Neuroepidemiology*, 36 (1), 2–18.

Warnier, N., Lambregts, S., and Port, I. Van De, 2019. Effect of virtual reality therapy on balance and walking in children with cerebral palsy: A systematic review. *Developmental Neurorehabilitation*, 1–17.

World Health Organization (WHO), 2001. *International Classification of Functioning, Disability and Health (ICF)*. Geneva: WHO.

Zwaigenbaum, L., Bauman, M.L., Choueiri, R., Kasari, C., Carter, A., Granpeesheh, D., Mailloux, Z., Roley, S.S., Wagner, S., Fein, D., Pierce, K., Buie, T., Davis, P.A., Newschaffer, C., Robins, D., Wetherby, A., Stone, W.L., Yirmiya, N., Estes, A., Hansen, R.L., McPartland, J.C., and Natowicz, M.R., 2015. Early Intervention for children with autism spectrum disorder under 3 years of age: Recommendations for practice and research. In: *Pediatrics*. Evanston: American Academy of Pediatrics, S60–S81.

Zwicker, J.G., Harris, S.R., and Klassen, A.F., 2013. Quality of life domains affected in children with developmental coordination disorder: A systematic review. *Child: Care, Health and Development*, 39 (4), 562–580.

7 The Use of Virtual Reality Environments in Cognitive Rehabilitation after Traumatic Brain Injury

E. Sorita
University of Bordeaux
EA 4136 HACS, France

P. Coignard, E. Guillaume, and J-L. Le Guiet
Centre mutualiste de rééducation et de réadaptation
fonctionnelles de Kerpape
France

E. Klinger and P-A. Joseph
University of Bordeaux
EA 4136 HACS
Institut Fédératif de Recherche sur le Handicap
France

CONTENTS

INTRODUCTION

In the 1990s, the wider dissemination of research outcomes combined with lower cost computer hardware encouraged researchers in cognitive rehabilitation to develop clinical virtual reality (VR) environments. It was hoped that VR would improve the standardization of traditional neuropsychological testing and enable larger amounts of data to be collected and compared. However, at the same time, one main criticism of traditional neuropsychological testing was that it did not reflect actual cognitive functioning in daily life (Rose, 1996). To address this limit, researchers sought to develop alternative neuropsychological tests which had "functional and predictive relationships" with natural performance (Sbordone, 1996). The concept of ecological validity (EV) appeared to be a key criterion in a new generation of neuropsychological tests aimed both to reflect and predict effects of cognitive disorders in an "open environment" (Franzen & Wilhelm, 1996). Shallice & Burgess's seminal work (Shallice & Burgess, 1991) concerning EV resulted in the creation of the Multiple Errands Test (MET) which assessed executive functioning in persons after traumatic brain injury (TBI). Indeed, the MET enabled the identification of impairments in daily functioning, in other words dysexecutive syndrome, reported by patients or families, that traditional neuropsychological test batteries could not identify (Burgess et al., 2006). The topic of EV generated a significant debate about the need for more realism and prediction of environmental behavior in neuropsychology testing (Franzen & Wilhelm, 1996).

In this context, it appeared that simulating naturalistic environments and activities using VR may have some advantages that could meet this need (Rizzo et al., 2004). Furthermore, computer systems enabled the development of objective behavioral indicators in these ecologically valid but also safe environments (Schultheis & Rizzo, 2001). Virtual environments could also benefit from experimental control, have multiple demand levels, deliver stimuli and opportunity for rehearsal that led to virtual cognitive rehabilitation (Rose et al., 2005). For Weiss et al. (2006) « *the ultimate goal of VR-based intervention is to enable clients to become more able to participate in their own real environments in as independent manner as possible* » (Weiss et al., 2006). Therefore, VR became increasingly more important in a naturalistic-orientated conception of CR, with its focus on the transferability and generalizability of compensation strategies into everyday life (Cicerone et al., 2019). However, although it is easier using VR to simulate realistic scenes, a key point is to structure VR for CR assessment or training on the basis of current cognitive theories.

VR GENERAL PRINCIPLES APPLIED TO CR AFTER TBI

It is challenging to face the complexity of cognitive and behavioral disorders in CR. Within acquired brain injury, TBI is a major concern throughout the world (Maas et al., 2017). Moderate to severe TBI leads to complex clinical situations that combine motor, cognitive, and behavioral disorders that have a major impact in daily functioning (Ponsford et al., 2014). This impact is of great concern, leading to stress, burden, and mood disorders for families (Poulin et al., 2019). For this population with TBI, impacts on activities and participation are related to instrumental activities of daily living (IADL) such as household chores, management of finances, diary management, shopping, and also outside navigation, transportation, and return to work (Dawson and Chipman, 1995). Daily life complexity may be defined in relation to the significant number and diversity of interactions with environmental contexts (Spector et al., 1987). Cognitive demand is notably higher when dealing with multitasking, problem-solving, and coping with environmental needs. Achieving goal-directed activities requires the mobilization of cognitive control resources (Cooper & Shallice, 2000) and to have good self-awareness in order to interact with the environment (Katz & Hartman-Maeir, 2005). In addition, this ability, which mobilizes cognitive resources to plan and organize actions and adapt voluntary behavior to the environment to achieve a goal, is called executive functioning (Lezak, 1982). Dysexecutive Syndrome has a tremendous impact on daily life and long-lasting community reintegration for patients with TBI (Caron et al., 2018).

VR clinical application in CR seeks to capture the activity demands of daily life in order to mobilize cognitive resources. Virtual technology makes it possible to interact in real time through simulated activities and so enables immersion and interaction (Fuchs, 2018). Available scripts can mirror a large variety of IADL (for instance, preparing a meal, doing errands in a mall, driving a car) or tasks included in IADL (e.g., moving around in a town, navigating to find a route). These scripts demand analytical and adaptation skills from the person in order that they can respond to the demands created in VR simulation. For persons with cognitive impairment, it requires them to use control resources and to process information by acting and carrying out several tasks simultaneously. Recording of behavioral data allows the clinician to develop clinical indicators and to use replay with clients to encourage self-awareness and efficiency of cognitive strategies. Virtual activity rehearsal in the same controlled conditions or multicontextual training (Toglia et al., 2011) promotes learning and generalization of efficient behavior (Klinger et al., 2010). Weiss et Col. proposed a cognitive model of the experience of patients using VR in CR (Weiss et al., 2006). They identified three levels of experience: the Interaction Space between the user and VR, where the user experiments and interacts with the virtual environment; the

Transfer Phase which refers to the process that leads to transfer and generalization of skills or strategy to the real-world, and the Real World Environment, where participation is targeted as the ultimate goal of CR. Clinicians who use VR in CR must first identify their CR goals: What has to be improved? What is the main goal of the intervention? How can they meet these goals making the best use of VR, the ultimate aim being to improve the client's participation in everyday life (Kizony, 2018)?

Many recent literature reviews have explored the usefulness and evidence of effectiveness of the use of VR in CR for persons with TBI (Pietrzak et al., 2014; Shin and Kim, 2015; Imhoff et al., 2016; Aida et al., 2018; Alashram et al., 2019; Banville et al., 2019; Maggio et al., 2019; Manivannan et al., 2019.) They strongly underline the issues mentioned above in relation to this specific population. At the same time, they underline the importance of developing VR to provide clinicians with effective tools that complement other more traditional CR approaches. All authors agree that the use of VR in CR for persons after TBI offers a significant potential in the assessment and treatment of people with TBI, whatever the stage of care, from the initial phase to long-term follow-up. However, despite the promising findings from the existing studies, the reviews underline the need to carry out high-powered and long-lasting follow-up experimental studies particularly in relation to CR and moderate to severe TBI (Imhoff et al., 2016; Alashram et al., 2019; Maggio et al., 2019).

Technological devices used in existing studies are essentially non immersive devices based on the use of both computer or laptop keyboard with mouse or joystick to interact and a computer screen for visual feedback (e.g., Besnard et al. 2016). Visual feedback can be from a video projection on a large screen to create a semi-immersive display (Figure 7.1). Finally, and contrary to common

FIGURE 7.1 Semi-immersive environment in cognitive rehabilitation. (Reprinted with permission – CMRF Kerpape – France.)

representations about VR, few clinical studies use fully immersive devices such as a head mounted display (HMD) (e.g., Dahdah et al., 2017). Despite many technological improvements in recent years (Mayor), there is still concern about the risk of cybersickness when using fully immersive devices, particularly with neurological patients. However, in studies with people with TBI, few problems have been reported (Cox et al., 2010; Dahdah et al., 2017; Robitaille et al., 2017).

In persons with moderate to severe TBI, dysexecutive syndrome is compounded when performing tasks that are not familiar (Poncet et al., 2017). Thus, the use of VR could increase task demand if the use of VR devices and virtual environment are not familiar. One way to control the familiarity of the technical use of VR is to begin any VR intervention with a familiarization time. This makes it possible to control both the technological variable and the functional features of the virtual environment, and thus avoid interpretation bias on the quality of performance (Klinger et al., 2010).

A variety of simulated activities of daily living are used with persons with TBI. For instance, cooking tasks such as preparing a hot beverage (Besnard et al., 2016), or preparing a simple meal (Zhang et al., 2003), or heating up a snack in a microwave oven (Yip & Man, 2009). Other tasks may concern navigating outside to learn or find a route (Caglio et al., 2012; Sorita et al., 2013). Navigation in simulated real environments is surprisingly little reported in VR literature while clinical literature reports significant problems of topographical orientation in TBI patients in their daily life (Boyd and Sautter, 1993). Other studies of VR have enabled a deeper understanding of the underlying cerebral processes involved in topographical disorders among the broad population of people with TBI (Cogné et al., 2017). Driving is a complex activity involving specific motor and cognitive skills. It is a major issue for people with TBI, who have a relatively young mean age. Several studies explore road safety and respect of driving rules in this population (Cox et al., 2010; Milleville-Pennel et al., 2010) or the underlying cognitive processes involved in driving (Lengenfelder et al., 2002). Considering the prevalence of executive disorders among people with TBI, the most important part of tasks in VR involve behavioral planning to reach a goal, time, and space organization and multitasking. Among these kinds of complex adaptative tasks are office task management (McGeorge et al., 2001), library management (Renison et al., 2012), or multiple errands tasks in the supermarket or mall (Jacoby et al., 2013; Canty et al., 2014). Multiple errands tasks in the supermarket are among the most widely used daily living tasks in VR. The supermarket is a natural, complex setting that mobilizes brain resources and notably executive functioning, which is impacted in numerous patients with brain injury (Josman et al., 2008). Finally, in several studies, the authors rely on commercial video games (Grealy et al., 1999; Caglio et al., 2012). The main objective of intervention is to improve memory or attention by stimulating cognitive processes.

VR ASSESSMENT OF COGNITIVE PERFORMANCE ON ACTIVITIES OF DAILY LIVING FOR PEOPLE WITH TBI

COMPARISON BETWEEN VIRTUAL AND REAL SETTINGS

Initial studies on the assessment of people with TBI have sought to measure whether performance was the same between VR and the real environment. Among earlier studies, Christiansen and Zhang (Christiansen et al., 1998; Zhang et al., 2003) introduced measures to compare virtual performance and real task performance when preparing a simple meal. The virtual environment provided a simulated kitchen environment on a computer screen with which patients can interact using the keyboard and the regular mouse. The main objective of these studies was to assess the impact of cognitive disorders after TBI on meal preparation and to compare virtual and real performance in this task. Task analysis was based on a step-by-step approach to fulfill the task and on the patients' abilities to spontaneously undertake every sequence, or with the need of cues. Totally, 54 patients with severe TBI were included in the study. The authors' findings showed strong correlations between performance result in the meal preparation and between real and virtual settings. Furthermore, they found correlations between the virtual meal preparation task and neuropsychological tests, highlighting that cognitive processes were associated with this kind of simple task.

In a recent study to compare virtual and real performance, our team compared two groups of people with moderate to severe TBI in a route-learning task (Sorita et al., 2013). The virtual district was simulated from a real urban district just next to the hospital of Bordeaux where the participants were inpatients. The virtual setting was semi-immersive with a video projection displayed on a large screen (Figure 7.1). Patients were seated in front of the screen and moved themselves in the virtual district using a joystick. The learning procedure was based on the route learning task (Barrash et al., 2000). The route distance was approximately 800 meters. The assessor showed the route once and subjects had to repeat the same route twice; immediately and after a delay of one day. After doing the route three times, subjects had to complete three spatial cognitive tasks, two map-like tasks, and one chronological arrangement of pictures from the route. Learning route scores did not show any differences between real or virtual learning settings. Furthermore, the learning curves were strictly parallel. The only task that differed between the real setting and VR was the picture arrangement task. We suggested that this discrepancy could be due to the lack of sensorimotor input during the virtual task compared to the real task, sensorimotor inputs being usually involved in these kinds of spatial representations.

According to the results of comparisons between performances carried out in virtual or real settings, a key question is whether the same cerebral networks are implicated in both environments. Slight differences have been shown when comparing performance in real and virtual settings in studies using functional magnetic resonance imaging (IRMf). Part of the cerebral networks involved is the same between settings, but virtual tasks involve specific cerebral networks due to the use of technology (see for instance Mellet et al., 2010). One suggestion from these studies is the importance of familiarizing subjects to the specific use of VR technology when doing a task, to reduce the bias of cognitive workload that could alter the task progress (Besnard et al., 2016).

COMPARISON BETWEEN PEOPLE AFTER TBI AND HEALTHY CONTROL SUBJECTS

Regarding the comparison between people after TBI and Healthy control subjects (HC), a study published in the early 2000s assessed a group of patients with dysexecutive syndromes and notably with planning disorders (McGeorge et al., 2001). The virtual environment was displayed on a 17″ computer screen and subjects interacted with a standard mouse. The subjects' performance was compared with the same task in a real setting using a video analysis. The authors designed an original task. Subjects had to plan and perform several subtasks that were part of an office workflow. Principles of the MET (Shallice and Burgess) were implemented to design this task. A TBI group of five patients was compared with a HC group. The comparison showed the power of the task to discriminate the quality of performance between groups. The TBI group performance was significantly lower than the HC performance due to dysexecutive syndrome. Performance in the virtual environment was significantly correlated with performance in the real environment for both groups.

Other studies focusing on the assessment of cognitive impact on behavioral performance have been published subsequently, with tasks designed on the basis of neuropsychological models. The main objective was to study the psychometric properties of virtual tasks and their sensitivity when comparing TBI and HC groups. For instance, Renison et al. (2012) included 20 participants in a post TBI group, with onset of TBI more than one year, and a HC group. Executive disorders were characteristic in the TBI group. The authors wanted to compare the performance of the groups in a real and virtually simulated library task. In this task subjects had to follow predefined rules. Instructions were displayed on the screen depicting the task to be performed. Tasks were designed on the basis of seven theoretical dimensions of executive functioning. The virtual environment was not immersive, using a computer screen and a gamepad. A conventional battery of neuropsychological paper and pencil tests was also administered, in addition to a questionnaire aiming to measure dysexecutive syndrome in daily life. Lastly, the virtual task was compared with a similar real library task with the purpose of assessing the EV of the virtual task. Results showed significant differences between groups, whether the task was done in VR or in the real setting. In the virtual task, four sub scores out of seven summarizing the executive functioning components were significantly different between groups. Components of prospective memory, inhibition capacity and multitasking were significantly different between groups. For the two groups, performance scores in real and virtual tasks were highly correlated with paper and pencil neuropsychological tests as well as for six of the seven executive functioning scores. Furthermore, the total score and three sub scores of the virtual task showed significant correlations with the daily living questionnaire. In this study, the authors demonstrated the EV of the virtual library task with a construct that reflects both the underlying neuropsychological processes implicated in the task and the behavior in daily life.

Correlations of virtual environments with measures aiming to capture daily living functioning are not sufficient in themselves. The development of virtual tasks also has to show convergent validity with neuropsychological tests intended to measure underlying injured cognitive processes. In a recent and interesting study, Besnard et al. (2016) compared the performance of TBI and HC groups in a virtual task of making a coffee in virtual and real kitchens. The TBI group included 19 subjects in a post-acute to chronic phase after TBI. The HC group included 24 volunteers matched with the TBI group for years of education and gender. All the participants received a familiarization training in the virtual environment. VR was non-immersive, using a 17″ laptop screen. VR performance indicators were the time to complete the task, percentage of completed sequences with or without errors, the total number of errors, with a qualitative identification of error-type classified into two main categories of omission errors (to forget an action), and commission errors (to do a non-expected action). These categories of error are commonly reported in the literature as a consequence of dysexecutive syndrome (Poncet et al., 2017). Comparison between HC and TBI groups showed significant differences between the two modalities, real and virtual, in the five task performance indicators. In the HC group, time to completion and total of errors correlated significantly between the two modalities. Interestingly, the authors noted that despite correlations, the number of errors as well as the time to completion was significantly higher in the virtual kitchen than in real kitchen for both groups. They suggested that VR was more sensitive to action impairment than the real environment. This discrepancy was due to the greater additional attentional demand in VR for persons with TBI, who have less resources than healthy subjects. Another hypothesis was that action schemas were not as intuitive as in the real setting, leading both groups to have a performance discrepancy. To conclude, the authors suggested that, despite the demonstrated EV of the virtual task, this particular task was more cognitively demanding than the same task in a real setting.

The development of cognitive assessments using virtual-simulated tasks can also be used to screen for specific neuropsychological diseases. For example, PM is a specific type of memory that can be particularly affected after TBI. PM can be described as memory for events occurring in the future and is therefore very important for optimal functioning in daily life. VR offers interesting potential in relation to the evaluation of PM. Canty et al. designed a shopping task test based on a cognitive model of PM. The task was based in a virtual shopping mall and was designed to be sensitive to PM. The authors tested the EV of VRST to demonstrate its close theoretical model construct and at the same time, they were interested in whether it could predict the impact of PM disorder on daily life functioning. Twenty-five patients with severe TBI, with an onset of less than two years were included in the study. The HC group consisted of 24 healthy adults. The VRST was presented on a laptop computer and commands were carried out using the keyboard keys. The shopping task involved purchasing 12 items in a prespecified chronological order, in a mall consisting of 20 shops. During the shopping task, subjects had to carry out three time-based PM tasks and three event-based PM tasks. The total duration of the task was 14 minutes overall. The results showed significantly lower scores for the TBI group compared to HC group for the overall scores on the evaluation battery, and notably for the time-based PM and event-based PM tasks. Moderate correlations were found between PM tasks and independent life skills, demonstrating the relative importance of PM in daily life. This study concluded that the virtual PM task was of clinical interest in a controlled virtual setting compared to real-life setting, but also was more sensitive than traditional PM assessment tasks. In addition, measures from the VR tasks were the strongest predictors of psychosocial functioning.

COGNITIVE VIRTUAL REHABILITATION PROGRAMS FOR PEOPLE WITH TBI

In addition to the development of virtual cognitive assessment, cognitive rehabilitation programs have begun to be evaluated. Cognitive rehabilitation includes both the restorative and functional approaches (Cicerone et al., 2011). The main objective of the restorative approach is to stimulate the restoration of the underlying damaged cognitive processes (Ben-Yishay & Diller, 1993) whereas the functional approach aims to teach patients to optimize their preserved cognitive information processing strategies to improve their effectiveness and safety (Goverover, 2018). Cognitive strategy disruptions when processing environmental information were specifically highlighted in the shopping tasks (Bottari et al., 2014), in household activities or high demand leisure activities (Goverover et al., 2017) or when driving (Bottari et al., 2012). These impaired cognitive strategies must be analyzed and understood to implement an effective rehabilitation program (Toglia et al., 2012; Bottari et al., 2014). Key principles of a cognitive functional rehabilitation program for people with TBI must consider four major aspects; the level of self-awareness of patients, their learning potential, their ability to abstract rules and generalize learning, and the efficiency of available cognitive strategies that can be used to improve efficiency and safety (Toglia, 1992; Dawson et al., 2009; Sohlberg & Turkstra, 2011).

For instance, several VR training studies looking at recovery of daily living abilities have focused on driving ability. Recovering the ability to drive requires on-road training (Ross et al., 2018). However, VR can be an additional useful alternative (Imhoff et al., 2016). However, there are currently few studies that focus on driving rehabilitation programs for people with TBI, compared to studies that focus on virtual driving assessment (Imhoff et al., 2016). Some authors have recently recommended that this kind of training program should be developed for people with brain injury (Dimech-Betancourt et al., 2019). VR driving simulators have the particular advantage of being able to control variables in order to promote the procedural training of the core abilities necessary for driving. Moreover, with VR the level of driving demand can be varied (e.g., urban or country driving, traffic intensity). Retraining of driving abilities is based on rehearsal practice in scenarios of increasing difficulty levels. In their study, Cox et al. (2010) compared a group of people with TBI following a virtual driving training program with a group of people with TBI taking part in a residential rehabilitation program.

The VR training simulator was a semi-immersive virtual environment displayed on a 180° field of vision screen with normal driving instruments. VR driving training phases were alternated with fun driving game time, aimed at mobilizing attentional resources and subjects' motivation. During the training phases, subjects were trained to respect driving rules (e.g., remain driving in own lane, braking in good time, indicating a turn, making regular safety checks). The VR driving simulator TBI group had four to six virtual driving sessions lasting 60–90 minutes. Significant improvement was shown with this VR driving program when comparing the results of the intervention group with the conventional residential rehabilitation TBI group. Number of errors was the main indicator. The residential TBI group made significantly more errors than the VR intervention group. Moreover, the driving behavior questionnaires showed significant differences between the two groups.

The generalizability of cognitive training is a pertinent and continuing question for researchers and clinicians using VR (Weiss et al., 2006; Joseph et al., 2014). Among the first studies aimed at improving daily living performance, those of Yip and Man (2009, 2013) have shown the advantage of using VR to re-acquire useful daily living skills in persons with brain injury.

In a pilot study, Yip and Man (2009) designed a program of ten training sessions of 35–40 minutes, three times a week, to develop Instrumental activities of Daily Living (IADL) abilities in tasks such as using road crossings, traveling by bus, shopping in a store, using a public phone, and meeting a friend. Subjects used a non-immersive 32″ wide LCD (liquid-crystal display) monitor with an interactive joystick. Out of the four subjects included in this pilot study, one was a person with a TBI. In a pre-post design, the outcome measures used were indicators of virtual performance (for instance, distance covered, task realization time). This study showed interesting learning effects despite the presence of cognitive impairment. Moreover, the authors asked subjects to perform VR learning tasks in a real urban setting with the aim of comparing performance before and after the training. Behavior was assessed using an expected behavior checklist that included 110 expected behaviors. The authors reported overall improvement of initial performance for all subjects. For the subject with TBI, the behavior checklist improvement was 15.6%. The authors suggested a transfer effect from VR training to real life settings.

More recently, Yip and Man (2013) designed a 12-session VR training program twice a week composed of a task using PM. Thirty-seven subjects were included in this study with a pre-posttest design. The origin of patients' brain injury is not specified in this study; however, PM disorders are frequently found in people after a TBI (Wilson et al., 2005). The main hypothesis of the study was that the group trained in the VR program would improve their PM performance in virtual and real situations compared to a non-trained control group. VR training was carried out in a non-immersive way using a LCD screen and joystick or keyboard with the choice left to the participant. Training tasks simulated daily living tasks (e.g., shopping, warming-up a meal in a microwave), and subjects had to respond to time-based or event-based stimuli to perform prospective tasks. Both groups had similar scores in the initial cognitive test battery. After VR training, the VR group showed a significant improvement in the indicators in virtual and real tasks compared to before training. Moreover, the prospective behaviors in real tasks were significantly better for the VR group compared to the control group, whereas the initial cognitive scores were similar between groups.

Beyond virtual training, it is of clinical importance to verify if behaviors that have been VR trained transfer and generalize in real situations. Thus, in a recent pilot randomized controlled trial (RCT) study, Jacoby et al. (2013) sought to verify the superiority of a VR intervention aiming to improve executive functioning compared to a traditional CR intervention in a pre-post design study. Both interventions were designed using the same CR principles that aimed to improve efficiency of information processing strategies in order to achieve an intentional goal. Twelve people who had sustained a TBI were included in the study and randomly divided in two groups. The VR intervention was based on a Virtual Mall (Vmall) designed by the same team few years earlier (Rand et al., 2005). The VR intervention consisted of ten sessions of 45 minutes, three to four times a week without any additional occupational therapy intervention. The traditional CR group received the same amount of time. Before the intervention, subjects were assessed in a real supermarket using the Multiple Errands Test – Simplified Version (MET-SV) assessment (Alderman et al., 2003) and with other conventional executive functioning assessments. At the end of intervention, scores in the MET-SV assessment were not significantly different between groups. However, the experimental group that had received the VR training significantly improved on the MET-SV score after training compared to the initial score. The authors concluded that an intervention using VR would improve the functional result of CR intervention. Moreover, they noted that performance trained in VR transferred into real settings, confirming the EV and transfer potential of VR training simulating the real environment.

CONCLUSION AND PERSPECTIVE

While the research concludes that VR has strong potential in CR for people with brain injuries, there is also agreement that further high-quality research is needed before it can be recommended in all cases (Pietrzak et al., 2014; Shin and Kim, 2015; Imhoff et al., 2016; Aida et al., 2018; Alashram et al., 2019; Banville et al., 2019; Maggio et al., 2019; Manivannan et al., 2019).

This chapter aimed to address a major clinical concern, which is the EV of VR and how it can be used to help people with TBI to further benefit from CR (Weiss et al., 2006). In the clinical development of VR, the gap between cognitive theories, growing neuroscience knowledge, and daily functioning is being gradually bridged (Parsons et al., 2017). VR appears to be one of the new technologies that could establish a possible link between cognition and daily functioning, whereas traditional CR approaches were marked by dissociations between clinical reality and the realities of daily life for patients. Burgess et al. (2006) underlined the need for CR to move towards a better understanding of everyday life actions when he stated: « *We know virtually nothing about how the brain allows us to organize simple, everyday activities like cooking, and interacts with environmental factors in doing so.* » (Burgess et al., 2006, p. 197). The study of the EV of VR to better understand how the brain works when interacting with environment was of great interest and enables a better understanding of large-scale spatial cognition (Cogné et al., 2017).

CR for persons with TBI is particularly complex and the effects of CR on daily living functioning can be relatively limited (Cicerone et al., 2019). In addition, there are still a lack of studies on CR that use daily living measures as outcomes and that show that interventions are relevant to daily life for people with TBI and their families. Introducing these kinds of outcome measures in studies that aim to demonstrate evidence for using VRCR to improve cognitive functioning should be self-evident, equally in relation to the assessment of underlying neuropsychological dysfunctions. CR models have shown that it was erroneous to think that improvement of single cognitive processes would automatically have an effect on functional behavior in everyday life. For instance, Campbell (2009) used VR to show that, for planning, one of the cognitive processes involved in executive functioning, the cerebral networks were not the same between a classic neuropsychological assessment of planning and the planning behavior used to actually plan a route in a virtual town. Thus, it is possible that intervention programs designed to improve the results of traditional nonecological tests have limited impact on real life. The need to develop new standardized and validated cognitive VR tests designed from simulated activities of daily living seems to be a key point in order to make CR more effective at helping people in their everyday life.

VR technologies are gradually embarking on a new phase (How et al., 2017). VR cognitive assessment has been well studied since the end of the nineties. Past and ongoing work enables us to see that VR has an interesting potential in cognitive evaluation and neuroscience research and can bridge a part of the gap between underlying cognitive processes and interactive behavior within the environment. CR training still needs to consolidate its effects and complementarity with traditional CR interventions.

Just as for people with TBI, there is a growing population of people with acquired brain injury or progressive brain disorders. This changing situation requires us to rethink the usual pathways for access to health care as well as the intervention and follow-up for these patients, because neurological disorders need lifelong care (Jackson et al., 2014). Thus, clinicians need to develop VR tools to enable the development of new practices of assessment and intervention in CR and a better access to care for more patients. To this end, it is time now to train students and clinicians in the use of VR technologies and to widely disseminate new research on VR environments that is theoretically and clinically well designed to meet the needs of patients (Dimech-Betancourt et al., 2019).

REFERENCES

Aida, J., Chau, B., & Dunn, J. (2018). Immersive virtual reality in traumatic brain injury rehabilitation: A literature review. *NeuroRehabilitation*, 42(4), 441–448.

Alashram, A. R., Annino, G., Padua, E., Romagnoli, C., & Mercuri, N. B. (2019). Cognitive rehabilitation post traumatic brain injury: A systematic review for emerging use of virtual reality technology. *Journal of Clinical Neuroscience*, 66, 209–219.

Alderman, N., Burgess, P. W., Knight, C., & Henman, C. (2003). Ecological validity of a simplified version of the multiple errands shopping test. *Journal of the International Neuropsychological Society*, 9(1), 31–44.

Banville, F., Nolin, P., Rosinvil, T., Verhulst, E., & Allain, P. (2019). Assessment and rehabilitation after traumatic brain injury using virtual reality: A systematic review and discussion concerning human-computer interactions. In A. "Skip". Rizzo, & S. Bouchard (eds.), *Virtual reality for psychological and neurocognitive interventions* (pp. 327–360). New York, NY: Springer.

Barrash, J., Damasio, H., Adolphs, R., & Tranel, D. (2000). The neuroanatomical correlates of route learning impairment. *Neuropsychologia*, 38(6), 820–836.

Ben-Yishay, Y., & Diller, L. (1993). Cognitive remediation in traumatic brain injury: Update and issues. *Archives of Physical Medicine and Rehabilitation*, 74(2), 204–213.

Besnard, J., Richard, P., Banville, F., Nolin, P., Aubin, G., Le Gall, D., ... & Allain, P. (2016). Virtual reality and neuropsychological assessment: The reliability of a virtual kitchen to assess daily-life activities in victims of traumatic brain injury. *Applied Neuropsychology: Adult*, 23(3), 223–235.

Bottari, C., Lamothe, M. P., Gosselin, N., Gélinas, I., & Ptito, A. (2012). Driving difficulties and adaptive strategies: The perception of individuals having sustained a mild traumatic brain injury. *Rehabilitation Research and Practice*. doi:10.1155/2012/837301.

Bottari, C., Shun, P. L. W., Le Dorze, G., Gosselin, N., & Dawson, D. (2014). Self-generated strategic behavior in an ecological shopping task. *American Journal of Occupational Therapy*, 68(1), 67–76.

Boyd, T. M., & Sautter, S. W. (1993). Route-finding: A measure of everyday executive functioning in the head-injured adult. *Applied Cognitive Psychology*, 7(2), 171–181.

Burgess, P. W., Alderman, N., Forbes, C., Costello, A., Coates, L. M., Dawson, D. R., ... & Channon, S. (2006). The case for the development and use of "ecologically valid" measures of executive function in experimental and clinical neuropsychology. *Journal of the International Neuropsychological Society*, 12(2), 194–209.

Caglio, M., Latini-Corazzini, L., D'Agata, F., Cauda, F., Sacco, K., Monteverdi, S., ... & Geminiani, G. (2012). Virtual navigation for memory rehabilitation in a traumatic brain injured patient. *Neurocase*, 18(2), 123–131.

Campbell, Z., Zakzanis, K. K., Jovanovski, D., Joordens, S., Mraz, R., & Graham, S. J. (2009). Utilizing virtual reality to improve the ecological validity of clinical neuropsychology: an FMRI case study elucidating the neural basis of planning by comparing the Tower of London with a three-dimensional navigation task. *Applied Neuropsychology*, 16(4), 295–306.

Canty, A. L., Fleming, J., Patterson, F., Green, H. J., Man, D., & Shum, D. H. (2014). Evaluation of a virtual reality prospective memory task for use with individuals with severe traumatic brain injury. *Neuropsychological Rehabilitation*, 24(2), 238–265.

Caron, E., Lesimple, B., Debarle, C., Lefort, M., Galanaud, D., Perlbarg, V. ... & Pradat-Diehl, P. (2018). Neuropsychological assessment of a long-term (LT) outcome after severe traumatic brain injury (TBI). *Annals of Physical and Rehabilitation Medicine*, 61, e230.

Christiansen, C., Abreu, B., Ottenbacher, K., Huffman, K., Masel, B., & Culpepper, R. (1998). Task performance in virtual environments used for cognitive rehabilitation after traumatic brain injury. *Archives of Physical Medicine and Rehabilitation*, 79(8), 888–892.

Cicerone, K. D., Langenbahn, D. M., Braden, C., Malec, J. F., Kalmar, K., Fraas, M., ... & Azulay, J. (2011). Evidence-based cognitive rehabilitation: Updated review of the literature from 2003 through 2008. *Archives of Physical Medicine and Rehabilitation*, 92(4), 519–530.

Cicerone, K. D., Goldin, Y., Ganci, K., Rosenbaum, A., Wethe, J. V., Langenbahn, D. M., ... & Trexler, L. (2019). Evidence-based cognitive rehabilitation: Systematic review of the literature from 2009 through 2014. *Archives of Physical Medicine and Rehabilitation*, 100(8), 1515–1533.

Cogné, M., Taillade, M., N'Kaoua, B., Tarruella, A., Klinger, E., Larrue, F., ... & Sorita, E. (2017). The contribution of virtual reality to the diagnosis of spatial navigation disorders and to the study of the role of navigational aids: A systematic literature review. *Annals of Physical and Rehabilitation Medicine*, 60(3), 164–176.

Cooper, R., & Shallice, T. (2000). Contention scheduling and the control of routine activities. *Cognitive Neuropsychology*, 17(4), 297–338.

Cox, D. J., Davis, M., Singh, H., Barbour, B., Nidiffer, F. D., Trudel, T., ... & Moncrief, R. (2010). Driving rehabilitation for military personnel recovering from traumatic brain injury using virtual reality driving simulation: A feasibility study. *Military Medicine*, 175(6), 411–416.

Dahdah, M. N., Bennett, M., Prajapati, P., Parsons, T. D., Sullivan, E., & Driver, S. (2017). Application of virtual environments in a multi-disciplinary day neurorehabilitation program to improve executive functioning using the Stroop task. *NeuroRehabilitation*, 41(4), 721–734.

Dawson, D. R., & Chipman, M. (1995). The disablement experienced by traumatically brain-injured adults living in the community. *Brain Injury*, 9(4), 339–353.

Dawson, D. R., Gaya, A., Hunt, A., Levine, B., Lemsky, C., & Polatajko, H. J. (2009). Using the cognitive orientation to occupational performance (CO-OP) with adults with executive dysfunction following traumatic brain injury. *Canadian Journal of Occupational Therapy*, 76(2), 115–127.

Dimech-Betancourt, B., Ross, P. E., Ponsford, J. L., Charlton, J. L., & Stolwyk, R. J. (2019). The development of a simulator-based intervention to rehabilitate driving skills in people with acquired brain injury. *Disability and Rehabilitation: Assistive Technology*, DOI: 10.1080/17483107.2019.1673835 1–12.

Franzen, M. D., & Wilhelm, K. L. (1996). Conceptual foundations of ecological validity in neuropsychology. In Sbordone, R. J., & Long, C. J. (eds.), *Ecological validity of neuropsychological testing* (pp. 91–112). Delray Beach, FL: GR Press/St. Lucie Press.

Fuchs, P. (2018). *Théorie de la réalité virtuelle. Les véritables usages (A theory of virtual reality: The real uses)*. Paris, France: Presses des Mines.

Goverover, Y. (2018). Cognitive rehabilitation: Evidence based interventions. In N. Katz, & J. Toglia (eds.), *Cognition, occupation, and participation across the lifespan: Neuroscience, neurorehabilitation, and models of intervention in occupational therapy* (pp. 51–66). Bethesda, MD: AOTA Press.

Goverover, Y., Genova, H., Smith, A., Chiaravalloti, N., & Lengenfelder, J. (2017). Changes in activity participation following traumatic brain injury. *Neuropsychological Rehabilitation*, 27(4), 472–485.

Grealy, M. A., Johnson, D. A., & Rushton, S. K. (1999). Improving cognitive function after brain injury: The use of exercise and virtual reality. *Archives of Physical Medicine and Rehabilitation*, 80(6), 661–667.

How, T. V., Hwang, A. S., Green, R. E., & Mihailidis, A. (2017). Envisioning future cognitive telerehabilitation technologies: A co-design process with clinicians. *Disability and Rehabilitation: Assistive Technology*, 12(3), 244–261.

Imhoff, S., Lavallière, M., Teasdale, N., & Fait, P. (2016). Driving assessment and rehabilitation using a driving simulator in individuals with traumatic brain injury: A scoping review. *NeuroRehabilitation*, 39(2), 239–251.

Jackson, D., McCrone, P., Mosweu, I., Siegert, R., & Turner-Stokes, L. (2014). Service use and costs for people with long-term neurological conditions in the first year following discharge from in-patient neurorehabilitation: A longitudinal cohort study. *PloS One*, 9(11), e113056.

Jacoby, M., Averbuch, S., Sacher, Y., Katz, N., Weiss, P. L., & Kizony, R. (2013). Effectiveness of executive functions training within a virtual supermarket for adults with traumatic brain injury: A pilot study. *IEEE Transactions on Neural Systems and Rehabilitation Engineering*, 21(2), 182–190.

Joseph, P. A., Mazaux, J. M., & Sorita, E. (2014). Virtual reality for cognitive rehabilitation: From new use of computers to better knowledge of brain black box? *International Journal on Disability and Human Development*, 13(3), 319–325.

Josman, N., Klinger, E., & Kizony, R. (2008). Performance within the virtual action planning supermarket (VAP-S): An executive function profile of three different populations suffering from deficits in the central nervous system. Maia, Portugal: Proc 7th ICDVRAT.

Katz, N., & Hartman-Maeir, A. (2005). Higher-level cognitive functions: Awareness and executive functions enabling engagement in occupation. In N. Katz, & J. Toglia (eds.), *Cognition and occupation across the life span: Models for intervention in occupational therapy*, 4th edition (pp. 3–25). Bethesda, MD: AOTA Press.

Kizony, R. (2018). Virtual reality for cognitive rehabilitation. In N. Katz, & J. Toglia (eds.), *Cognition, occupation and participation across the life span: Neuroscience, neurorehabilitation, and models of intervention in occupational therapy*, 4th edition (pp. 231–242). Bethesda, MD: AOTA Press.

Klinger, E., Weiss, P. L., & Joseph, P. A. (2010). Virtual reality for learning and rehabilitation. In J-P. Didier, & E. Bigand (eds.), *Rethinking physical and rehabilitation medicine* (pp. 203–221). Paris: Springer.

Lengenfelder, J., Schultheis, M. T., Al-Shihabi, T., Mourant, R., & DeLuca, J. (2002). Divided attention and driving: A pilot study using virtual reality technology. *The Journal of Head Trauma Rehabilitation*, 17(1), 26–37.

Lezak, M. D. (1982). The problem of assessing executive functions. *International Journal of Psychology*, 17(1–4), 281–297. doi:10.1080/00207598208247445

Maas, A. I., Menon, D. K., Adelson, P. D., Andelic, N., Bell, M. J., Belli, A., ... & Citerio, G. (2017). Traumatic brain injury: Integrated approaches to improve prevention, clinical care, and research. *The Lancet Neurology*, 16(12), 987–1048.

McGeorge, P., Phillips, L. H., Crawford, J. R., Garden, S. E., Sala, S. D., Milne, A. B., ... & Callender, J. S. (2001). Using virtual environments in the assessment of executive dysfunction. Presence. *Teleoperators & Virtual Environments*, 10(4), 375–383.

Maggio, M. G., De Luca, R., Molonia, F., Porcari, B., Destro, M., Casella, C., ... & Calabro, R. S. (2019). Cognitive rehabilitation in patients with traumatic brain injury: A narrative review on the emerging use of virtual reality. *Journal of Clinical Neuroscience*, 61, 1–4.

Manivannan, S., Al-Amri, M., Postans, M., Westacott, L. J., Gray, W., & Zaben, M. (2019). The effectiveness of virtual reality interventions for improvement of neurocognitive performance after traumatic brain injury: A systematic review. *The Journal of Head Trauma Rehabilitation*, 34(2), E52–E65.

Mellet, E., Laou, L., Petit, L., Zago, L., Mazoyer, B., & Tzourio-Mazoyer, N. (2010). Impact of the virtual reality on the neural representation of an environment. *Human Brain Mapping*, 31(7), 1065–1075.

Milleville-Pennel, I., Pothier, J., Hoc, J. M., & Mathé, J. F. (2010). Consequences of cognitive impairments following traumatic brain injury: Pilot study on visual exploration while driving. *Brain Injury*, 24(4), 678–691.

Parsons, T. D., Carlew, A. R., Magtoto, J., & Stonecipher, K. (2017). The potential of function-led virtual environments for ecologically valid measures of executive function in experimental and clinical neuropsychology. *Neuropsychological Rehabilitation*, 27(5), 777–807.

Pietrzak, E., Pullman, S., & McGuire, A. (2014). Using virtual reality and videogames for traumatic brain injury rehabilitation: A structured literature review. *Games for Health: Research, Development, and Clinical Applications*, 3(4), 202–214.

Poncet, F., Swaine, B., Dutil, E., Chevignard, M., & Pradat-Diehl, P. (2017). How do assessments of activities of daily living address executive functions: A scoping review. *Neuropsychological Rehabilitation*, 27(5), 618–666.

Ponsford, J. L., Downing, M. G., Olver, J., Ponsford, M., Acher, R., Carty, M., & Spitz, G. (2014). Longitudinal follow-up of patients with traumatic brain injury: Outcome at two, five, and ten years post-injury. *Journal of Neurotrauma*, 31(1), 64–77.

Poulin, V., Dawson, D. R., Bottari, C., Verreault, C., Turcotte, S., & Jean, A. (2019). Managing cognitive difficulties after traumatic brain injury: A review of online resources for families. *Disability and Rehabilitation*, 41(16), 1955–1965.

Rand, D., Katz, N., Shahar, M., Kizony, R., & Weiss, P. L. (2005). The virtual mall: A functional virtual environment for stroke rehabilitation. *Annual Review of Cybertherapy and Telemedicine: A Decade of VR*, 3, 193–198.

Renison, B., Ponsford, J., Testa, R., Richardson, B., & Brownfield, K. (2012). The ecological and construct validity of a newly developed measure of executive function: the virtual library task. *Journal Of The International Neuropsychological Society: JINS*, 18(3), 440–450.

Rizzo, A. A., Schultheis, M., Kerns, K. A., & Mateer, C. (2004). Analysis of assets for virtual reality applications in neuropsychology. *Neuropsychological Rehabilitation*, 14(1–2), 207–239.

Robitaille, N., Jackson, P. L., Hébert, L. J., Mercier, C., Bouyer, L. J., Fecteau, S., ... & McFadyen, B. J. (2017). A virtual reality avatar interaction (VRai) platform to assess residual executive dysfunction in active military personnel with previous mild traumatic brain injury: Proof of concept. *Disability and Rehabilitation: Assistive Technology*, 12(7), 758–764.

Rose, F. D. (1996). Virtual reality in rehabilitation following traumatic brain injury. In Proceedings of the European Conference on Disability, Virtual Reality and Associated Technology (pp. 5–12).

Rose, F. D., Brooks, B. M., & Rizzo, A. A. (2005). Virtual reality in brain damage rehabilitation: Review. *CyberPsychology & Behavior*, 8(3), 241–262.

Ross, P. E., Di Stefano, M., Charlton, J., Spitz, G., & Ponsford, J. L. (2018). Interventions for resuming driving after traumatic brain injury. *Disability and Rehabilitation*, 40(7), 757–764.

Sbordone, R. J. (1996). Ecological validity: Some critical issues for the neuropsychologist. In R. J. Sbordone, & C. J. Long (eds.), *Ecological validity of neuropsychological testing* (pp. 15–41). Delray Beach, FL: GR Press/St. Lucie Press.

Schultheis, M. T., & Rizzo, A. A. (2001). The application of virtual reality technology in rehabilitation. *Rehabilitation Psychology*, 46(3), 296–311.

Shallice, T. I. M., & Burgess, P. W. (1991). Deficits in strategy application following frontal lobe damage in man. *Brain*, 114(2), 727–741.

Shin, H., & Kim, K. (2015). Virtual reality for cognitive rehabilitation after brain injury: A systematic review. *Journal of Physical Therapy Science*, 27(9), 2999–3002.

Sohlberg, M. M., & Turkstra, L. S. (2011). *Optimizing cognitive rehabilitation: Effective instructional methods*. New York, NY: Guilford Press.

Sorita, E., N'Kaoua, B., Larrue, F., Criquillon, J., Simion, A., Sauzéon, H., ... & Mazaux, J. M. (2013). Do patients with traumatic brain injury learn a route in the same way in real and virtual environments? *Disability and Rehabilitation*, 35(16), 1371–1379.

Spector, W. D., Katz, S., Murphy, J. B., & Fulton, J. P. (1987). The hierarchical relationship between activities of daily living and instrumental activities of daily living. *Journal of Chronic Diseases*, 40(6), 481–489.

Toglia, J. P. (1992). *A dynamic interactional approach to cognitive rehabilitation. In N. Katz (ed.), Cognitive rehabilitation: Models for intervention in occupational therapy* (pp. 104–143). Boston, MA: Andover Medical.

Toglia, J., Goverover, Y., Johnston, M. V., & Dain, B. (2011). Application of the multicontextual approach in promoting learning and transfer of strategy use in an individual with TBI and executive dysfunction. *Occupation, Participation and Health*, 31(1_suppl), S53–S60.

Toglia, J. P., Rodger, S. A., & Polatajko, H. J. (2012). Anatomy of cognitive strategies: A therapist's primer for enabling occupational performance. *Canadian Journal of Occupational Therapy*, 79(4), 225–236.

Weiss, P. L., Kizony, R., Feintuch, U., & Katz, N. (2006). Virtual reality in neurorehabilitation. In M. E. Selzer, L. Cohen, F. H. Gage, S. Clarke, & P. W. Duncan (eds.), *Textbook of neural repair and neurorehabilitation*, 2, (pp. 182–197). Cambridge, UK: Cambridge University Press

Wilson, B. A., Emslie, H., Foley, J., Shiel, A., Watson, P, Hawkins, K. et al. (2005). *The Cambridge prospective memory test*. London, UK: Harcourt Assessment.

Yip, B. C., & Man, D. W. (2009). Virtual reality (VR)-based community living skills training for people with acquired brain injury: A pilot study. *Brain Injury*, 23(13–14), 1017–1026.

Yip, B. C., & Man, D. W. (2013). Virtual reality-based prospective memory training program for people with acquired brain injury. *Neurorehabilitation*, 32(1), 103–115.

Zhang, L., Abreu, B. C., Seale, G. S., Masel, B., Christiansen, C. H., & Ottenbacher, K. J. (2003). A virtual reality environment for evaluation of a daily living skill in brain injury rehabilitation: reliability and validity. *Archives of Physical Medicine and Rehabilitation*, 84(8), 1118–1124.

8 Effect of Virtual Reality-Based Training of the Ankle, Hip, and Stepping Strategies on Balance after Stroke

Roberto Llorens
Universitat Politècnica de València
NEURORHB. Servicio de Neurorrehabilitación de Hospitales Vithas
València, Spain

CONTENTS

BACKGROUND

Despite its widespread use and common understanding, there is no universally accepted definition of human balance (Pollock et al. 2000). Definition of this term is complex and may arise from different fields of knowledge. From the biomechanical point of view, balance can be defined as the human ability to maintain the center of pressure, the vertical projection of the center of body mass onto the support surface, within the area of the base of support (Spaulding 2008). A systematic and outstanding body of research has been done on human balance, most of it addressing the motor control mechanisms that allow maintaining standing balance. Although from the eighties, successive studies have challenged and redefined the inverted pendulum model, it identified the role of the ankles and the hip during unbalancing and rebalancing periods (Winter 1995). The ankle and hip strategies encompass the motor mechanisms that lead to swaying from the ankles or the hip to maintain balance without modifying the base of support. The stepping strategy, in contrast, is used against larger perturbations to realign the base of support under the center of body mass (Horak 1987).

Balance is particularly vulnerable to some of the harmful consequences of a stroke, as it can cause multiple biomechanical alterations, such as muscle weakness, exaggerated reflex activity, and impaired coordination (Tyson et al. 2006), and harm multisensory integration (Bonan et al. 2004) and cognitive processing (Brown, Sleik, and Winder 2002). In addition, impairments associated with ageing and the higher incidence of stroke among the elderly (Cuevas-Trisan 2017) make balance disorders one of the most frequent impairments of individuals who have sustained a stroke (Geurts et al. 2005).

All these deleterious stroke- and age-related factors can lead to an impaired use of the balance strategies in stroke survivors (Chen et al. 2000). However, as other motor skills, balance strategies can improve with training and practice (Pollock et al. 2000). An increasing body of evidence supports that interventions based on motor learning principles can promote both clinical and neural changes (Maier, Ballester, and Verschure 2019). According to this, motor interventions should include repetitive, intensive, task-oriented, and motivating exercises with adaptive difficulty while providing multisensory feedback in enriched environments. The capacity of virtual reality (VR) to recreate safe and controlled simulated environments while potentially increasing motivation through engaging narratives with adjusted goal setting, rewards, and challenges has motivated its use in neurorehabilitation (Bermúdez i Badia et al. 2016). In the last years, multiple reviews and meta-analyses have summarized the findings of VR-based interventions on balance post-stroke (Chen et al. 2016; de Rooij, van de Port, and Meijer 2016; Laver et al. 2017; Iruthayarajah et al. 2017; Corbetta, Imeri, and Gatti 2015; Darekar et al. 2015; Dominguez-Tellez et al. 2019). Rather than an effort at summarizing the main existing literature or at building upon it just to add slight variations to the reviews, the aim of this book chapter is to describe the main findings of a series of studies conducted by our research group on the effectiveness and acceptability of a VR-based training of the balance strategies in adults with chronic stroke and their clinical implications. New analyses on aggregated data collected during all the studies and on unpublished data have been performed retrospectively to investigate the consistency of the published results.

Specifically, this chapter comprises a longitudinal follow-up study and a randomized controlled trial that examined the effectiveness and acceptability of the VR-based training of the ankle and hip strategies (Llorens et al. 2014; Gil-Gómez et al. 2011) and a subsequent and larger controlled study that investigated the influence of time since on the effectiveness of the training (Llorens et al. 2018). Furthermore, the chapter encompasses three longitudinal follow-up studies and a randomized controlled trial on the effectiveness and acceptability of the VR-based training of the stepping strategy (Llorens et al. 2011, 2012, 2013; Llorens, Gil-Gómez, Alcañiz, et al. 2015). An additional randomized control trial that explored the feasibility of a telerehabilitation program involving the stepping exercise (Llorens, Noé, Colomer, et al. 2015) and an exploratory study on the subjective perceptions of stroke survivors and physical therapists after using different motion tracking technologies have been accounted for (Llorens, Noé, Naranjo, et al. 2015).

All the studies were conducted in the clinical facilities of the Neurorehabilitation Service of Vithas Hospitals in Spain ("NEURORHB. Servicio de Neurorrehabilitación de Hospitales Vithas" 2020). The greatest success of all this research has been, beyond the scientific outcomes, the integration of the customized VR-based exercises, presented further in the text, into daily clinical practice. The exercises are used on a daily basis in ten workstations distributed throughout the hospital network, according to the particular care needs of each center (Figure 8.1). The overwhelming

Hospital El Consuelo   Hospital Sevilla Aljarafe   Hospital Aguas Vivas
(València, Spain)   (Sevilla, Spain)   (Carcaixent, Spain)

FIGURE 8.1   Individuals with a brain injury training with the stepping exercise in a different clinical facility of the hospital network.

interest and unconditional support of both clinical and managing staff have motivated and encouraged all the studies presented hereunder.

TRAINING OF THE ANKLE AND HIP STRATEGIES

INSTRUMENTATION

Since its launch in 2006, the capacity of the Nintendo Wii (Nintendo, Kyoto, Japan) game console to enable interaction with actual body movements drew the attention of the scientific community (Laver et al. 2017; Cheok et al. 2015). The Nintendo Wii Balance Board (WBB), a force platform bundled with the Wii Fit software, released one year later, was especially interesting for researchers interested in balance, as it is able to estimate the center of pressure of the users from the stress caused by their body weight distribution on four built-in strain gauges. However, the most interesting features of this device are, in contrast to laboratory-grade posturography systems, its low cost, portability, and widespread availability. Dedicated software for the WBB requires users to interact through weight shifts resulting from adjusting the body posture while (mostly) maintaining the soles in contact with the surface of the platform, which necessarily involves the use of the ankle and hip strategies. Although different studies have investigated the effects of balance training with the WBB using off-the-self video games, these applications are likely to present motor and cognitive demands that individuals post-stroke cannot meet. Consequently, to benefit from the advantages of the WBB yet covering the wide range of possible motor and cognitive sequelae after stroke, we designed customized exercises that could be adapted to fit the motor condition and needs of each individual while providing simple feedback of their motor performance and the task to accomplish without distractors. Interaction with all the exercises required participants to continuously use the ankle and hip strategies to control their center of pressure (Figure 8.2).

The Table 8.1 shows four exercises that have been used in different studies. Besides the parameters that modulated the difficulty of the exercises, individuals registered their limits of stability in

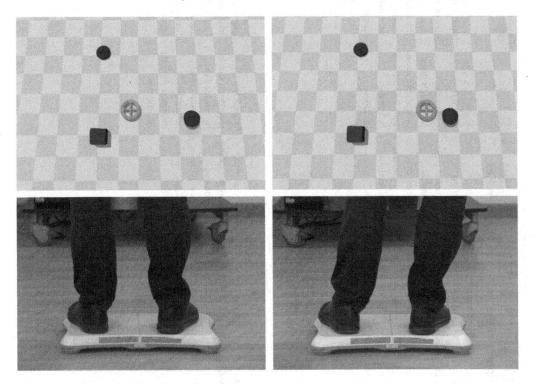

FIGURE 8.2 Real movement (bottom frame) and their representation in the virtual environment (top frame) during the training of the ankle and hip strategy.

TABLE 8.1

Description of the Ankle and Hip Exercises

Exercise	Virtual Environment	Interaction	Objective	Input Parameters	Output Parameters	In-Game Screenshot
Playdough blast	A checkered pattern with a crosshair, which represents the current position of the center of pressure. Colored playdough shapes appear and disappear a few seconds later	Users control the crosshair by displacing the center of pressure in all directions	To blow the playdough shapes by reaching them with the crosshair before they disappear	Area of appearance, size, time to reach the shapes, maximum number of simultaneous shapes, movement (yes/no), speed	Correct answers and omissions at each area	
Catch the playdoughs	A checkered pattern with a multicolored cube, which represents the current position of the center of pressure. Colored playdough shapes move from the top to the bottom of the virtual environment	Users control the cube by displacing the center of pressure in the medial-lateral axis	To catch the playdough shapes by reaching them with the cube before they disappear from the environment	Area of appearance, trajectory (vertical, with an angle), time between shapes, speed	Correct answers and omissions at each area	

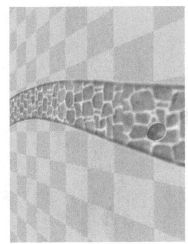

Simon	A virtual replica of the original Simon game and a crosshair, which represents the current position of the center of pressure. A series of sequences of tones and lights are displayed	Users control the crosshair by displacing the center of pressure in all directions	To repeat the sequences in the same order. As the user succeeds, the sequences progressively become longer	Time to reach the buttons, maximum length of the sequence	Correct answers, omissions, commissions, maximum sequence length
Follow the path	A checkered pattern with a winding road and a colored ball that represents the position of the center of pressure in the medial-lateral axis and progresses along the path	Users control the ball by displacing the center of pressure in the medial-lateral axis	To remain within the path	Width of the path, forward speed	Percentage of time within the path

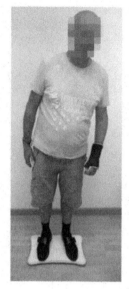

FIGURE 8.3   An individual post-stroke registering his limits of stability in the medial–lateral and anterior–posterior axis.

the medial–lateral and anterior–posterior axis prior their interaction (Figure 8.3). This allowed to estimate the maximum excursion of the center of pressure that individuals could perform and adjust the motor requirements of the exercises to each particular case. Interestingly, the WBB has shown comparable accuracy to laboratory-grade platforms to estimate the limits of stability after stroke (Llorens et al. 2016). In practice, exercises required displacements of up to 80% of the registered values.

Effectiveness

Different studies have investigated the effectiveness of the VR-based training of the ankle and hip strategies in individuals with chronic stroke (Gil-Gómez et al. 2011; Llorens et al. 2014, 2018). Interventions consisted of 20 sessions administered from three to five days a week and combined 30–40 minutes of the VR-based exercises with 20–30 minutes of conventional physical therapy. Studies so far involved adults with different age ranges with sufficient motor and cognitive condition to enable successful interaction with the exercises.

All existing studies, either longitudinal (Llorens et al. 2014) or controlled (Gil-Gómez et al. 2011; Llorens et al. 2018), exploring the effects of an intervention with the described exercises have reported a significant improvement in different aspects of balance, which, in some cases, has additionally shown certain transference to functional gait (Table 8.2) (Llorens et al. 2014; Gil-Gómez et al. 2011).

Interestingly, the benefits of the VR-based intervention detected in the Berg Balance Scale (Berg et al. 1992) and the Functional Reach Test (Duncan et al. 1990) have been consistent in all studies (Gil-Gómez et al. 2011; Llorens et al. 2014, 2018). The effect on these instruments is also evident when analyzing aggregated data of all the individuals with stroke who participated in the studies. The scores in the Berg Balance Scale and the Functional Reach Test of 60 participants, 17 women and 43 men, who had suffered either an ischemic ($n = 41$) or an hemorrhagic stroke ($n = 19$), either in their left ($n = 31$) or right hemisphere ($n = 29$), and had a mean age of 58.2 ± 8.1 years and a mean time since the injury of 846.2 ± 1004.7 days[1] were analyzed. A significant improvement was detected after the intervention in the Berg Balance Scale (40.7 ± 8.9 to 44.8 ± 7.5; $t(59) = -12.438$, $p < 0.001$)

TABLE 8.2

Effects of the Training of the Ankle and Hip Strategies Detected by Balance and Gait Instruments in the Existing Studies

Scale	Gil-Gómez et al. (2011)	Llorens et al. (2014)	Llorens et al. (2017)
Berg balance scale	✓	✓	✓
Brunel balance assessment	✓		
Functional reach test – Standing	✓	✓	✓
Functional reach test – Sitting	✗		
Stepping test – Paretic side	✓	✓	✗
Stepping test - Non-paretic side	✓	✓	✗
30-Second chair-to-stand	✓	✓	✓
Timed up-and-down test	✗		✗
1-Minute walking test	✓		
10-Meter walk test	✗		✗
Timed up-and-go	✓	✓	
Sway speed			✓
Limits of stability			✓

and the Functional Reaches Test (24.4 ± 6.8 to 27.2 ± 6.7 cm; $t(58) = -5.983$, $p < 0.001$) (Figure 8.4). Interestingly, the effect of the intervention was not only maintained but also enhanced one month after in both the Berg Balance Scale (44.8 ± 7.5 to 46.1 ± 7.0; $t(59) = -4.053$, $p < 0.001$) and the Functional Reach Test (26.8 ± 7.4 to 27.6 ± 7.4 cm; $t(59) = -2.157$, $p = 0.035$). It should be noted that all the participants returned to conventional physical therapy after the VR-based intervention.

The improvement exhibited in these instruments after the VR-based training of the ankle and hip strategies was also higher than that provided by conventional physical therapy (Gil-Gómez et al. 2011), which support the efficacy of the intensive customized task-oriented training mediated by VR. These results also provide partial evidence of the specificity of the training, as the training effects were exhibited by the Functional Reach Test in standing but not in sitting position. While an improved control of the ankle and hip strategies, presumably derived from the VR-based intervention, could have had a positive influence on forward reaching in standing,[2] it would not have a direct effect on postural control in sitting, which is mainly driven by the tonic activation of trunk muscles (Masani et al. 2009).

The clinical relevance of the results should be highlighted, as all the participants in all studies were chronic and the improvement in the Berg Balance Scale exceeded the minimally detectable

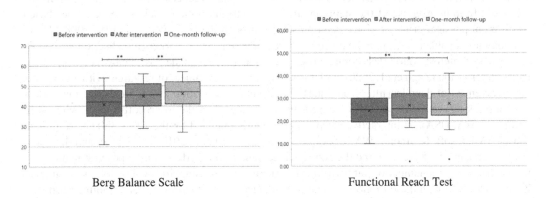

FIGURE 8.4 Aggregated time effect of the training of the ankle and hip strategies in the berg balance scale and the functional reach test.

change of the scale in this population (Liston and Brouwer 1996; Flansbjer, Blom, and Brogårdh 2012). In addition, the Functional Reach Test has shown to be sensitive to training effects in acute (Brooks, Davis, and Naglie 2006) and subacute stroke (Weiner et al. 1993) and also risk of falls in the elderly (Duncan et al. 1990). Consequently, the improvement detected in this scale could also reflect an enhancement of the balance abilities of the participants. A specific investigation of the influence of time since injury on the effectiveness of the VR-based training of the ankle and hip strategies on balance compared the performance of three groups of subjects who had had a stroke within less than one year, between one and two years, and more than two years before the enrollment. Results showed that all groups improved their balance condition, although participants with higher time since injury had more difficulty to maintain the gains (Llorens et al. 2018).

ACCEPTABILITY

The acceptability of the VR-based exercises has been explored using self-reported measures of their usability and motivation. Individual scores of participants with stroke who were enrolled in the controlled study (Gil-Gómez et al. 2011) rated the usability of the exercises with a mean score of 53.2 ± 5.6 over 65 in the Short Feedback Questionnaire (Kizony and Katz 2006), and only one participant reported not being able to control the exercises. Analysis of aggregated data of 54 participants poststroke, including those enrolled in the different studies (Llorens et al. 2014, 2018) and unpublished data, showed that the exercises were found to be very usable, as evidenced by a mean score of 77.1 ± 9.3 over 100 to the System Usability Scale (Brooke 1996), well above the average usability threshold of 68 (Czaja et al. 2019). The subjective experience after the intervention was also assessed using the pressure/tension, value/usefulness, perceived competence, and interest/enjoyment items of the Intrinsic Motivation Inventory (McAuley, Duncan, and Tammen 1989). According to unpublished aggregated reports of the same 54 participants, the intervention was reported as being enjoyable and useful, as described by scores of 5.5 ± 1.0 and 5.7 ± 1.0 over 7, respectively. Participants also felt competent and not pressured during the interaction, as described by scores of 5.3 ± 1.0 and 2.0 ± 0.8[3] over 7. All these results evidence that the VR-based interventions were well accepted, and the exercises were usable and motivating.

TRAINING OF THE STEPPING STRATEGY

INSTRUMENTATION

Opposite to the ankle and hip strategies, which require maintaining the feet in contact with the support surface without modifying the base of support, the stepping strategy requires active movement of the feet with the objective of increasing the base of support. Consequently, real-time tracking of the feet in the real world was necessary to provide visual feedback of their movements in the VR-based training exercise of the stepping strategy. Different motion tracking technologies based on different physical principles have been presented through the years (Zhou and Hu 2008). These technologies are able to estimate the position of wearable sensors or markers attached to the specific body parts of interest. Markerless human pose estimation has been made possible by recent advances in computer vision (Zago et al. 2020), and it is still a very active subject of research, with an increasing involvement of deep learning techniques (Gu, Zhang, and Kamijo 2020). Real-time markerless motion tracking was worldwide available for the first time when the Microsoft Kinect (Microsoft, Redmond, WA, USA), an interactive peripheral originally intended for the Xbox 360 entertainment console (Microsoft, Redmond, WA, USA), broke into the market in 2010. The Kinect and their successive versions, the Kinect v2 and the Azure Kinect, enable markerless human pose estimation and motion tracking of a number of joints from single depth images using different techniques (Lun and Zhao 2015). Although these motion sensing input devices were intended for entertaining purposes, a significant body of research has investigated the reliability of the skeleton tracking to assess different motor skills, from upper limb function (Cai et al. 2019) to standing balance (Clark et al. 2015) and gait (Latorre et al. 2019), with promising results. These features, along with its portability and low-cost, have motivated its use for rehabilitation (Knippenberg et al. 2017). However, analogously to the Nintendo Wii and other previous motion-enabled entertainment systems, off-the-shelf video games for the Kinect are likely to have limited adaptability to motor and cognitive impairments post-stroke.

To overcome these limitations of the entertainment software, we have designed different versions of a customized VR-based exercise of the stepping strategy over the years, which have been interacted using different motion technologies existing at the time (Table 8.3). The implications of using optical, electromagnetic, and skeleton tracking on the subjective perceptions and clinical assistance during interaction with VR applications are discussed further into this chapter.

TABLE 8.3

In-Game Screenshots and Tracking Technologies of the Different Versions of the Virtual Reality-Based Stepping Exercise

In-game Screenshots	Interactive Technology
 2007	• Optical tracking using two OptiTrack FLEX:C120 cameras (NaturalPoint, Corvallis, OR)
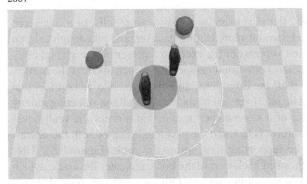 2011	• Electromagnetic tracking using the G4 (Polhemus, Colchester, VT, USA) • Skeleton tracking using the Kinect and Kinect v2
 2016	• Skeleton tracking using the Kinect v2 and Azure Kinect

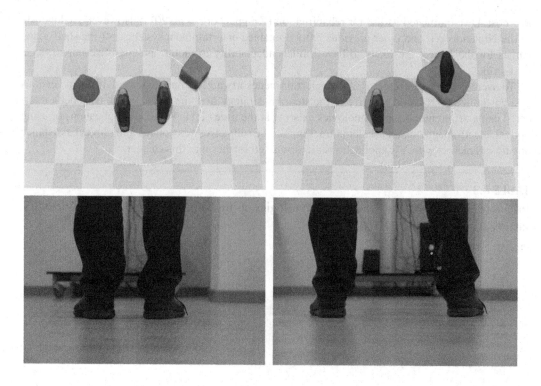

FIGURE 8.5   Real movement (bottom frame) and their representation in the virtual environment (top frame) during the training of the stepping strategy.

The virtual environment of the VR-based stepping exercise consists, essentially, of a checkered floor, whose center is indicated by a circle, an avatar representing the user, and playdough colored shapes that appear around the circle. The first versions of the exercise represented the virtual environment from an overhead and slightly backward one-point perspective, which allowed the participants to perceive their peripersonal space in all the directions, and displayed two virtual feet that mimicked the movements of the real feet. The latest version of the exercise allows selecting among three perspectives and displays a silhouette of a sex-matched avatar that mimics the real movements of the whole body. The goal of the exercise is to reach the playdough shapes with the nearest foot while maintaining the other foot, i.e. the supporting foot, within the circle (Figure 8.5). After reaching the item, the extended leg must be moved back toward the body and enter the circle, which is indicated in the virtual environment by some flashing arrows. Otherwise, the exercise does not allow reaching new shapes. The exercise can be configured by modifying the region of appearance, distance, size, lifetime, and number of simultaneous shapes to customize the task and its difficulty to the particular needs of each user.

The VR-based exercise of the stepping strategy trains the control of the trunk and lower extremities and the movement patterns initiated by these components in a standing position. Movement patterns initiated by the upper trunk result in postural adjustments to maintain balance. Movement patterns initiated by the lower extremities result in a weight shift between the legs, a weight shift to one leg, or in the movement of one leg in the space. Specifically, anterior–posterior reaches require flexion and extension of the hip, pelvis, knees and spine, and plantarflexion and dorsiflexion of the ankles. Medial–lateral reaches require abduction–adduction of the hip and pelvis, extension of the knees, and curvature of the spine. In general, the exercise trains trunk control, variations in the base of support, one leg stance, and some phases of the gait cycle (Ryerson and Levit 1997).

EFFECTIVENESS

The effectiveness of the VR-based training of the stepping strategy has been investigated in successive studies (Llorens et al. 2011, 2012, 2013; Llorens, Gil-Gómez, Alcañiz, et al. 2015; Llorens, Noé, Colomer, et al. 2015). Studies have included 15–20 sessions that combined conventional physical therapy with 20–45 minutes of training with the VR-based stepping exercise, administered from three to five days a week. Studies involved adults with chronic stroke and a motor and cognitive condition that allowed successful interaction with the exercise.

Existing reports of the effects of the VR-based exercise of the stepping strategy have consistently shown improvement of balance, evidenced by overall measures of static balance (Llorens et al. 2011, 2012, 2013; Llorens, Gil-Gómez, Alcañiz, et al. 2015; Llorens, Noé, Colomer, et al. 2015), but also a transfer of the benefits to locomotion, similarly to the training of the ankle and hip strategies, evidenced by a decrease in the gait speed (Llorens, Gil-Gómez, Alcañiz, et al. 2015) (Table 8.4). Specifically, benefits were observed in the scores of the Berg Balance Scale and the level in the Brunnel Balance Assessment (Tyson, and DeSouza 2004). Interestingly, these scales also exhibited greater benefits derived from a training with the stepping exercises than from that after a conventional physical therapy program. The comparison to conventional physical therapy also showed an improvement in gait speed during the 10 Meter Walk Test (Bowden et al. 2008), which could have benefitted from the repetitive use of the load–unloading sway strategy at the hip during the VR-based intervention. Consequently, the training of the stepping strategy could have had a positive effect on the one-leg support phase of the gait cycle (Mercer et al. 2009), which support the specificity of the intervention.

TABLE 8.4
Effects of the Training of the Stepping Strategy Detected by Balance and Gait Instruments in the Existing Studies

Scale	Llorens et al. (2011)	Llorens et al. (2012)	Llorens et al. (2013)	Llorens et al. (2015)[a]	Llorens et al. (2015)[b]
Berg balance scale	✓	✓	✓	✓	✓
Tinetti performance oriented mobility assessment – balance subscale	✓	✗	✓	✗	✓
Tinetti performance oriented mobility assessment – gait subscale	✓			✗	✓
Brunel balance assessment		✓		✓	✓
10 Meter walk test				✓	
Limits of stability	✓[c]		✓		

[a] Llorens, Roberto, José Antonio Gil-Gómez, Mariano Alcañiz, Carolina Colomer, and Enrique Noé. 2015. "Improvement in Balance Using a Virtual Reality-Based Stepping Exercise: A Randomized Controlled Trial Involving Individuals with Chronic Stroke." *Clinical Rehabilitation.* 29 (3): 261–68.

[b] Llorens, Roberto, Enrique Noé, Carolina Colomer, and Mariano Alcañiz. 2015. "Effectiveness, Usability, and Cost-Benefit of a Virtual Reality-Based Telerehabilitation Program for Balance Recovery after Stroke: A Randomized Controlled Trial." *Archives of Physical Medicine and Rehabilitation.* 96 (3): 418–25.

[c] Significant effects were found on the limits of stability in the medial–lateral and anterior–posterior axis, but not in the total score.

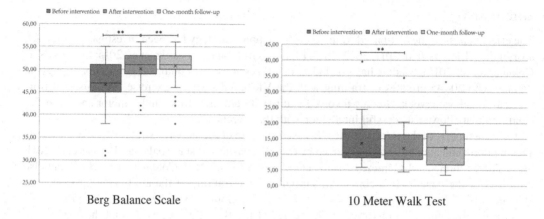

FIGURE 8.6 Aggregated time effect of the training of the stepping strategy in the berg balance scale and the 10 meter walk test.

Analyses of aggregated data collected through the years, including published and unpublished data, support these findings. Data from 131 individuals, 56 women and 75 men, with a mean age of 57.3 ± 11.8 years, who had suffered either an ischemic ($n = 97$) or hemorrhagic stroke ($n = 34$), affecting the left ($n = 37$), right ($n = 88$) or both hemispheres ($n = 6$), a mean of 361.9 ± 229.8 days before their participation in an intervention with the VR-based stepping exercises were examined. Results in the Berg Balance Scale exhibited a significant improvement after the training (46.6 ± 6.0 to 50.1 ± 4.3; $t(130) = -11.746$, $p < 0.001$), which were even improved one month after (50.2 ± 4.4 to 50.7 ± 4.0; $t(110) = -5.375$, $p < 0.001$) (Figure 8.6). Results in the 10-Meter Walk Test also evidenced a significant improvement after the intervention (13.6 ± 6.6 to 12.0 ± 5.5 cm; $t(94) = 8.396$, $p < 0.001$) and were maintained in the follow-up assessment (12.0 ± 5.9 to 12.2 ± 6.0 cm; $t(70) = -0.390$, $p = 0.698$) (Figure 8.6).

Regarding the clinical relevance of these findings, as for the training of the ankle and hip strategies, the improvement detected in the Berg Balance Scale should be highlighted given the time since injury of the participants. Additionally, special mention should be made of the improvement in the gait speed, provided that it reached the minimally clinically importance difference established for subjects post-stroke (Perera et al. 2006), even for the acute phase (Kuo and Donelan 2010). This improvement not only supports the relevance of the gains but could have also mandated a change in the clinical management of the participants (Jaeschke, Singer, and Guyatt 1989). It is also important to highlight that the benefits derived from the intervention did not differ when the VR-based training was administered at home (Llorens, Noé, Colomer, et al. 2015), which endorses the feasibility of a telerehabilitation program including the stepping exercise (Figure 8.7). Additionally, the telerehabilitation paradigm entailed a reduction of the costs, mainly caused by a decrease in the number of trips to the clinic. These results support that, in some particular cases and during certain stages of the rehabilitation process, part of the therapy could be conducted at home under supervision from a physical therapist. This could be specially interesting for those subjects who either are on a low income, live in remote or rural areas, or are homebound.

ACCEPTABILITY

The acceptability of the VR-based stepping exercises has been explored using the same self-reported measures of usability and motivation as in the studies of the ankle and hip strategy. In line with the reports of the published studies (Llorens et al. 2013; Llorens, Gil-Gómez, Alcañiz, et al. 2015), analysis of aggregated unpublished data of usability including self-reports of 72 individuals post-stroke on the Short Feedback Questionnaire showed that the interaction with the VR-based exercise

FIGURE 8.7 Subject with stroke training with the stepping exercise at home under supervision from his son.

was enjoyable and successful, as described by a mean total score in this instrument of 53.7 ± 3.9 over 65. These results are in accordance with the average score of the 30 participants of the study on the effectiveness and usability of a telerehabilitation program with the stepping exercise to the System Usability Score (Llorens, Noé, Colomer, et al. 2015), which was 86.5 ± 5.1 over 100, far above the minimum threshold of 68. According to the reports of the same participants in the Intrinsic Motivation Inventory, the experience was perceived as being enjoyable and useful, with average scores of 6.1 ± 0.3 and 6.1 ± 0.6 over 7, respectively. In addition, participants felt competent but no pressured, as described by average scores of 5.0 ± 0.3 and 1.2 ± 0.4 over 7, respectively (Llorens, Noé, Colomer, et al. 2015).

Although participants did not reported significant difficulties at using interfacing devices, as evidenced by mean scores of 4.3 ± 0.7 and 4.0 ± 1.0 over 5 to that item in the Short Feedback Questionnaire in the published studies (Llorens et al. 2013; Llorens, Gil-Gómez, Alcañiz, et al. 2015), we conducted an exploratory study to determine the influence of motion tracking technologies on subjective perceptions of healthy subjects, individuals with stroke, and physical therapists (Llorens, Noé, Naranjo, et al. 2015). Optical, electromagnetic, and skeleton tracking, enabled with OptiTrack FLEX:C120 cameras, the Polhemus G4, and the Microsoft Kinect (Table 8.3), were examined. A group of 19 healthy subjects and 22 individuals with stroke interacted with the stepping exercise in three 15-minute sessions with the three different motion tracking solutions, in counterbalanced order. A group of 14 physical therapists supervised 15 rehabilitation sessions with each tracking technology. The subjective perceptions of all the participants were collected using ad-hoc questionnaires. The use of markers and sensors had a negative effect on the configuration of the device, as evidenced by the reports of healthy subjects and physical therapists, who gave the best scores to the Kinect. This technology was also identified as being the most comfortable solution. However, it was reported as being the most difficult to calibrate. Therapists provided the highest scores to the skeleton tracking, followed by the optical and the electromagnetic tracking, on the ease of fixation, insensibility to changes, maintenance, working range, integration in the clinical setting and value for money. In contrast, ease of assistance with the three motion tracking devices was rated in inverse order. Although the Kinect was identified as being difficult to calibrate, it is important to highlight that this process was improved in newer versions of the software development kit, which only require performing slight movement to start the tracking. Moreover, the provision of assistance also improved as therapists interacted with the Kinect and learned its fundamentals. These factors, together with its low-cost and portability led physical therapists to choose this device over the other alternatives (Llorens, Noé, Naranjo, et al. 2015). As a proof, the skeleton tracking, enabled by the successive versions of the Kinect, was adopted in all the workstations distributed throughout the clinical network of the neurorehabilitation service, which are being updated to the Azure Kinect after discontinuation of the Kinect v2 (Figure 8.1).

CONCLUSIONS

All the longitudinal studies on the effectiveness of the VR-based training of the ankle, hip, and stepping strategies exhibited measurable improvements over time in balance (Llorens et al. 2011, 2012, 2013, 2014). Controlled studies showed that programs that combined VR-based training of the balance strategies with conventional physical therapy promoted greater improvements in balance compared to conventional physical therapy programs alone (Gil-Gómez et al. 2011; Llorens, Gil-Gómez, Alcañiz, et al. 2015). Interestingly, the effectiveness of the VR-based training of the stepping exercise was not different when the intervention was administered at home, which supports the feasibility of telerehabilitation programs including this exercise (Llorens, Noé, Colomer, et al. 2015).

The improvements detected in balance instruments, often higher than their corresponding minimally detectable change, are remarkable despite the fact that time since injury to participation was longer than the six-month period, traditionally considered as the period of maximum recovery (Jørgensen et al. 1995; Ferrarello et al. 2011). An in-depth study of the influence of time since injury on the effect of the training of the ankle and hip strategies showed that, while it was possible to obtain significant improvements, the intensity of these improvements was inversely proportional to time since injury (Llorens et al. 2018). Moreover, this effect was even more evident in the maintenance of gains.

The most relevant benefits were detected in skills that were specifically trained. However, certain transference to other abilities was also observed. While the training of the ankle and hip strategies, in comparison to conventional physical therapy, provided consistent improvement in the Functional Reach Test, no significant improvements were detected in functional gait (Gil-Gómez et al. 2011). In contrast, the training of the stepping strategy did not have detectable effects on the Functional Reach Test compared to conventional physical therapy, but promoted an improvement in the gait speed detected by the 10 Meter Walk Test (Llorens, Gil-Gómez, Alcañiz, et al. 2015). Consequently, it is possible to conclude that the VR-based training of the balance strategies should be reinforced with exercises that promote the improvement of other abilities that are not specifically trained.

All the existing reports and new analyses of aggregated and unpublished data evidence that VR-based exercises that encompass the fundamentals of motor learning can promote improvement in balance, allowing for controlled, safe and gradual training of intensive and repetitive exercises, as well as observation, practice, and representation of meaningful activities. Results also evidence that VR-based interventions are very well accepted, and the exercises were usable and motivating. Although all the interventions that used the VR-based exercises of the ankle, hip, and stepping strategies were well accepted, and the exercises were found usable and motivating, the enabling technology should consider the particular needs and subjective perceptions of both individuals with stroke (clients) and clinicians (professionals), to facilitate the integration of VR into clinical practice.

NOTES

1 The high time since injury is caused by the characteristics of the participants in the study by Llorens et al. (2017), which investigated the effect of time since injury in the improvement.
2 The improvement of the limits of stability after the virtual reality-based intervention reported by Llorens et al. (2018) also supports this hypothesis. However, the absence of a control group prevents identification of the source of improvement.
3 Lower scores to the pressure/tension item represent better experience, as they are a favorable predictor of intrinsic motivation.

REFERENCES

Berg, K. O., S. L. Wood-Dauphinee, J. I. Williams, and B. Maki. 1992. "Measuring Balance in the Elderly: Validation of an Instrument." *Canadian Journal of Public Health*. 83. https://doi.org/10.1016/j.archger.2009.10.008.

Bermúdez i Badia, Sergi, Gerard G. Fluet, Roberto Llorens, and Judith E. Deutsch. 2016. *"Virtual Reality for Sensorimotor Rehabilitation Post Stroke: Design Principles and Evidence."* Neurorehabilitation Technology. 2nd ed. 573–603. Springer. https://doi.org/10.1007/978-3-319-28603-7_28.

Bonan, Isabelle V., Florence M. Colle, Jean P. Guichard, Eric Vicaut, Martine Eisenfisz, P. Tran Ba Huy, and Alain P. Yelnik. 2004. "Reliance on Visual Information after Stroke. Part I: Balance on Dynamic Posturography." *Archives of Physical Medicine and Rehabilitation*. 85 (2): 268–273. https://doi.org/10.1016/j.apmr.2003.06.017.

Bowden, Mark G., Chitralakshmi K. Balasubramanian, Andrea L. Behrman, and Steven A. Kautz. 2008. "Validation of a Speed-Based Classification System Using Quantitative Measures of Walking Performance Poststroke." *Neurorehabilitation and Neural Repair*. https://doi.org/10.1177/1545968308318837.

Brooke, John. 1996. "SUS - A Quick and Dirty Usability Scale." *Usability Evaluation in Industry*. 189 (194): 4–7. https://doi.org/10.1002/hbm.20701.

Brooks, Dina, Aileen M. Davis, and Gary Naglie. 2006. "Validity of 3 Physical Performance Measures in Inpatient Geriatric Rehabilitation." *Archives of Physical Medicine and Rehabilitation*. https://doi.org/10.1016/j.apmr.2005.08.109.

Brown, Lesley A., Ryan J. Sleik, and Toni R. Winder. 2002. "Attentional Demands for Static Postural Control after Stroke." *Archives of Physical Medicine and Rehabilitation*. https://doi.org/10.1053/apmr.2002.36400.

Cai, Laisi, Ye Ma, Shuping Xiong, and Yanxin Zhang. 2019. "Validity and Reliability of Upper Limb Functional Assessment Using the Microsoft Kinect V2 Sensor." *Applied Bionics and Biomechanics*. https://doi.org/10.1155/2019/7175240.

Chen, I. C., P. T. Cheng, A. L. Hu, M. Y. Liaw, L. R. Chen, W. H. Hong, and M. K. Wong. 2000. "Balance Evaluation in Hemiplegic Stroke Patients." *Chang Gung Medical Journal*. 23 (6): 339–47.

Chen, Ling, Wai Leung Ambrose Lo, Yu Rong Mao, Ming Hui Ding, Qiang Lin, Hai Li, Jiang Li Zhao, Zhi Qin Xu, Rui Hao Bian, and Dong Feng Huang. 2016. "Effect of Virtual Reality on Postural and Balance Control in Patients with Stroke: A Systematic Literature Review." *BioMed Research International*. https://doi.org/10.1155/2016/7309272.

Cheok, Gary, Dawn Tan, Aiying Low, and Jonathan Hewitt. 2015. "Is Nintendo Wii an Effective Intervention for Individuals with Stroke? A Systematic Review and Meta-Analysis." *Journal of the American Medical Directors Association*. https://doi.org/10.1016/j.jamda.2015.06.010.

Clark, Ross A., Yong Hao Pua, Cristino C. Oliveira, Kelly J. Bower, Shamala Thilarajah, Rebekah McGaw, Ksaniel Hasanki, and Benjamin F. Mentiplay. 2015. "Reliability and Concurrent Validity of the Microsoft Xbox One Kinect for Assessment of Standing Balance and Postural Control." *Gait and Posture*. https://doi.org/10.1016/j.gaitpost.2015.03.005.

Corbetta, Davide, Federico Imeri, and Roberto Gatti. 2015. "Rehabilitation That Incorporates Virtual Reality Is More Effective than Standard Rehabilitation for Improving Walking Speed, Balance and Mobility after Stroke: A Systematic Review." *Journal of Physiotherapy*. https://doi.org/10.1016/j.jphys.2015.05.017.

Cuevas-Trisan, Ramon. 2017. "Balance Problems and Fall Risks in the Elderly." *Physical Medicine and Rehabilitation Clinics of North America*. https://doi.org/10.1016/j.pmr.2017.06.006.

Czaja, Sara J., Walter R. Boot, Neil Charness, and Wendy A. Rogers. 2019. *Designing for Older Adults: Principles and Creative Human Factors Approaches*. 3rd ed. Boca Raton, Florida: CRC Press.

Darekar, Anuja, Bradford J. McFadyen, Anouk Lamontagne, and Joyce Fung. 2015. "Efficacy of Virtual Reality-Based Intervention on Balance and Mobility Disorders Post-Stroke: A Scoping Review." *Journal of NeuroEngineering and Rehabilitation*. 12 (1). https://doi.org/10.1186/s12984-015-0035-3.

Dominguez-Tellez, P., J. A. Moral-Munoz, E. Casado-Fernandez, A. Salazar, and D. Lucena-Anton. 2019. "Effects of Virtual Reality on Balance and Gait in Stroke: A Systematic Review and Meta-Analysis." *Revista de Neurologia*. https://doi.org/10.33588/rn.6906.2019063.

Duncan, Pamela W., Debra K. Weiner, Julie Chandler, and Stephanie Studenski. 1990. "Functional Reach: A New Clinical Measure of Balance." *Journal of Gerontology*. 45 (6): M192–97. https://doi.org/10.1093/geronj/45.6.M192.

Ferrarello, Francesco, Marco Baccini, Lucio Antonio Rinaldi, Maria Chiara Cavallini, Enrico Mossello, Giulio Masotti, Niccolò Marchionni, and Mauro Di Bari. 2011. "Efficacy of Physiotherapy Interventions Late after Stroke: A Meta-Analysis." *Journal of Neurology, Neurosurgery and Psychiatry*. https://doi.org/10.1136/jnnp.2009.196428.

Flansbjer, Ulla Britt, Johanna Blom, and Christina Brogårdh. 2012. "The Reproducibility of Berg Balance Scale and the Single-Leg Stance in Chronic Stroke and the Relationship between the Two Tests." *PM and R*. https://doi.org/10.1016/j.pmrj.2011.11.004.

Geurts, Alexander C.H., Mirjam De Haart, Ilse J.W. Van Nes, and Jaak Duysens. 2005. "A Review of Standing Balance Recovery from Stroke." *Gait and Posture*. https://doi.org/10.1016/j.gaitpost.2004.10.002.

Gil-Gómez, José Antonio, Roberto Llorens, Mariano Alcñiz, and Carolina Colomer. 2011. "Effectiveness of a Wii Balance Board-Based System (EBaViR) for Balance Rehabilitation: A Pilot Randomized Clinical Trial in Patients with Acquired Brain Injury." *Journal of NeuroEngineering and Rehabilitation*. 8 (1). https://doi.org/10.1186/1743-0003-8-30.

Gu, Yanlei, Huiyang Zhang, and Shunsuke Kamijo. 2020. "Multi-person Pose Estimation Using an Orientation and Occlusion Aware Deep Learning Network." *Sensors (Switzerland)*. 20 (6). https://doi.org/10.3390/s20061593.

Horak, F. B. 1987. "Clinical Measurement of Postural Control in Adults." *Physical Therapy*. https://doi.org/10.1093/ptj/67.12.1881.

Iruthayarajah, Jerome, Amanda McIntyre, Andreea Cotoi, Steven Macaluso, and Robert Teasell. 2017. "The Use of Virtual Reality for Balance among Individuals with Chronic Stroke: A Systematic Review and Meta-Analysis." *Topics in Stroke Rehabilitation*. https://doi.org/10.1080/10749357.2016.1192361.

Jaeschke, Roman, Joel Singer, and Gordon H. Guyatt. 1989. "Measurement of Health Status. Ascertaining the Minimal Clinically Important Difference." *Controlled Clinical Trials*. 10 (4): 407–15. https://doi.org/10.1016/0197-2456(89)90005-6.

Jørgensen, Henrik S., Hirofumi Nakayama, Hans O. Raaschou, Jørgen Vive-Larsen, Mogens Støier, and Tom S. Olsen. 1995. "Outcome and Time Course of Recovery in Stroke. Part II: Time Course of Recovery. The Copenhagen Stroke Study." *Archives of Physical Medicine and Rehabilitation*. 76 (5): 406–12. https://doi.org/10.1016/S0003-9993(95)80568-0.

Kizony, Rachel, and Noomi Katz. 2006. "Short Feedback Questionnaire (SFQ) to Enhance Client-Centered Participation in Virtual Environments." *CyberPsychology & Behavior*. 9 (6): 687–88. http://scholar.google.com/scholar?hl=en&btnG=Search&q=intitle:Short+Feedback+Questionnaire+(SFQ)+to+enhance+client-centered+participation+in+virtual+environments#0.

Knippenberg, Els, Jonas Verbrugghe, Ilse Lamers, Steven Palmaers, Annick Timmermans, and Annemie Spooren. 2017. "Markerless Motion Capture Systems as Training Device in Neurological Rehabilitation: A Systematic Review of their Use, Application, Target Population and Efficacy." *Journal of NeuroEngineering and Rehabilitation*. https://doi.org/10.1186/s12984-017-0270-x.

Kuo, Arthur D., and J. Maxwell Donelan. 2010. "Dynamic Principles of Gait and their Clinical Implications." *Physical Therapy*. 90 (2): 157–74. https://doi.org/10.2522/ptj.20090125.

Latorre, Jorge, Carolina Colomer, Mariano Alcañiz, and Roberto Llorens. 2019. "Gait Analysis with the Kinect v2: Normative Study with Healthy Individuals and Comprehensive Study of its Sensitivity, Validity, and Reliability in Individuals with Stroke." *Journal of NeuroEngineering and Rehabilitation*. 16 (1). https://doi.org/10.1186/s12984-019-0568-y.

Laver, Kate E., Belinda Lange, Stacey George, Judith E. Deutsch, Gustavo Saposnik, and Maria Crotty. 2017. "Virtual Reality for Stroke Rehabilitation." *Cochrane Database of Systematic Reviews*. 2017 (11). https://doi.org/10.1002/14651858.CD008349.pub4.

Liston, Rebecca A.L., and Brenda J. Brouwer. 1996. "Reliability and Validity of Measures Obtained from Stroke Patients Using the Balance Master." *Archives of Physical Medicine and Rehabilitation*. https://doi.org/10.1016/S0003-9993(96)90028-3.

Llorens, Roberto, Sergio Albiol, José Antonio Gil-Gómez, Mariano Alcañiz, Carolina Colomer, and Enrique Noé. 2014. "Balance Rehabilitation Using Custom-Made Wii Balance Board Exercises: Clinical Effectiveness and Maintenance of Gains in an Acquired Brain Injury Population." *International Journal on Disability and Human Development*. 13 (3): 327–32. https://doi.org/10.1515/ijdhd-2014-0323.

Llorens, Roberto, Mariano Alcañiz, Carolina Colomer, and Maria Dolores Navarro. 2012. "Balance Recovery through Virtual Stepping Exercises Using Kinect Skeleton Tracking: A Follow-up Study." *Annual Review of CyberTherapy and Telemedicine*. 181: 108–12. https://doi.org/10.3233/978-1-61499-121-2-108.

Llorens, Roberto, Carolina Colomer, Mariano Alcañiz, and Enrique Noé. 2013. "BioTrak Virtual Reality System: Effectiveness and Satisfaction Analysis for Balance Rehabilitation in Patients with Brain Injury." *Neurología (English Edition)*. 28 (5): 268–75. https://doi.org/10.1016/j.nrleng.2012.04.016.

Llorens, Roberto, José Antonio Gil-Gómez, Mariano Alcañiz, Carolina Colomer, and Enrique Noé. 2015. "Improvement in Balance Using a Virtual Reality-Based Stepping Exercise: A Randomized Controlled Trial Involving Individuals with Chronic Stroke." *Clinical Rehabilitation*. 29 (3): 261–68. https://doi.org/10.1177/0269215514543333.

Llorens, Roberto, José Antonio Gil-Goméz, Patricia Mesa-Gresa, Mariano Alcañiz, Carolina Colomer, and Enrique Noé. 2011. "BioTrak: A Comprehensive Overview." In *2011 International Conference on Virtual Rehabilitation, ICVR 2011*. https://doi.org/10.1109/ICVR.2011.5971843.

Llorens, Roberto, Jorge Latorre, Enrique Noé, and Emily A. Keshner. 2016. "Posturography Using the Wii Balance Board. A Feasibility Study with Healthy Adults and Adults Post-Stroke." *Gait and Posture*. 43: 228–32. https://doi.org/10.1016/j.gaitpost.2015.10.002.

Llorens, Roberto, Enrique Noé, Mariano Alcañiz, and Judith E. Deutsch. 2018. "Time since Injury Limits but Does Not Prevent Improvement and Maintenance of Gains in Balance in Chronic Stroke." *Brain Injury*. 32 (3): 303–09. https://doi.org/10.1080/02699052.2017.1418905.

Llorens, Roberto, Enrique Noé, Carolina Colomer, and Mariano Alcañiz. 2015. "Effectiveness, Usability, and Cost-Benefit of a Virtual Reality-Based Telerehabilitation Program for Balance Recovery after Stroke: A Randomized Controlled Trial." *Archives of Physical Medicine and Rehabilitation*. 96 (3): 418–25. https://doi.org/10.1016/j.apmr.2014.10.019.

Llorens, Roberto, Enrique Noé, Valery Naranjo, Adrián Borrego, Jorge Latorre, and Mariano Alcañiz. 2015. "Tracking Systems for Virtual Rehabilitation: Objective Performance vs. Subjective Experience. A Practical Scenario." *Sensors (Switzerland)*. 15 (3): 6586–606. https://doi.org/10.3390/s150306586.

Lun, Roanna, and Wenbing Zhao. 2015. "A Survey of Applications and Human Motion Recognition with Microsoft Kinect." *International Journal of Pattern Recognition and Artificial Intelligence*. https://doi.org/10.1142/S0218001415550083.

Maier, Martina, Belén Rubio Ballester, and Paul F.M.J. Verschure. 2019. "Principles of Neurorehabilitation after Stroke Based on Motor Learning and Brain Plasticity Mechanisms." *Frontiers in Systems Neuroscience*. https://doi.org/10.3389/fnsys.2019.00074.

Masani, Kei, Vivian W. Sin, Albert H. Vette, T. Adam Thrasher, Noritaka Kawashima, Alan Morris, Richard Preuss, and Milos R. Popovic. 2009. "Postural Reactions of the Trunk Muscles to Multi-Directional Perturbations in Sitting." *Clinical Biomechanics*. 24 (2): 176–82. https://doi.org/10.1016/j.clinbiomech.2008.12.001.

McAuley, Edward D., Terry Duncan, and Vance V. Tammen. 1989. "Psychometric Properties of the Intrinsic Motivation Inventory in a Competitive Sport Setting: A Confirmatory Factor Analysis." *Research Quarterly for Exercise and Sport*. https://doi.org/10.1080/02701367.1989.10607413.

Mercer, Vicki Stemmons, Janet Kues Freburger, Shuo-Hsiu Chang, and Jama L. Purser. 2009. "Measurement of Paretic–Lower-Extremity Loading and Weight Transfer after Stroke." *Physical Therapy*. 89 (7): 653–64. https://doi.org/10.2522/ptj.20080230.

"NEURORHB. Servicio de Neurorrehabilitación de Hospitales Vithas." 2020. Website. 2020. www.neurorhb.com.

Perera, Subashan, Samir H. Mody, Richard C. Woodman, and Stephanie A. Studenski. 2006. "Meaningful Change and Responsiveness in Common Physical Performance Measures in Older Adults." *Journal of the American Geriatrics Society*. 54 (5): 743–49. https://doi.org/10.1111/j.1532-5415.2006.00701.x.

Pollock, A. S., B. R. Durward, P. J. Rowe, and J. P. Paul. 2000. "What Is Balance?" *Clinical Rehabilitation*. 14 (4): 402–6. https://doi.org/10.1191/0269215500cr342oa.

de Rooij, Ilona J.M., Ingrid G.L. van de Port, and Jan-Willem G. Meijer. 2016. "Effect of Virtual Reality Training on Balance and Gait Ability in Patients with Stroke: Systematic Review and Meta-Analysis." *Physical Therapy*. 96 (12): 1905–18. https://doi.org/10.2522/ptj.20160054.

Ryerson, Susan, and Kathryn Levit. 1997. *Functional Movement Reeducation: A Contemporary Model for Stroke Rehabilitation*. Churchill Livingstone. http://books.google.fr/books?id=H_FrAAAAMAAJ.

Spaulding, Sandi J. 2008. *Basic Biomechanics. Ergonomics for Therapists*. 6th ed. McGraw-Hill. https://doi.org/10.1016/B978-032304853-8.50009-3.

Tyson, Sarah F, and Lorraine H DeSouza. 2004. "Development of the Brunel Balance Assessment: A New Measure of Balance Disability Post Stroke." *Clinical Rehabilitation*. 18 (7): 801–10. https://doi.org/10.1191/0269215504cr744oa.

Tyson, Sarah F, Marie Hanley, Jay Chillala, Andrea Selley, and Raymond C Tallis. 2006. "Balance Disability after Stroke." *Physical Therapy*. 86 (1): 30–38. https://doi.org/16386060.

Weiner, Debra K., Dennis R. Bongiorni, Stephanie A. Studenski, Pamela W. Duncan, and Gary G. Kochersberger. 1993. "Does Functional Reach Improve with Rehabilitation?" *Archives of Physical Medicine and Rehabilitation*. 74 (8): 796–800. https://doi.org/10.1016/0003-9993(93)90003-S.

Winter, D. A. 1995. "Human Balance and Posture Control during Standing and Walking." *Gait and Posture*. https://doi.org/10.1016/0966-6362(96)82849-9.

Zago, Matteo, Matteo Luzzago, Tommaso Marangoni, Mariolino De Cecco, Marco Tarabini, and Manuela Galli. 2020. "3D Tracking of Human Motion Using Visual Skeletonization and Stereoscopic Vision." *Frontiers in Bioengineering and Biotechnology*. 8. https://doi.org/10.3389/fbioe.2020.00181.

Zhou, Huiyu, and Huosheng Hu. 2008. "Human Motion Tracking for Rehabilitation – A Survey." *Biomedical Signal Processing and Control*. https://doi.org/10.1016/j.bspc.2007.09.001.

9 Engaging Stroke Survivors with Virtual Neurorehabilitation Technology

Marcus King
Callaghan Innovation
Christchurch, New Zealand

Holger Regenbrecht
University of Otago
Dunedin, New Zealand

Simon Hoermann
University of Canterbury
Christchurch, New Zealand

Chris Heinrich and Leigh Hale
University of Otago
Dunedin, New Zealand

CONTENTS

INTRODUCTION

For stroke survivors to regain upper limb motor function and recovery, it is important for them to engage in high intensity, functional, and meaningful exercise. Professionally trained physiotherapists and occupational therapists, amongst others, frequently work with patients in one-on-one supervised training sessions to achieve this (French et al., 2016). Research with animal models has demonstrated that high intensity practice of movement can drive neural change. Unfortunately, the amount of practice provided during post-stroke rehabilitation in humans is small compared with that which can be attained in animal-based research. It is possible that current doses of task-specific practice during rehabilitation are not adequate to drive the neural reorganization needed to promote optimal functional recovery post-stroke (Lang et al., 2009). Insufficient intensity of therapy appears to be primarily caused by lack of time and funding (French et al., 2016). Methods where patients can increase the intensity of therapeutic training by themselves or semi-supervised are thus in high demand.

Virtual reality (VR) and augmented reality (AR) technology—computer generated worlds which can be explored in interactive real-time—have potential to support the physical rehabilitation process in a safe and engaging environment. With VR technology, the therapeutic intensity can be adjusted so that patients can practice at a level that is neither too difficult nor too easy, known as the "state of flow" (Csíkszentmihályi, 1990). Patients can be supported to successfully complete tasks beyond their comfort level through the guidance of the VR system or a clinical professional. Working in this way is postulated to more effective and engaging than traditional therapy thereby enhancing motivation to undertake the training for longer.

Neuroplasticity is the process by which neural circuits in the brain are modified by experience, learning, and/or injury (cf. Nudo, 2003), and this process currently underpins rehabilitation principles of motor-relearning and recovery after stroke. A high number of repetitions of a movement pattern, with feedback, forms the physiological basis of motor learning and is an essential component of motor-relearning (Butefisch et al., 1995).

Neuroplasticity effects can be used in combination with VR techniques to provide many forms of extrinsic feedback (Regenbrecht et al., 2014) to motivate and engage patients in repeated exercise and support their implicit motor learning (Subramanian et al., 2010). Studies have reported increased participant motivation, enjoyment, or perceived improvement in physical ability following the inclusion of VR into stroke rehabilitation (Broeren et al., 2008; Yavuzer et al., 2008; Housman et al., 2009). Importantly, computer games can improve compliance with prescribed rehabilitation exercises (Kwakkel et al., 2008). Combining elements of research and practice in neuroplasticity, physical rehabilitation, VR and AR, and gameplay lead to the new field of virtual neurorehabilitation:

> Virtual neurorehabilitation is the field of practice and research to achieve effective physical rehabilitation by using virtual reality techniques to generate neuroplastic change.

Unfortunately, to date, shortcomings in a systematic approach to progress virtual rehabilitation have included (a) a lack of systematic development and evaluation of virtual neurorehabilitation systems based on sound theoretical and practical principles; (b) combining this technology with tailored therapeutic protocols that can lead to effective gains in motor and cognitive function; and (c) effectively engaging patients in a sustainable way. With our work reported here, we address these shortcomings.

UPPER LIMB STROKE NEUROREHABILITATION

Globally, stroke is the leading cause of complex disability (Adamson et al., 2004). Recovery of upper limb function after stroke is critical to a person's independence and self-care, yet up to 85% of patients never regain upper limb function and remain dependent on caregivers (Harris & Eng, 2007; Alia et al., 2017). Loss of upper limb function accounts for most of the poor subjective well-being after stroke (Wyller et al., 1997; Singam et al., 2015).

Current best practice for the rehabilitation of upper limb function post-stroke suggests the repetitive training of functional upper limb motor tasks, intensity of practice, and the functional relevance of the motor tasks are critical components (Van Peppen et al., 2004; Han et al., 2013; Veerbeek et al., 2014; French et al., 2016). Practice is postulated to advance neuroplastic changes in the damaged neurological pathways augmenting adaptation and enhancing recovery. The functional relevance of the task boosts motivation and cognitive involvement and encourages the neuroplastic changes to be focused on movements that are important for the individual (French et al., 2016). The amount of repetition required to induce these neuroplastic changes is not exactly known (Han et al., 2013; French et al., 2016). Han et al. (2013), however, found that between 30 and 90 hours (1–3 hours/day for 5 days/week over 6 weeks) of repetitive, task-specific training lead to improved arm motor function in stroke survivors.

Upper limb rehabilitation appears to thus require concentrated and intense training which in turn necessitates commitment and motivation from both stroke survivors and highly skilled health professionals providing therapy. Availability of skilled health professionals and the costs of their intense involvement can be prohibitive. Machine-assisted physiotherapy could help in alleviating therapist shortages and costs (Amirabdollahian et al., 2001) and has the potential to improve the outcome of upper limb rehabilitation (Islam et al., 2006). While robot-assisted training is a means of delivering intensive, repeatable, task-specific training, the evidence for its effectiveness to improve upper limb function post-stroke is ambiguous (Alia et al., 2017). In VR-assisted therapy, patients appear to engage in machine-assisted exercise for longer, seemingly motivated by the inclusion into the machine by a mixture of sensory and VR systems (Broeren et al., 2008; Alia et al., 2017).

VR has been defined as the "use of interactive simulations created with computer hardware and software to present users with opportunities to engage in environments that appear and feel similar to real world objects and events" (Weiss et al., 2006). From a technological point of view, VR has been classified on a spectrum between immersive (which is usually delivered via head-mounted displays (HMDs)) and non-immersive VR (usually delivered via personal computer screen and controlled via joystick, mouse, or other form of control device) (Rose et al., 1999).

Low-immersive VR can often be acquired at low cost as it does not require complex hardware; it can be delivered simply via a standard computer and monitor (Holden & Dyar, 2002). The user can simultaneously experience movement in the real world and receive feedback from the system of their movement. Immersive VR requires sophisticated hardware making it expensive and it is often not well tolerated for long durations, with some users reporting nausea, eye-strain, and dizziness or "simulator sickness" (Kennedy et al., 1993).

The potential benefits of VR are seen to arise from: improved patient motivation during rehabilitation (Flores et al., 2008), its ability to be combined with rehabilitation exercise devices (Guberek et al., 2008), and to assist movement, visually capture and assess movement, and provide movement feedback (King et al., 2010). Transcranial magnetic stimulation testing has revealed that this "action-observation" training can enhance the effects of post-stroke motor training by increasing the magnitude of motor memory formation and differential corticomotor excitability change (Celnik et al., 2008). Functional magnetic resonance imaging studies show that VR hand avatars (i.e., the image of a hand in a computer game) can create an "observation condition" that can be mimicked (Adamovich et al., 2009).

Evidence for the use of VR in upper limb stroke rehabiliation is emerging, with positive effects for low-immersive VR systems reported by Holden and Dyar (2002) and a review of VR by Henderson et al. (2007) concluded that the evidence was encouraging and warranted further investigation. Dosage of VR therapy is also emerging, with Laver et al. (2017) reporting that it should be provided for at least 15 hours total therapy time and is most effective when used in the first six months after stroke.

The 2017 Australian Clinical Guidelines for Stroke Management (updated 21/11/2019) (InformMe, n.d.) for upper limb rehabilitation following stroke states that for stroke survivors (a) with at least some voluntary movement of the arm and hand, repetitive task-specific training may be used to improve arm and hand function (French et al., 2016) and (b) with mild to moderate arm impairment, VR and interactive games may be used to improve upper limb function. These guidelines claim that research into upper limb stroke rehabilitation is ongoing, thus the recommendation, albeit weak, for the impact of VR therapy is encouraging.

THE POTENTIAL OF VIRTUAL REALITY TECHNIQUES

VR and related techniques are gaining popularity due to the availability of affordable and robust hardware and software, which makes VR useable in stroke rehabilitation clinics and not only in research environments. Besides its technical characteristics of being a computer-generated, interactive, three-dimensional environment, the distinctive aspect of VR is its ability to stimulate a sense of presence. Regardless of whether the VR environment is experienced on a computer monitor or by way of stereoscopic projection systems, including HMDs, the user should feel they are part of the virtual environment and they should experience the virtual environment in a similar way as a real environment. This experience is the premise of both VR's potential effectiveness and is potential transferability into the real world.

In a wider sense, VR can be extended, or brought back to the real world by combining real and virtual elements in a coherent way, leading to mixed reality concepts of AR or augmented virtuality. In such scenarios, the same concept of presence applies: the users should experience the environment as plausible, develop a sense of place, and experience the virtual objects present. Both virtual and mixed reality environments exhibit the same potential characteristics, which make them applicable to the virtual neurorehabilitation context:

- Users are able to initiate training in a simplified environment so that they can easily learn the basics
- Complexity of training can be increased at a pace that the user can manage
- Users can challenge themselves to practice complex movements or simulate risky situations within a controlled and safe environment and without risk of damage to self or equipment
- Users can experience accurate and realistic simulations
- Systems can cater for large numbers of users, even over various locations
- Virtual environments provide a highly visual approach to aid learning
- Systems can allow for peer review, feedback, and ongoing assessment
- VR can deconstruct complex data into manageable chunks and visualize complex concepts and theories
- VR systems can be very cost effective
- VR ensures that learning is motivating, fun, and enjoyable where appropriate
- Users can explore virtual scenarios in preparation for real-world scenarios.

VR has shown effective in areas such as training and learning, data analytics, and in particular, psychotherapy. For instance, in the treatment of special phobias, like fear of heights, fear of flying, or fear of spiders, clients are gradually exposed to fear-evoking stimuli and develop coping mechanisms, which then can be transferred to their real-world behavior. Again, central to the effectiveness of the therapy is the concept of presence and with this the ability of the VR system to evoke the same psychological and physiological responses as in real life, for example, exhibited avoidance behavior or an increasing heart rate. VR therapy environments are safe for the client and the therapist is able to control stimuli, measure responses, and systematically repeat exercises (Rizzo & Koenig, 2017).

Of particular interest in rehabilitation is VR's potential to enhance learning. For instance, in studies of serious gaming in women's health care, Knowles (1970) described four elements important to adult learning:

- Adults are autonomic and want independence in their learning. Gaming promotes an active form of learning and allows independence.
- Adults use their past experience. Gaming facilitates this by offering different scenarios according to experience.
- Adults are goal orientated. Gaming is designed around completion of a level or task.
- Adults tend to be problem-based learners and not content-oriented learners. Gaming provides learning experiences that players can relate to realistic clinical problems.

Hence, in combining VR with elements of gamification, we create highly interactive and engaging environments suitable for virtual neurorehabilitation.

VR is possibly the "ultimate neuroplasticity vehicle." Neuroplasticity, the brain's ability to adapt its functions and activities in response to environmental and psychological change, is mediated by sensations, perceptions, emotions, and, finally, beliefs (Regenbrecht et al., 2014). Through its interactive feedback loop of providing meaningful stimuli to be perceived by and responded to by its users, VR changes the emotional state of the users, and eventually, if effective, alters their beliefs. For example, by visually amplifying goal-oriented movements during exercises, the learned nonuse of the impaired limbs can be counteracted (Regenbrecht et al., 2012, 2014; Ballester et al., 2015). While this belief altering can be exploited and abused in certain scenarios, in virtual neurorehabilitation this ability is directed toward neuroplastic change to regain motor function.

The potential for VR to change beliefs solves a problem that occurs when using traditional mirror therapy: the issue of the patient's potential disbelief. In traditional upper limb mirror therapy, the patient's impaired limb is placed behind an optical mirror; the patient knows that the impaired limb is unable to move and so has to develop a suspension of disbelief when observing their unimpaired limb moving in the mirror for the therapy to take effect. They have to believe that what they see in the mirror is their impaired limb moving. VR is less obvious in its inner workings and can be seen more as "magic" by the patient when they observe their impaired limb "moving" and therefore requires much less belief.

In the following, we present three examples of where virtual and AR techniques have been used for upper limb stroke rehabilitation. All three examples use virtual neurorehabilitation as their main principle, applied in different settings. All aim at maximizing patient engagement and promoting the interplay of neuroplasticity, motivation, and repeated exercise and were developed based on sound technological, scientific, and clinical theory, knowledge, and practice.

EXAMPLE 1: AUGMENTED REFLECTION TECHNOLOGY

While the use of fully immersive, HMD-based VR for physical rehabilitation is still in its infancy, the use of semi-immersive VR and variants of augmented realty systems have been studied for close to two decades (Burdea & Thalmann, 2003; Weiss et al., 2004; Cameirao et al., 2008).

Many of these systems visualize patients' movement through the computer system. Systems like the Rehabilitation Gaming System (Cameirão et al., 2009) and YouRehab (Eng et al., 2007) represent patients' hands with computer generated virtual limbs, whereas Weiss et al. (2004) researched a system that used video capture to integrate the video of the patient's actual body into the visualization.

Similar to therapist-led therapeutic interventions, VR systems aim to optimize the frequency and intensity of a given exercise and to provide patients with task-oriented practice and feedback. These systems both stimulate movement execution and provide visual feedback. VR systems have, however, the capability to provide visual feedback beyond what is possible in traditional therapist-led rehabilitation (Schüler et al., 2015): they enable movement visualization, performance feedback, and context information. Movement visualization includes not only the movement pattern of the limb but also a representation of the limb. A wide range of representations are available, from video images of the real hand or limb to computer generated limbs to abstract representations (Schüler, 2012; Schuler et al., 2013).

Presence in virtual rehabilitation interventions is postulated as important (Schüler et al., 2014), but so too is the use of visual feedback beyond just the accurate representation of the physical reality. For example, use of visual illusions of movement to extend the visual feedback beyond the actual capabilities of the patient are thought to further stimulate and enhance rehabilitation outcomes (Ballester et al., 2015, 2016).

Augmented reflection technology (ART) (Figure 9.1) is a computerized system which was developed based on the principles of mirror therapy, a form of therapy shown in a systematic review of 62 studies to be effective for improving upper limb function and motor impairment after stroke (Thieme et al., 2018). These principles include the provision of mirror visual feedback through the ART system while performing systematic and repetitive practice of standardized hand exercises (Morkisch & Dohle, 2015).

The ART has been studied in several empirical evaluations. In these experiments with unimpaired participants, the "fooling" potential of the ART was demonstrated—participants were fooled about their limb ownership and laterality. For example, participants believe that a hand displayed on the left side of a display screen (see Figure 9.1) is their own left hand when in fact it was just a video of their right hand (Regenbrecht et al., 2011, 2012, 2014).

Clinical studies with the ART have also been conducted: application of the ART with six individuals with chronic stroke in a physiotherapy clinic demonstrated that all six were able to use the system with sustained engagement and motivation (Hoermann et al., 2014). Two additional studies showed that the ART was feasible for use in an inpatient setting as an adjunct rehabilitation intervention in the early phase post-stroke (Hoermann et al., 2015, 2017).

Early evidence of the ART's capability to provide autonomous instructions and feedback without the direct involvement of a therapist has also been shown in a nonclinical study with 28 unimpaired

FIGURE 9.1   Augmented reflection technology system (version 4) in action.

volunteers. In this study, participants were able to carry out a subset of a traditional mirror therapy movement protocol via computerized feedback and instructions only. The system instructed participants to execute motor gestures, assessed the accuracy of the execution, and progressed when the gestures were completed adequately. The system provided continuous as well as summative feedback to participants (Pinches & Hoermann, 2016).

EXAMPLE 2: NEUROREHABILITATION INVOLVING SPECIALISED PHYSICAL DEVICES

An integrated rehabilitation system was created to enable upper limb rehabilitation post-stroke comprising of a set of computer games that could be played using three devices: the Able-B (Sampson et al., 2011); the Able-M (Jordan et al., 2014); and the Able-X (Hijmans et al., 2011) (Figure 9.2). The devices were aligned to the degree of impairment, or weakness presented by the patient and encouraged movements appropriate for their recovery. Figure 9.2 shows the relationship of the components of the system to the patient's strength as measured by the Oxford scale of muscle strength (Parkinson, 2000).

The Able-B provides a mechanical linkage between the hemiparetic and the unaffected arm so that the unaffected arm powered a mirrored motion in the affected arm. The Able-M is an "arm-skate" (an arm version of a skateboard) which enables the stroke survivor to interact with the computer while fully supported on a mobile device on a flat surface (e.g., their kitchen table at home), thus eliminating the impact of gravity and friction to movement.

The Able-X comprises an air-mouse, which coupled both arms so that the hemiparetic arm can be guided through space via gravity resistive movement of the unaffected arm, enabling the hand held air mouse to interact with the computer.

Oxford scale of movement

Grade 0 no movement is observed

Grade 1 only a trace or flicker of movement

Grade 2 Can move only if the resistance of gravity is removed

Grade 3 Can move against gravity if other resistance is removed

Grade 4 Strength is reduced

Grade 5 Muscle contracts normally

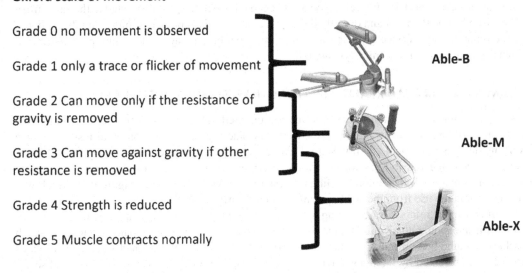

Able-B

Able-M

Able-X

FIGURE 9.2 Able-B, Able-M, and Able-X devices for upper limb rehabilitation and the relationship of the components of the system to the patient's strength.

Computer games were developed using user-centric design principles, as described in King et al. (2010) and Hale et al. (2012), and user perspectives (Lewis et al., 2011). Several factors were identified that motivated stroke survivors to play the games during rehabilitation and these were then included in the game designs:

- Intellectual stimulation during game play
- Feedback (e.g., game scores)
- Physical benefits from the exercise
- Tolerance for disabilities (e.g., game levels suitable for a range of abilities)
- Connecting to the game, i.e., participant understands and relates to the game
- Social interaction during group play.

Furthermore, the games were graded to anticipate the need that some older adults with stroke might require guidance on how to play computer games. Hence, the suite of games started with "easy to achieve" large stationary target reaching games in which random motion of the device would achieve success; graduated to smaller and later, moving targets; then to targets that required manipulation and placement on icons; then to strategic target games; and finally to a choice of sports or mind-challenging games.

As well as the exploratory studies above, the equipment was provided to a small cohort of stroke survivors for unsupervised use in their homes (King et al., 2012). Participants practiced for 4.5–5.5 sessions per week over the 55-day duration of the trial, each averaging 33.5 hours of exercise. This level of exercise was significantly higher than the 16 hours suggested by Kwakkel et al. (2004), or the 15 hours suggested by Laver et al. (2015).

Galea et al. (2016) used the Able-M and Able-B in a 92-participant clinical trial which showed significant improvement in arm function and strength, muscle tone (as assessed by the Modified Ashworth Scale), the Wolf Motor Function Test, the Functional Independence Measure (locomotion, mobility, and psychosocial subscales), quality of life (as measured by the EQ-5D), and overall health. This trial was particularly interesting as it was conducted using a clinical practice improvement approach (Horn & Gassaway, 2007) in a busy clinical practice and it not only improved patient outcomes but it also built capacity in the provision of subacute rehabilitation services. The "Hand Hub," as the site of the trial was called, typically enabled five patients to be treated at the same time, supervised by one therapist and one allied health assistant. This approach is improving the response of the rehabilitation service to their waiting list by providing a way of streaming patients toward intensive therapy for upper limb dysfunction.

EXAMPLE 3: VIRTUAL REALITY IN HOME-BASED MIRROR THERAPY

VR systems for home rehabilitation have mainly targeted patient motivation and engagement for repetitive exercises by incorporating game elements. These games often consist of tasks that have varying degrees of clinical validity. With the recent emergence of affordable immersive VR hardware, it opens the possibilities for new ways to target patients' home rehabilitation. We present one such immersive VR home rehabilitation system, the ART VR Home (Figure 9.3), which was adapted for home use from the ART system presented earlier.

Immersive VR refers to surrounding a person in virtual content by wearing a HMD as opposed to semi-immersion with a computer monitor. Immersive VR systems have many advantages which include an embodied virtual experience (e.g., mirror therapy illusion), can block out the distractions to exercise presented by the home environment, and hardware that is easy to transport and setup. It is also an approach often considered as novel and exciting by patients, and this novelty has the potential to motivate the person to engage in their rehabilitation.

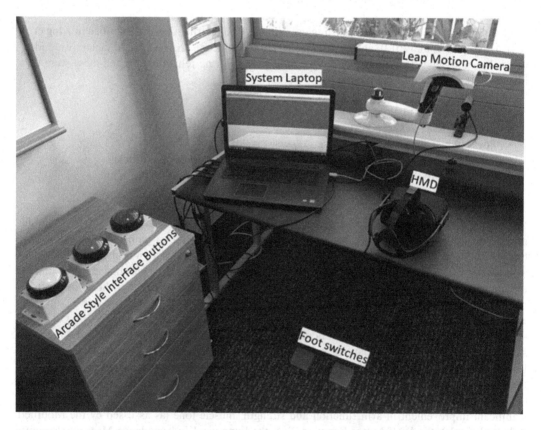

FIGURE 9.3 The ART VR Home system consists of six main components: Laptop to run system, Leap Motion camera for hand tracking, HMD for the patient to experience the virtual environment, arcade style buttons to interact with system outside VR, foot pedals to interact with system inside VR, and a height adjustable desk to place patient's hands while interacting inside VR.

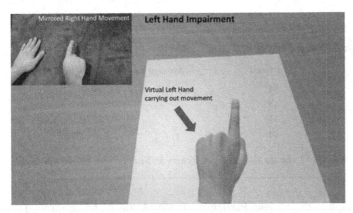
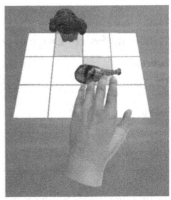

FIGURE 9.4 ART VR home system patient tasks: rehabilitation task (a) and TheraMem memory game (b). This presented scenario is for a patient with a left hand impairment. Our system captures their right (non-affected) hand movement and mirrors this and presents it to the patient in VR such that their left (impaired) virtual hand is performing their mirrored hand movement.

We aimed to provide a home-based VR mirror therapy system suitable for use by stroke survivors with unilateral upper limb impairment that was intuitive and easy to use and that delivered standardized exercises. We worked with the clinicians who had developed a previously used mirror therapy protocol (BeST-ART) and adapted it for use for immersive virtual environments (Figure 9.4). The adapted protocol, the BeST-ART VR protocol, comprised making the numbers 1 through 5 with hand movements, in both the palm down and palm up hand orientations. We also included in the protocol an easy to play memory game (*TheraMem*) that required hand movements to activate virtual tiles (Figure 9.4b). In the rehabilitation task, a virtual computer monitor is placed in the virtual environment and displays an image of the BeST-ART VR hand position. The patient is then asked to copy the position with their hand. In the *TheraMem* task, virtual tiles can be activated to reveal matching pairs of food items.

The clinical ART VR base system comprises four components: a laptop computer to run the VR system, an HMD to experience immersive VR, a hand-tracking camera to bring patient's hand movements into VR, and a height-adjustable table to place their hand on while interacting in the VR environment. As our targeted population are often part of a demographic group that can lack confidence in using computers and technology, an easy-to-use, intuitive interface that patients could feel confident operating on their own was necessary. We thus designed an interface that consisted of three arcade style buttons to operate the system outside of VR and two foot pedals to interact with once inside VR (while wearing an HMD). The three arcade buttons corresponded with starting the VR system, changing the VR task to *TheraMem* and changing the VR task back to rehabilitation. The two foot pedals provide patients with the options of moving onto the next hand exercise and showing a virtual training hand (help) performing the assigned hand exercise.

The ART VR Home system runs on a standard (computer gaming) laptop and the hand-tracking camera is a Leap Motion depth sensing camera that is attached to a desk mount arm angled down toward the desk. The desk is a 70 cm × 70 cm height adjustable desk with wheels. The system creates a patient log file every time the system is started which consists of: system time usage, hand exercises completed, how much time spent on each hand exercise/task, and an early prototype of a machine learning classifier to output which BeST ART VR hand exercise the patient was performing.

We conducted a feasibility study with four people with stroke exploring acceptance and usability of the system. The ART VR Home system was placed in participants' homes for four weeks, and they were asked to use the system for a minimum of 15 minutes a day for five days a week (20 days total) and log their use. After a pre-assessment (Upper Limb subsection of Motor Assessment Scale), participants were provided with a 30-minute demonstration of the system and

TABLE 9.1

Participant System Usage over a 4-Week Period

Participant	Days Used	Average Time Used/Day (min:sec)
P1	24	12:00
P2	15	16:37
P3	4	17:22
P4	20	21:56

It was recommended that participants spend at least 15 minutes a day using the system for 5 days a week (20 days in total).

the protocol before having it installed in their home. Participant feedback was gathered through a semi-structured interview at the conclusion of the study. The participant logs were also analyzed to determine how long and often they used the system. Table 9.1 shows that participants were able to use the system for the requested time length; however, the number of days the system was used varied across participants as most reported life events interrupting their schedule (family travel and medical events/procedures).

All four participants were able to complete the study and did not report any adverse events with using the system nor any complaints about having the system in their home. All reported that they were easily able to use the interface (arcade buttons and foot pedals), albeit three classified themselves as not competent with computers. For the rehabilitation task, participants reported they were able to follow the assigned hand exercises shown on the virtual computer monitor within the VR environment; however, they wished there was more variety in the hand positions assigned to them. For the *TheraMem* memory game, all participants found it fun to play but wanted more variety of games to play. Lastly, three participants reported some form of improvement in their impaired limb after using the system (e.g., capable of more fingers movements, able to open fingers more widely, and improvement in shoulder movement).

CONCLUSIONS AND FUTURE WORK

For this chapter, we brought together contributions from clinical and technological virtual neuro-rehabilitation specialists. VR techniques have been demonstrated to be effective in the rehabilitation of upper limb impairment post-stroke and have potential not only in stroke rehabilitation but also in other therapeutic and rehabilitation areas. Virtual environments are highly controllable and can give immediate and progressive feedback to patients. These environments have the potential for providing unsupervised and semi-supervised rehabilitation and potentially optimizing clinician productivity. We have been able to demonstrate that virtual neurorehabilitation systems are sufficiently motivating for patients to engage in tailored and repetitive movement practice.

In particular, VR can be used to amplify visual simulation and, combined with game elements, potentially enhancing engagement in therapeutic exercises and thus improved outcomes. Tailored applications are preferred over the "black box" applications of commercial games, as individualizing the approach to the patient may result in sustained use after the "wow" effect of the technology wears off.

The biggest challenge is how to implement VR technology into clinical practice. Studies investigating VR have mostly concentrated on chronic stroke populations with marginal results, possibly due to the chronicity of the population with limited recovery potential. More clinical trials are required in acute stroke rehabilitation to demonstrate effectiveness for clinical implementation.

There is significant potential to combine VR techniques with other technical components. Unfortunately, much research in this field is focused on the more complex field of robotics, with associated instrumentation and capital investment. Low-cost and low-complexity systems, such as computer–game interface exercise devices and customized off-the-shelf VR devices tailored to patients' needs will encourage uptake by busy clinical practitioners or by patients in home-based settings. A future focus can be how to integrate rehabilitation technologies with activities of daily living and to make it more meaningful for patients and clinicians.

REFERENCES

Adamovich, S. V., August, K., Merians, A., & Tunik, E. (2009). A virtual reality-based system integrated with FMRI to study neural mechanisms of action observation-execution: A proof of concept study. *Restorative Neurology and Neuroscience, 27*(3), 209–223. https://doi.org/10.3233/RNN-2009-0471

Adamson, J., Beswick, A., & Ebrahim, S. (2004). Is stroke the most common cause of disability? *Journal of Stroke and Cerebrovascular Diseases, 13*(4), 171–177. https://doi.org/10.1016/j.jstrokecerebrovasdis.2004.06.003

Alia, C., Spalletti, C., Lai, S., Panarese, A., Lamola, G., Bertolucci, F., Vallone, F., Di Garbo, A., Chisari, C., Micera, S., & Caleo, M. (2017). Neuroplastic changes following brain ischemia and their contribution to stroke recovery: Novel approaches in neurorehabilitation. *Frontiers in Cellular Neuroscience, 11.* https://doi.org/10.3389/fncel.2017.00076

Amirabdollahian, F., Loureiro, R., Driessen, B., & Harwin, W. (2001). Error correction movement for machine assisted stroke rehabilitation. *Integration of Assistive Technology in the Information Age, 9,* 60–65.

Ballester, B. R., Maier, M., San Segundo Mozo, R. M., Castañeda, V., Duff, A., & M. J. Verschure, P. F. (2016). Counteracting learned non-use in chronic stroke patients with reinforcement-induced movement therapy. *Journal of NeuroEngineering and Rehabilitation, 13*(1), 74. https://doi.org/10.1186/s12984-016-0178-x

Ballester, B. R., Nirme, J., Duarte, E., Cuxart, A., Rodriguez, S., Verschure, P., & Duff, A. (2015). The visual amplification of goal-oriented movements counteracts acquired non-use in hemiparetic stroke patients. *Journal of NeuroEngineering and Rehabilitation, 12*(1), 50. https://doi.org/10.1186/s12984-015-0039-z

Broeren, J., Claesson, L., Goude, D., Rydmark, M., & Sunnerhagen, K. S. (2008). Virtual rehabilitation in an activity centre for community-dwelling persons with stroke. *Cerebrovascular Diseases, 26*(3), 289–296. https://doi.org/10.1159/000149576

Butefisch, C., Hummelsheim, H., Denzler, P., & Mauritz, K. H. (1995). Repetitive training of isolated movements improves the outcome of motor rehabilitation of the centrally paretic hand. *Journal of the Neurological Sciences, 130*(1), 59–68.

Burdea, G., & Thalmann, D. (2003). Editorial. *The Journal of Visualization and Computer Animation, 14*(5), i. https://doi.org/10.1002/vis.328

Cameirao, M. S., Badia, S. B., & Verschure, P. F. M. J. (2008). Virtual reality based upper extremity rehabilitation following stroke: A review. *Journal of Cyber Therapy and Rehabilitation, 1*(1), 63–75.

Cameirão, M. S., Bermúdez I Badia, S., Duarte Oller, E., & Verschure, P. F. M. J. (2009). The rehabilitation gaming system: A review. *Studies in Health Technology and Informatics, 145,* 65–83.

Celnik, P., Webster, B., Glasser, D. M., & Cohen L. G. (2008). Effects of action observation on physical training after stroke. *Stroke, 39*(6), 1814–1820. https://doi.org/10.1161/STROKEAHA.107.508184

Csíkszentmihályi, M. (1990). *Flow: The Psychology of Optimal Experience.* Harper Perennial Modern Classics.

Eng, K., Siekierka, E., Pyk, P., Chevrier, E., Hauser, Y., Cameirao, M., Holper, L., Hägni, K., Zimmerli, L., Duff, A., Schuster, C., Bassetti, C., Verschure, P., & Kiper, D. (2007). Interactive visuo-motor therapy system for stroke rehabilitation. *Medical & Biological Engineering & Computing, 45*(9), 901–907. https://doi.org/10.1007/s11517-007-0239-1

Flores, E., Tobon, G., Cavallaro, E., Cavallaro, F. I., Perry, J. C., & Keller, T. (2008). Improving patient motivation in game development for motor deficit rehabilitation. *Proceedings of the 2008 International Conference on Advances in Computer Entertainment Technology,* 381–384. https://doi.org/10.1145/1501750.1501839

French, B., Thomas, L. H., Coupe, J., McMahon, N. E., Connell, L., Harrison, J., Sutton, C. J., Tishkovskaya, S., & Watkins, C. L. (2016). Repetitive task training for improving functional ability after stroke. *Cochrane Database of Systematic Reviews, 11.* https://doi.org/10.1002/14651858.CD006073.pub3

Galea, M. P., Khan, F., Amatya, B., Elmalik, A., Klaic, M., & Abbott, G. (2016). Implementation of a technology-assisted programme to intensify upper limb rehabilitation in neurologically impaired participants: A prospective study. *Journal of Rehabilitation Medicine, 13,* 48 (6), 522–528. doi: 10.2340/16501977-2087. PMID: 27068229.

Guberek, R., Schneiberg, S., Sveistrup, H., McKinley, P., & Levin, M. F. (2008). Application of virtual reality in upper limb rehabilitation. *2008 Virtual Rehabilitation*, 65–65. https://doi.org/10.1109/ICVR.2008.4625128

Hale, L. A., Satherley, J. M., McMillan, N. J., Milosavljevic, S., Hijmans, J. M., & King, M. J. (2012). Participant perceptions of the use of an adapted CyWee Z as an adjunct to rehabilitation of upper limb function following stroke. *Journal of Rehabilitation Research and Development*, 49 (4), 623–634.

Han, C., Wang, Q., Meng, P., & Qi, M. (2013). Effects of intensity of arm training on hemiplegic upper extremity motor recovery in stroke patients: A randomized controlled trial. *Clinical Rehabilitation*, 27(1), 75–81. https://doi.org/10.1177/0269215512447223

Harris, J. E., & Eng, J. J. (2007). Paretic upper-limb strength best explains arm activity in people with stroke. *Physical Therapy*, 87(1), 88–97. https://doi.org/10.2522/ptj.20060065

Henderson, A., Korner-Bitensky, N., & Levin, M. (2007). Virtual reality in stroke rehabilitation: A systematic review of its effectiveness for upper limb motor recovery. *Topics in Stroke Rehabilitation*, 14(2), 52–61. https://doi.org/10.1310/tsr1402-52

Hijmans, J. M., Hale, L. A., Satherly, J. A., McMillan, N. J., & King, M. J. (2011). Bilateral upper-limb rehabilitation after stroke using movement-based game controller. *Journal of Rehabilitation Research and Development*, 48 (8), 1005–1014.

Hoermann, S., Ferreira dos Santos, L., Morkisch, N., Jettkowski, K., Sillis, M., Cutfield, N., Schmidt, H., Hale, L., Krüger, J., Regenbrecht, H., & Dohle, C. (2015, June 9). Computerized mirror therapy with augmented reflection technology for stroke rehabilitation—A feasibility study in a rehabilitation center. *Proceedings of the 2015 International Conference on Virtual Rehabilitation*. International Conference on Virtual Rehabilitation (ICVR), Valencia, Spain.

Hoermann, S., Ferreira dos Santos, L., Morkisch, N., Jettkowski, K., Sillis, M., Devan, H., Kanagasabai, P. S., Schmidt, H., Krüger, J., Dohle, C., Regenbrecht, H., Hale, L., & Cutfield, N. J. (2017). Computerised mirror therapy with augmented reflection technology for early stroke rehabilitation: Clinical feasibility and integration as an adjunct therapy. *Disability and Rehabilitation*, 39 (15), 1503–1514. https://doi.org/10.1080/09638288.2017.1291765

Hoermann, S., Hale, L., Winser, S. J., & Regenbrecht, H. (2014). Patient engagement and clinical feasibility of augmented reflection technology for stroke rehabilitation. *International Journal on Disability and Human Development*, 13 (3), 355–360. https://doi.org/10.1515/ijdhd-2014-0328

Holden, M. K., & Dyar, T. (2002). Virtual environment training-a new tool for neurorehabilitation? *Neurology Report*, 26 (2), 62–71.

Horn, S. D., Gassaway, J. (2007). Practice-based evidence study design for comparative effectiveness research. *Medical Care*, 45(10 Supl 2), S50–7. doi: 10.1097/MLR.0b013e318070c07b. PMID: 17909384

Housman, S., Scott, K. M., & Reinkensmeyer, D. J. (2009). A randomized controlled trial of gravity-supported, computer-enhanced arm exercise for individuals with severe hemiparesis. *Neurorehabil Neural Repair*, 23, 505–514

InformMe. (n.d.). *Clinical Guidelines for Stroke Management*. Retrieved April 21, 2020, from https://informme.org.au/Guidelines/Clinical-Guidelines-for-Stroke-Management-2017

Islam, N., Harris, N. D., & Eccleston, C. (2006). Does technology have a role to play in assisting stroke therapy? A review of practical issues for practitioners. *Quality in Ageing and Older Adults*, 7(1), 49.

Jordan, K., Sampson, M., & King, M. (2014). Gravity-supported exercise with computer gaming improves arm function in chronic stroke. *APMR*, 95 (8), 1484–1489. doi:https://doi.org/10.1016/j.apmr.2014.02.028

Kennedy, R. S., Lane, N. E., Berbaum, K. S., & Lilienthal, M. G. (1993). Simulator sickness questionnaire: An enhanced method for quantifying simulator sickness. *The International Journal of Aviation Psychology*, 3(3), 203–220.

King, M., Hale, L., Pekkari, A., Persson, M., Gregorsson, M., & Nilsson, M. (2010). An affordable, computerised, table-based exercise system for stroke survivors. *Disability and Rehabilitation: Assistive Technology*, 5(4), 288–293. https://doi.org/10.3109/17483101003718161

King, M., Hijmans, J. M., Sampson, M., Satherley, J., & Hale, L. (2012). Home-based stroke rehabilitation using computer gaming. *New Zealand Journal of Physiotherapy*, 40(3), 129–135.

Knowles, M. S. (1970). *The modern practice of adult education. Andragogy versus pedagogy*. Englewood Cliffs, NJ: Prentice Hall.

Kwakkel, G., Kollen, B.J., & Krebs, H.I. (2008). EGects of robot-assisted therapy on upper limb recovery a-er stroke: a systematic review. *Neurorehabilitation and Neural Repair*, 22, 111–121. doi:10.1177/1545968307305457

Kwakkel, G., Van Peppen, R., Wagenaar, R. C., Dauphinee, S. W., Richards, C., Ashburn, A., Miller, K., Lincoln, N., Partridge, C., Wellwood, I., & Langhorne, P. (2004). Effects of augmented exercise therapy time after stroke: a meta-analysis. *Stroke, 35,* 2529–2539.

Lang, C. E., MacDonald, J. R., Reisman, D. S., Boyd, L., Jacobson Kimberley, T., Schindler-Ivens, S. M., Hornby, T. G., Ross, S. A., & Scheets P. L. (2009) Observation of amounts of movement practice provided during stroke rehabilitation. *Archives of Physical Medicine and Rehabilitation, 90* (10), 1692–1698.

Laver, K. E., George, S., Thomas, S., Deutsch, J. E., & Crotty, M. (2015). Virtual reality for stroke rehabilitation. *Cochrane Database of Systematic Reviews.* https://doi.org/10.1002/14651858.CD008349.pub3

Laver, K. E., Lange, B., George, S., Deutsch, J. E., Saposnik, G., & Crotty, M. (2017). Virtual reality for stroke rehabilitation. *Cochrane Database of Systematic Reviews* (11), 1–170. CD008349.

Lewis, G.N., Woods, C., Rosie, J.A., & McPherson, K. M. (2011). Virtual reality games for rehabilitation of people with stroke: perspectives from the users. *Disability Rehabilitation Assistance Technology, 6* (5):453–463. doi: 10.3109/17483107.2011.574310. Epub 2011 Apr 17. PMID: 21495917.

Morkisch, N., & Dohle, C. (2015). *BeST - Berliner Spiegeltherapieprotokoll: Ein Wissenschaftlich Evaluiertes Manual Zur Durchführung der Spiegeltherapie.* Hippocampus.

Nudo R. J. (2003). Adaptive plasticity in motor cortex: implications for rehabilitation after brain injury. *Journal of Rehabilitation Medicine Supplement, 41,* 7–10.

Pinches, J., & Hoermann, S. (2016). Evaluating automated real time feedback and instructions during computerized mirror therapy for upper limb rehabilitation using augmented reflection technology. In P. M. Sharkey & A. A. Rizzo (Eds.), Proceedings 11th International Conference on Disability, Virtual Reality & Associated Technologies, 65–72. ICDVRAT archive.

Regenbrecht, H., Franz, E. A., McGregor, G., Dixon, B. G., & Hoermann, S. (2011). Beyond the looking glass: Fooling the brain with the augmented mirror box. *Presence: Teleoperators and Virtual Environments, 20* (6), 559–576. https://doi.org/10.1162/PRES_a_00082

Regenbrecht, H., Hoermann, S., McGregor, G., Dixon, B., Franz, E., Ott, C., Hale, L., Schubert, T., & Hoermann, J. (2012). Visual manipulations for motor rehabilitation. *Computers & Graphics, 36*(7), 819–834. https://doi.org/10.1016/j.cag.2012.04.012

Regenbrecht, H., Hoermann, S., Ott, C., Müller, L., & Franz, E. (2014). Manipulating the experience of reality for rehabilitation applications. *Proceedings of the IEEE, 102*(2), 170–184. https://doi.org/10.1109/JPROC.2013.2294178

Rizzo, A. "S." &Koenig, S. T. (2017). Is clinical virtual reality ready for primetime? *Neuropsychology, 31*(8), 877–899. https://doi.org/10.1037/neu0000405

Rose, D., Brooks, F., Attree, B. M., Parslow, E. A., Leadbetter, D. M., McNeil, A. G., Jayawardena, J. E., Greenwood, S., & Potter, J. (1999). A preliminary investigation into the use of virtual environments in memory retraining after vascular brain injury: Indications for future strategy?. *Disability and Rehabilitation, 21*(12), 548–554.

Sampson, M., Shau, Y. -W., & King, M. J. (2012). Bilateral upper limb trainer with virtual reality for post-stroke rehabilitation: case series report. *Disability Rehabilitation Assistance Technology, 7,* 55–62

Schüler, T. (2012). Generative design as a method to foster explorative behaviour in virtual motor rehabilitation. *Proceedings of the 9th International Conference on Disability, Virtual Reality and Associated Technologies,* 495–498.

Schüler, T., dos Santos, L. F., & Hoermann, S. (2015, June 9). Designing virtual environments for motor rehabilitation: Towards a framework for the integration of best-practice information. Proceedings of International Conference on Virtual Rehabilitation 2015. ICVR 2015, Valencia, Spain.

Schuler, T., Drehlmann, S., Kane, F., & von Piekartz, H. (2013). Abstract virtual environment for motor rehabilitation of stroke patients with upper limb dysfunction. A pilot study. *2013 International Conference on Virtual Rehabilitation (ICVR),* 184–185. https://doi.org/10.1109/ICVR.2013.6662117

Schüler, T., Ferreira dos Santos, L., & Hoermann, S. (2014). Harnessing the experience of presence for virtual motor rehabilitation: Towards a guideline for the development of virtual reality environments. *Proceedings 10th International Conference on Disability, Virtual Reality & Associated Technologies,* 373–376. http://centaur.reading.ac.uk/37397/

Singam, A., Ytterberg, C., Tham, K., & von Koch, L. (2015). Participation in complex and social everyday activities six years after stroke: Predictors for return to pre-stroke level. *PLoS One, 10*(12). https://doi.org/10.1371/journal.pone.0144344

Subramanian, S. K., Massie, C. L., Malcolm, M. P., & Levin, M. F. (2010). Does provision of extrinsic feedback result in improved motor learning in the upper limb poststroke? a systematic review of the evidence. *Neurorehabilitation and Neural Repair, 24*(2), 113–124. https://doi.org/10.1177/1545968309349941

Thieme, H., Morkisch, N., Mehrholz, J., Pohl, M., Behrens, J., Borgetto, B., & Dohle, C. (2018). Mirror therapy for improving motor function after stroke. *Cochrane Database of Systematic Reviews, 7*. https://doi.org/10.1002/14651858.CD008449.pub3

Van Peppen, R. P., Kwakkel, G., Wood-Dauphinee, S., Hendriks, H. J., Van der Wees, P. J., & Dekker, J. (2004). The impact of physical therapy on functional outcomes after stroke: What's the evidence? *Clinical Rehabilitation, 18*(8), 833–862. https://doi.org/10.1191/0269215504cr843oa

Veerbeek, J. M., van Wegen, E., van Peppen, R., van der Wees, P. J., Hendriks, E., Rietberg, M., & Kwakkel, G. (2014). What is the evidence for physical therapy poststroke? A systematic review and meta-analysis. *PLoS One, 9*(2). https://doi.org/10.1371/journal.pone.0087987

Weiss, P. L., Kizony, R., Feintuch, U., & Katz, N. (2006). Virtual reality in neurorehabilitation. *Textbook of Neural Repair and Rehabilitation, 51*(8), 182–197.

Weiss, P. L., Rand, D., Katz, N., & Kizony, R. (2004). Video capture virtual reality as a flexible and effective rehabilitation tool. *Journal of NeuroEngineering and Rehabilitation, 1*, 12. https://doi.org/10.1186/1743-0003-1-12

Williams, M. (1956). Manual muscle testing, development and current use. *Physical Therapy Rev. 36*(12), 797–805. doi: 10.1093/ptj/36.12.797. PMID: 13378993.

Wyller, T. B., Sveen, U., Sødring, K. M., Pettersen, A. M., & Bautz-Holter, E. (1997). Subjective well-being one year after stroke. *Clinical Rehabilitation, 11*(2), 139–145. https://doi.org/10.1177/026921559701100207

Yavuzer, G., Senel, A., Atay, M. B., & Stam, H. J. (2008). "Playstation EyeToy games" improve upper extremity-related motor functioning in subacute stroke: A randomized controlled clinical trial. *European Journal of Physical Rehabilitation Medicine, 44*(3), 237–44. [PMID: 18469735]

Section III

Virtual Reality for Health Education

10 Simulation-Based Training for Ultrasound Practice

May Almestehi
King Saud University
Riyadh, Saudi Arabia
University College Dublin
Belfield, Dublin 4, Ireland

CONTENTS

INTRODUCTION

Ultrasonography has become one of the most commonly requested imaging utilities for both diagnostic and therapeutic purposes. Several reasons contribute to the extensive usage of ultrasound; for example, ultrasound applies non-ionising radiation which is considered reasonably safe for both patients and users. Moreover, ultrasound devices are relatively small compared to other imaging devices, which make them effortlessly portable, and their lower cost facilitates their availability in clinics (Cavanagh and Smith, 2017). However, ultrasound is highly operator-dependent, and the accuracy of scanning examinations relies fundamentally on various complex skills (SCoR/BMUS, 2019). For example, the UK based Consortium for the Accreditation of Sonographic Education (UK-CASE) reported that 'the acquisition of suitable images and assessment of them is entirely operator-dependent at the time of the scan. Deficiencies in acquisition cannot be rectified by involving a more skilled practitioner at a later stage' (Harrison and Dolbear, 2018, p. 6). Achieving competency in ultrasound practice requires adequate efforts in the provision of optimum training methods for ultrasound-novice healthcare practitioners. The European Federation of Societies for Ultrasound in Medicine and Biology (EFSUMB) highlighted the requirements needed to effectively deliver ultrasound training, including the acknowledgement of the training courses that must be completed and the time that must be put in by the trainee in clinical settings (EFSUMB, 2005).

Some challenges have emerged in recent years that could prove to be obstacles for the implementation of optimum ultrasound training programmes in clinical settings. For example, the shortage of available supervisors for training, busy departments being non-learning-friendly environments and the necessity of assuring patient safety are the main concerns that surround clinical-based training. These challenges have contributed to the incorporation of alternative training methods that are mainly based on virtual reality (VR) simulation.

VR simulation has provided a number of positive impacts on ultrasound training in recent years. According to the current literature, ultrasound simulators provide a safe learning environment that contributes to the development and enhancement of both cognitive and psychomotor skills which are essential for performing competent ultrasound practice. Moreover, VR simulators, computer-based especially, usually provide immediate automatic feedback to learners, offering a suitable solution to the absence of direct clinical supervision. Simulators contribute to ensuring patient safety by decreasing scanning time for cases that are sensitive to the prolonged scans such as transvaginal ultrasound in early pregnancy (Tolsgaard et al., 2017). To justify the significance of employing VR simulation in ultrasound training, this chapter starts with a brief overview of the concerns associated with clinical-based training and then discusses the positive influence of VR simulation on the cognitive and psychomotor skills of ultrasound practitioners.

CONCERNS WITH CLINICAL-BASED TRAINING

It is a general practice to train healthcare practitioners to perform independent ultrasound scanning, for which training is based on clinical life and real patients. However, many concerns have been raised in recent years that could constrain the provision of an optimum educational programme based on live clinical training. The following section outlines the main challenges that relate to performing optimum clinical-based training in ultrasound, namely, a shortage of available ultrasound trainers, the environment of clinical practice and the necessity for ensuring patient safety.

SUPERVISION SHORTAGE

With the recent rise of the international population, workloads in hospitals have increased, impacting the training capacity in clinical sites as the demands of qualified practitioners to complete their non-teaching duties have intensified. The issue is further complicated within ultrasound practice due to the decreasing chance of finding an expert sonographer who could be available to train novice practitioners. The last statistics report of the UK National Health Service (NHS) indicates a 21% rise in non-obstetric ultrasound examinations over the previous five years (Baker, 2019). Crucially, one-third of UK sonographers are over 50 years of age, and thus many could be retiring over the following ten years (Thomson, 2014). This was one of the rationales that promoted UK-CASE to update the sonographic education standards to improve the future workforce in the field of ultrasound (Harrison and Dolbear, 2018). Consequently, it has become crucial that the healthcare system considers the incorporation of alternative training methods that reduce the reliance on receiving clinical supervision and therefore enhance the process of ultrasound learning and skills acquisition.

CLINICAL ENVIRONMENT

Receiving clinical training in a busy environment is another challenge that could impede the ultrasound learning process and therefore delay skills acquisition. Clinical-based training is not necessarily a learning-friendly environment for novice practitioners because it exerts time pressures on learners that mostly allows only for passive observation, limited active engagement and limited constructive feedback (Spencer, 2003). Learning ultrasound skills, on the other hand, is better when subjected to experiential learning, requiring active interaction with the learning process, playing a role and receiving a reflective response that assists in achieving optimum learning efficiency (Fanning and Gaba, 2007). In clinical practice, novice practitioners expect to receive a well-oriented programme that permits them to practice what they have learned in theory. At the same time, patients expect to receive a high quality of care from practitioners. However, when the novice practitioners fail to meet educational expectations, their stress level elevates, a frustrating relationship with supervisors occurs and the potential to creating error increases; as a result, the quality of patient care reduces (Haskvitz and Koop, 2004). According to the new standards for providing ultrasound service in the UK, which are recommended by the Society and College of Radiographers (SCoR) and the Royal College of Radiology (RCR), experiencing appropriate training is one of the factors that fundamentally affects the quality of ultrasound examinations (Cavanagh and Smith, 2017). As the learning environment profoundly impacts on the resulting skills, and as the clinical setting could induce more pressure on ultrasound learners, establishing an alternative environment which is suitable for learning and developing ultrasound skills should therefore be mandatory.

PATIENT SAFETY

Ensuring patient safety in ultrasound practice is a superior aim to training for skills acquisition, as some sensitive procedures cannot be undertaken during for long scans, and passive observations are thus sometimes performed with such scans rather than hands-on training. Even with the safe nature of ultrasound beams, as a non-ionising radiation, employing the ALARA principle is still crucial to ensure patient safety. ALARA stands for 'As Low As Reasonably Achievable', and refers to the minimisation of acoustic output with shortest scanning time while achieving optimum diagnosis (Joy et al., 2006). The international ultrasound societies emphasise the importance of occupying safe ultrasound practice in clinical fields. For example, both the American Institution for Ultrasound in Medicine (AIUM) and the British Medical Ultrasound Society (BMUS) provide the highest evidence-based guidelines that recommend ultrasound users to ascertain the endorsed measures of exposure time and thermal/mechanical index for sensitive scans such as obstetric, neonatal, trans-cranial and vascular ultrasound (BMUS, 2009; AIUM, 2016; SCoR/BMUS, 2019). Moreover, The International Society of Ultrasound in Obstetrics and Gynaecology (ISUOG) recommends that practitioners follow the ALARA principle in early pregnancy ultrasound and do not include pulsed Doppler ultrasound in routine scanning during the first trimester. If used at all, the thermal index (TI) should be ≤1.0 and the exposure time should be minimised as far as possible with no longer than 5–10 minutes' exposure (Salvesen et al., 2011, 2016).

According to ISUOG, healthcare educators must follow the ALARA principle when setting up training programmes for ultrasound practice (ISUOG, 2014; Salvesen et al., 2016). Additionally, BMUS states that scanning time should be taken into account when teaching or training is in process (BMUS, 2017). However, monitoring the length of time taken by scans used for sensitive examinations by novice practitioners can be a challenge. For example, one study showed that ultrasound-novice participants spent five times longer on VR scans than the participating experts in measuring foetal crown-rump length (CRL) and also spent around 11 minutes scanning for foetal biometry (FB) while the experts spent five minutes (Burden et al., 2013). Based on the ALARA principle, spending so much time on obstetric ultrasound could be acceptable only if the TI of ultrasound equipment is less than two, meaning that ultrasound equipment induces a tissue-temperature elevation of less than 2°C (BMUS, 2009). However, the study demonstrates that novice practitioners could take a longer scanning time that could negatively impact safe ultrasound practice (Burden et al., 2013). Consequently, preparing novice practitioners through the performance of simulation-based training before starting clinical practice can contribute to assuring patient safety. Indeed, the literature has highlighted the significant decrease in time of ultrasound examinations performed after completing simulation-based training (Burden et al., 2013; Tolsgaard et al., 2017).

As discussed above, the challenges faced by any novice ultrasound practitioner during clinical-based training could delay skills acquisition and promote longer training sessions. Due to the high workloads of ultrasound departments, trainees could receive passive observations on cases that require practical practice, as meeting the scheduled patients' examinations is the highest priority. Trainees, additionally, could experience limited supervision and inadequate feedback due to a shortage of ultrasound experts who are available for training. Importantly, the obligation to declare patient safety by keeping scanning time as short as possible could impact on the efficiency of ultrasound training. These challenges encourage healthcare educators to invest in alternative training methods based on VR simulation, especially when the literature has examined the effects of simulation-based training on skills acquisition, showing positive impacts for ultrasound practice (Sidhu et al., 2012; Blum et al., 2013).

ULTRASOUND SKILLS ACQUISITION USING VR SIMULATION

Skills acquisition is a core principle in competent ultrasound practice that contributes to maintaining high-quality services with minimum diagnostic errors (EFSUMB, 2005; Konge et al., 2014). Unlike X-ray, computed tomography (CT) or magnetic resonance imaging (MRI) technology, ultrasound highly depends on the practitioner's competency during the scan, and the generation of any error cannot be resolved later by another more competent practitioner (Harrison and Dolbear, 2018). Thus, the EFSUMB highlights the importance of adequate ultrasound training for novice practitioners to eliminate any potential diagnostic errors (EFSUMB, 2005). Moreover, the literature shows that as long as the practitioner proceeds through several varied ultrasound examinations, their competency in the practice of ultrasound improves. However, ultrasound imaging requires a variety of complex skills that involve both cognitive and psychomotor skills, for which acquisition can ensure a competent scanning and contribute to efficient image recognition and interpretation. Such skills require a systematic training programme that permits maximum learning proficiency in a safe environment to provide optimum maintenance of care services. Interestingly, the recent literature has examined the impact of embedding simulation practice within ultrasound training programmes on practitioners' skills, showing positive influences on both cognitive and psychomotor ultrasound skills. Simulation-based training has therefore been shown to assist in improving learning efficiency and knowledge retention (cognitive skills) and scanning competence (psychomotor skills).

COGNITIVE SKILLS: LEARNING EFFICIENCY AND KNOWLEDGE RETENTION

The best description of the term 'cognitive skills' does not refer to gifted intelligence, but rather refers to the acquisition of an accumulative knowledge which can reach to the actual performance. For example, playing chess requires learning the rules of the game or at least having experienced it before, even highly intelligent individuals cannot play chess without learning its rules (Bloom, 1956; Perkins and Salomon, 1989). Developing cognitive skills for a task involves building knowledge including recalling and remembering, and therefore, developing mental abilities including reflection on this knowledge, i.e. critical thinking (Bloom, 1956). In ultrasound imaging, the primary step in developing cognitive skills is to build a basic knowledge in anatomy, pathology and ultrasound principles relating to instrumentations, scanning techniques, ultrasound physics and safety. Recalling knowledge as a skill in ultrasound can contribute to identifying internal structures, recognising anatomical plans and detecting image artefacts. For example, the attenuation artefact is a common ultrasound artefact which can be caused by using an inappropriate probe or performing an inadequate scan, i.e. keeping some anatomical structures unclear in the image (Feldman et al., 2009). The recognition of such artefacts not only requires knowledge acquisition based on a conventional learning process (e.g. lectures, reading, etc.) but further experiencing the hands-on practice of a number and variety of scans. Additionally, to master the cognitive skills in ultrasound practice that exceeds knowledge retention, the practitioner requires an improvement in their critical thinking skills as vital skills to attaining best practice. Critical thinking, in general, refers to a construction of a state linked with a rationale that is based on an accumulation of knowledge to provide a conclusion or a decision (Sternberg et al., 2007). For example, when a patient complains of an upper-right abdominal pain with the existence of a gallstone, good critical thinking in ultrasound practice would occur when the practitioner performs a systematic liver scan to exclude any other lesions, rather than limiting the scan to the gallbladder (Edwards, 2006).

The recent literature has emphasised the capability of simulation-based training to improve the learning efficiency that contributes to developing cognitive skills for ultrasound practice. For example, a study conducted in 2018 showed that employing VR simulation in teaching foetal ultrasound can contribute to better learning efficiency and knowledge retention than conventional teaching methods. The participants for this study were a group of junior doctors and medical master's students who had no previous ultrasound experience ($n = 51$) and were recruited from different universities and hospitals in Switzerland and Germany. The participants were randomly allocated into a control group ($n = 30$) who received a traditional teaching method of video lectures and electronic articles to learn normal and abnormal sonographic foetal anatomy. The rest of the group ($n = 21$) was allocated into an intervention of simulation-based learning, in addition to the conventional methods, that utilised online software, Pocket Brain VR objects (Tutschek and Pilu, 2017). The knowledge of both groups was tested using online multiple-choice questions (MCQs), and the results showed that the intervention group answered more questions correctly than the control group with a statistically significant shorter time (13 minutes for the intervention group vs. 16 minutes for the control group, $p < 0.001$) (Ebert and Tutschek, 2019). The results indicate that applying VR simulation for ultrasound education can assist in improving learning efficiency and knowledge retention as cognitive skills.

Another study highlighted the benefit of adding simulation-based training in improving ultrasound knowledge for medical students in emergency departments. The study embedded four hours per week of simulation practice (using a mannequin and SonoSim ultrasound training solution device) with one hour per week of theoretical learning over a period of four weeks of ultrasound training in emergency departments (SonoSim, 2019). The study recruited 96 final year medical students who attended the emergency department in the Kirikkale University Faculty of Medicine in Turkey between 2015 and 2016. The participants were examined theoretically and practically before and after the intervention. The results showed a significant increase ($p < 0.001$) in post-intervention test mean scores from 7.9 ± 2.2 to 17.14 ± 1.6 in 20 MCQs in ultrasound practice in emergency departments, including focused assessment with sonography for trauma (FAST) and rapid ultrasound for shock and hypotension (RUSH). Additionally, the results demonstrated a significant increase ($p < 0.001$) in the post-intervention practical test from a mean score of 1.1 ± 0.9 to 10.9 ± 0.2 in 12 different questions regarding selecting the probe, improving image quality, recognising internal structures and defining lesions and injured organs (Eroglu and Coskun, 2018). The study indicates that simulation-based training could contribute to an efficient learning process of ultrasound in emergency medicine that involves developing better knowledge in a shorter time compared with the conventional learning methods displayed in the previous literature (Favot et al., 2015; Turner et al., 2015).

A recent study has underlined the efficiency of simulation-based learning on acquiring accurate ultrasound measurements of kidneys. Approximately 66 medical students were recruited in Germany in 2016 and were allocated into a control group ($n = 33$) and an intervention group ($n = 33$). Both groups were given 60 minutes to learn about kidney ultrasound using textbooks; however, the intervention group had additional access to an ultrasound simulator app. The students were asked to obtain kidney images by scanning the tutor, who facilitated the learning session. The results showed that the intervention group provided more accurate measurements for kidneys compared to the control group ($p < 0.001$) (Ebner et al., 2019). It is known that the normal length of an adult kidney is 11.2 cm for the left kidney and 10.9 cm for the right kidney (Emamian et al., 1993). In the study, the intervention group median measure for the left kidney was 10.03 cm, and for the right kidney was 10.53 cm, while the control group measures were 8.5 for the left kidney and 9.2 for the right kidney (Ebner et al., 2019). This finding shows that simulation-based training improves the quality of learning in ultrasound, which impacts positively on practitioner's practice skills.

The perspectives of learners who had a chance to practice ultrasound using VR simulation show positive insights about the effect of this technology on the quality of learning. The literature showed two studies used a qualitative approach to collect opinions from master's level ultrasound students and their mentors about using the ScanTrainer as a transvaginal ultrasound VR Simulator, and both studies demonstrated various views about the usefulness of simulation training in learning ultrasound (Gibbs, 2014, 2015). For example, one of the students said: 'Scanning the pathology cases was much better than just reading it in a textbook because you had to work your way through the step by step assessment'. Another student said: 'I was struggling with understanding 2D anatomy and orientation in clinical practice but the diagrams and instructions as you move the simulator transducer made it suddenly click with me' (Gibbs, 2014, p. 175). Moreover, one of the students' mentors said: 'I found that after she worked on the simulator, my student has taught me a few things about orientation which I'd never really thought about before!'; while another mentor said: 'This year my student has grasped the anatomy and relationship of organs very quickly – not sure if she's just spent more time learning this than my previous students or whether it was the simulator that helped' (Gibbs, 2015, p. 206).

Furthermore, the most recent study has collected the views of 41 healthcare practitioners who experienced four hours of training on the ultrasound simulator, Canadian Aviation Electronics-Vimedix, involving modules in adult and paediatric cardiac ultrasound, emergency medicine and obestetric and gynecology ultrasound (Healthcare, 2019). The results showed that 98% ($n = 40$) agreed that the simulator facilitates training and learning pathology, while 100% ($n = 41$) agreed that VR simulators should be part of the curriculum either in medical schools or during residency (Hani et al., 2019). All these perspectives support the notion that VR simulators could be efficient learning tools that assist in developing ultrasound cognitive skills.

Psychomotor Skills: Scanning Competence

Psychomotor learning is generally related to the skills that involve physical movement, manual tasks and the operation of equipment (Rovai et al., 2009, Simpson, 1966). The definition of psychomotor skills in the healthcare domain varies from one discipline to another (Nicholls et al., 2014). For example, psychomotor skills in surgery refer to a pattern of mental and motor skills that contribute to performing a manual task; however, in physiotherapy, psychomotor skills are more related to performance that is accomplished correctly, efficiently and safely (Kovacs, 1997, Rose and Best, 2005). Regarding ultrasound practice, Nicholls et al. defined psychomotor skills as 'the unique mental and motor activities required to execute a manual task safely and efficiently for each clinical situation' (Nicholls et al., 2014, p. 1350).

Psychomotor skills in ultrasound are divided into visuomotor skills and visuospatial skills. *Visuomotor skills* are hand-eye coordination skills that rely on a limb movement to guide the probe to create a visual image on the screen. For example, when the image is unclear or the target organ is obscured with shadow, a motor task would be performed to change the probe pressure or angle. On the other hand, *visuospatial skills* relate to developing 3D mental vision of an organ or pathology from 2D images and depend on the recognition of the different anatomical projections that facilitate the practitioner to view an abnormality from all its angles. For example, recognising a retroverted uterus in the sagittal plane is a practice of visuospatial skills (Nicholls et al., 2014).

Mastering psychomotor skills in ultrasound practice requires experiencing a hands-on practice that involves unlimited access to various cases for an adequate length of time in training (SCoR/BMUS, 2019). The literature shows encouraging impacts of simulation-based training on developing psychomotor skills for ultrasound practice (Blum et al., 2013). Training using VR simulators provides an opportunity for the learner to practically practice ultrasound scans in a realistic manner that contributes to better performance within the first clinical scan on real patients. For example, a student said in Gibbs's qualitative study 'I found it (using the simulator) more useful at the beginning of my scanning because, although it wasn't completely life-like, it did prepare me for when I had to scan my first real patient on my own' (Gibbs, 2015, p. 206). The reason for performing enhanced scans in clinical practice after simulation-based training attributes to the rise of the learner's confidence, as acquiring skills related to ultrasound technical aspects and image perceptions are significant factors for enhancing confidence in ultrasound performance (Tolsgaard et al., 2014). Based on the literature, many medical doctors who had a chance to experience VR simulation for transvaginal ultrasound training reported that simulation practice could influence their scanning confidence (Chao et al., 2015, Williams et al., 2013). Furthermore, the theme which was generated from Gibbs's qualitative study based on interviewing 12 postgraduate ultrasound students influences the capability of VR simulators in providing opportunities to build confidence in a safe environment (Gibbs, 2014).

The literature not only shows the efficiency of developing psychomotor skills for ultrasound scanning based on developing the self-confidence of the simulation learner, it further highlights the actual effects of simulation-based training by examining learners' skills following simulation intervention. For example, the most recent study that explored this aspect and recruited 20 radiology residents from different hospitals in Denmark concludes that simulation-based training can improve practical ultrasound skills (Østergaard et al., 2019). The participants in this study were randomly allocated into an intervention group ($n = 11$), who received up to 17 hours of abdominal ultrasound practice on the Schallware simulator before starting clinical training (Schallware, 2019), and a control group ($n = 9$) who started their clinical training without this intervention. The abdominal ultrasound scans performed in the first six weeks of clinical training were scored for both groups involving scores on knowledge of equipment, image optimisation, systematic examination and image interpretation. The results demonstrated that within the first 29 clinical scans, the performance scores of the intervention group were significantly higher than those of the control group ($p < 0.001$) (Østergaard et al., 2019).

Furthermore, another study examined the efficiency of ultrasound training in developing the practical skills of point-of-care ultrasound for medical students. Nineteen medical students were recruited from McGill University in Canada and were assigned into two groups: control and intervention groups. The intervention group ($n = 10$) received four hours of thoracic and abdominal ultrasound sessions on the Vimedix simulator (Healthcare, 2019) in addition to the four weeks of clinical training during emergency rotation; however, the control group ($n = 9$) completed their rotation without any additional practice. The ultrasound practical skills for both groups were examined before starting the emergency rotation (pre-test) and in the fourth week after completing the rotation (post-test). While there was no difference in the pre-test scores of the groups (73.6% vs. 61.4%, $p = 0.212$), the post-test scores of the intervention group were significantly higher than those of the control group (77.9% vs. 56.8%, $p = 0.006$) (Le et al., 2019). Consequently, both studies indicate the usefulness of VR simulation-based training in developing psychomotor skills that are essential to starting clinical ultrasound practice.

Simulation-based training does not only improve the psychomotor skills of novice practitioners in ultrasound but can also extend to strengthen the skills of those with previous practical experience. For example, a recent study has examined the effect of simulation-based training on the practical skills of performing an accurate obstetric ultrasound. The study followed 70 obstetricians with different experience levels in ultrasound who were randomly allocated in an intervention group that had around three hours of simulation training to perform foetal weight scans and a control group who had no intervention. Experts assessed the scanning performances for both groups before and after the intervention. The results showed that the FB images of the experienced participants in the intervention group were significantly more accurate than those in the control group ($p = 0.03$). The accuracy in foetal measurements increased by 45% in the intervention group and by 18% in the control group (Andreasen et al., 2019). This finding shows that practicing using simulation can improve the psychomotor skills of even experienced ultrasound healthcare practitioners.

FUTURE DEVELOPMENTS

Although the literature provides real insights into employing simulation-based training in ultrasound practice that indicates likely impacts on healthcare practitioners' skills, simulation-based training still cannot replace clinical-based training. The need for other crucial skills that assist in maintaining a competent ultrasound practice, i.e. communication skills, renders a sole reliance on simulation-based training inapplicable. For example, an ultrasound practitioner is responsible for building a personal connection with the patient once he/she arrives in the room using simple language that can lead to a successful examination process (Duffy et al., 2004, Pillai et al., 2018).

Starting in this way shows how highly the patient is regarded and how they are involved in their clinical examination, ultimately assisting in relieving any anxiety or discomfort during the ultrasound scan. Ensuring patient confidentiality when another person is present in the room, being aware of the susceptibility of the patient when receiving the results of ultrasound scans, and dealing with any emotional reactions are examples of communication skills that represent competent ultrasound practice (Pillai et al., 2018). Such skills require hands-on training in a clinical environment, which has not been implanted within VR simulation to date. Incorporating virtual scenarios with ultrasound simulators may help to enhance the communication skills that proceed the performance of an optimum ultrasound practice.

Despite the necessity consolidate communication skills development with VR ultrasound simulation, incorporating new technologies with simulators could provide a promising future in improving ultrasound practice. For example, utilising gamification-based learning could become an upcoming trend in ultrasound education. The programme 'SonoGame' has been proposed in the meetings of various different European societies as a VR gaming tool for ultrasound practice, and participants report having appreciated the entertaining part of the device, encouraging them to develop their basic skills for ultrasound practice. Furthermore, Dr. Teistler, a professor of computer science at the Flensburg University of Applied Science in Germany, reported that this university plans to employ candidates' skills in programming, game design, computer graphics and VR to develop the SonoGame to involve more mini games with mobile versions for Android and iPhones (EFSUMB, 2019a).

At the same time, the future could see development in the practice of engaging handled ultrasound devices (pocket-size) with simulation training. Based on statement from EFSUMB, it appears that these devices have a lower cost than conventional ultrasound scanners and contain probes with plugs that can be connected to smartphones to start scanning at any time (Nielsen et al., 2019). A significant decrease in the cost of such devices could encourage healthcare educators to invest in such tools, especially within countries with low resources, such as seen in the use of devices to train doctors in Uganda to diagnose pneumonia (EFSUMB, 2019b). Such technology has been in use for some time at point-of-care ultrasound but due to its limitation of small screens with low resolution that could cause inaccurate detection of small lesions, a reliance on such devices for clinical practice has been restricted (Nielsen et al., 2019). However, integrating these simple and inexpensive pocket-sized ultrasound devices with simulation for educational purposes could provide a promising step forward in developing basic skills for ultrasound practice.

CONCLUSION

Simulation-based training provides a unique opportunity for a healthcare practitioner to acquire and develop the complex skills required for optimum ultrasound practice involving both cognitive and psychomotor skills. Several scientific studies highlight the benefits of virtual-reality simulators in enhancing learning efficiency and knowledge retention as well as learners' confidence and practical skills. Simulation-based training could overcome the drawbacks related to clinical-based training such as supervision limitations, busy environments and limiting the training for patient safety assurance; however, sole reliance on simulation-based training is ineffective and therefore unsafe. The fact that simulators lack live elements restricts the development of practitioners' communication skills, for example. Such limitations could be mitigated in the future by incorporating live scenarios with simulation settings.

REFERENCES

AIUM. 2016. *Recommended Maximum Scanning Times for Displayed Thermal Index (TI) Values* [online]. Available at: https://www.aium.org/officialStatements/65 [Accessed 12/11/2019].

Andreasen, L. A. et al. 2019. Is simulation training only for inexperienced trainees? A multicenter randomized trial exploring the effects of simulation-based ultrasound training on obstetricians' diagnostic accuracy. *Ultrasound in Obstetrics & Gynecology*.doi: 10.1002/uog.20362.

Baker, C. 2019. *NHS Key Statistics: England October 2019* [online]. Available at: https://researchbriefings. parliament.uk/ResearchBriefing/Summary/CBP-7281 [Accessed 02/10/2019].

Bloom, B. S. 1956. *Taxonomy of Educational Objectives: Cognitive Domain/the Classification of Educational Goals.* London: Longmans.

Blum, T. et al. 2013. A review of computer-based simulators for ultrasound training. *Simulation in Healthcare,* 8, pp. 98–108.

BMUS. 2009. *Guidelines for the Safe Use of Diagnostic Ultrasound Equipment* [online]. Available at: https://www.bmus.org/ [Accessed 15/11/2019].

BMUS. 2017. *Guidelines for the Management of Safety When Using Volunteers & Patients for Practical Training in Ultrasound Scanning* [online]. Available at: https://www.bmus.org/ [Accessed 15/11/2019].

Burden, C. et al. 2013. Usability of virtual-reality simulation training in obstetric ultrasonography: a prospective cohort study. *Ultrasound in Obstetrics & Gynecology*, 42, pp. 213–217.

Cavanagh, P. & Smith, K, 2017. *Standards for Provision Ultrasound Service* [online]. Available at: https://www.rcr.ac.uk/publication/standards-provision-ultrasound-service [Accessed 17/11/2019].

Chao, C. et al. 2015. Randomized clinical trial of virtual reality simulation training for transvaginal gynecologic ultrasound skills. *Journal of Ultrasound in Medicine*, 34, pp. 1663–1667.

Duffy, F. D. et al. 2004. Assessing competence in communication and interpersonal skills: the Kalamazoo II report. *Academic Medicine: Journal of the Association of American Medical Colleges*, 79, pp. 495–507.

Ebert, J. & Tutschek, B. 2019. Virtual reality objects improve learning efficiency and retention of diagnostic ability in fetal ultrasound. *Ultrasound in Obstetrics & Gynecology*, 53, pp. 525–528.

Ebner, F. et al. 2019. Effect of an augmented reality ultrasound trainer app on the motor skills needed for a kidney ultrasound: prospective trial. *Journal of Medical Internet Research Serious Games*, 7, e12713.

Edwards, H. 2006. Critical thinking and the role of the clinical ultrasound tutor. *Radiography*, 12, pp. 209–214.

EFSUMB. 2005. *Minimum Training Recommendations for the Practice of Medical Ultrasound* [online]. Available at: https://www.efsumb.org/blog/archives/1687 [Accessed 7/10/2019].

EFSUMB. 2019a. Learning by playing – using computer games to enhance ultrasound education. *Ultraschall in der Medizin - European Journal of Ultrasound*, 40(01), pp. 93–94.

EFSUMB. 2019b. Butterfly network. *Ultraschall in der Medizin - European Journal of Ultrasound*, 40(04), pp. 528–529.

Emamian, S. A., Nielsen, M. B., Pedersen, J. F. & Ytte, L. 1993. Kidney dimensions at sonography: correlation with age, sex, and habitus in 665 adult volunteers. *American Journal of Roentgenology*, 160, pp. 83–86.

Eroglu, O. & Coskun, F. 2018. Medical students' knowledge of ultrasonography: effects of a simulation-based ultrasound training program. *The Pan African Medical Journal*, 30, p. 122.

Fanning, R. M. & Gaba, D. M. 2007. The role of debriefing in simulation-based learning. *Simulation in Healthcare*, 2, pp. 115–125.

Favot, M., Courage, C., Mantouffel, J. & Amponsah, D. 2015. Ultrasound training in the emergency medicine clerkship. *Western Journal of Emergency Medicine*, 16, p. 938.

Feldman, M. K., Katyal, S. & Blackwood, M. S. 2009. US artifacts. *Radiographics: A Review Publication of the Radiological Society of North America, Inc*, 29, pp. 1179–1189.

Gibbs, V. 2014. An investigation into sonography student experiences of simulation teaching and learning in the acquisition of clinical skills. *Ultrasound*, 22, pp. 173–178.

Gibbs, V. 2015. The role of ultrasound simulators in education: an investigation into sonography student experiences and clinical mentor perceptions. *Ultrasound*, 23, pp. 204–211.

Hani, S., Chalouhi, G., Lakissian, Z. & Sharara-Chami, R. 2019. Introduction of ultrasound simulation in medical education: exploratory study. *Journal of Medical Internet Research Medical Education*, 5, e13568.

Harrison, G. & Dolbear, G. 2018. *Standards for Sonographic Education. Consortium for the Accreditation of Sonographic Education* [online]. Available at: http://www.case-uk.org [Accessed: 07/11/2018].

Haskvitz, L. M. & Koop, E. C. 2004. Students struggling in clinical? A new role for the patient simulator. *Journal of Nursing Education*, 43, pp. 181–184.

Healthcare 2019. *CAE Vimedix Ultrasound Simulation* [online]. Available at: https://caehealthcare.com/ultrasound-simulation/vimedix/ [Accessed 08/11/2019].

ISUOG. 2014. ISUOG education committee recommendations for basic training in obstetric and gynecological ultrasound: ISUOG recommendations. *Ultrasound in Obstetrics & Gynecology*, 43, pp. 113–116.

Joy, J., Cooke, I. & Love, M. 2006. Is ultrasound safe? *The Obstetrician & Gynaecologist*, 8, pp. 222–227.

Konge, L., Albrecht-Beste, E. & Nielsen, M. B. 2014. Virtual-reality simulation-based training in ultrasound. *Ultraschall in der Medizin-European Journal of Ultrasound*, 35, pp. 95–97.

Kovacs, G. 1997. Procedural skills in medicine: linking theory to practice. *The Journal of Emergency Medicine*, 15, pp. 387–391.

Le, C. K., Lewis, J., Steinmetz, P., Dyachenko, A. & Oleskevich, S. 2019. The use of ultrasound simulators to strengthen scanning skills in medical students: a randomized controlled trial. *Journal of Ultrasound in Medicine*, 38, pp. 1249–1257.

Nicholls, D., Sweet, L. & Hyett, J. 2014. Psychomotor skills in medical ultrasound imaging. *Journal of Ultrasound in Medicine*, 33, pp. 1349–1352.

Nielsen, M. B. et al. 2019. The use of handheld ultrasound devices – an efsumb position paper. *Ultraschall in der Medizin - European Journal of Ultrasound*, 40(01), pp. 30–39.

Østergaard, M. L. et al. 2019. Simulator training improves ultrasound scanning performance on patients: a randomized controlled trial. *European Radiology*, 29, pp. 3210–3218.

Perkins, D. N. & Salomon, G. 1989. Are cognitive skills context-bound? *Educational Researcher*, 18, pp. 16–25.

Pillai, M., Briggs, P. & Bridson, J.-M. 2018. *Ultrasound in Reproductive Healthcare Practice*. Cambridge: Cambridge University Press.

Rose, M. L. & Best, D. L. 2005. *Transforming Practice through Clinical Education, Professional Supervision and Mentoring*. Philadelphia, USA: Elsevier Health Sciences.

Rovai, A. P., Wighting, M. J., Baker, J. D. & Grooms, L. D. 2009. Development of an instrument to measure perceived cognitive, affective, and psychomotor learning in traditional and virtual classroom higher education settings. *The Internet and Higher Education*, 12, pp. 7–13.

Salvesen, K. et al. 2011. ISUOG statement on the safe use of Doppler in the 11 to 13+ 6-week fetal ultrasound examination. *Ultrasound in Obstetrics & Gynecology*, 37, pp. 628–628.

Salvesen, K. et al. 2016. *ISUOG statement on ultrasound exposure in the first trimester and autism spectrum disorders* [online]. Available at: https://www.isuog.org/resource/isuog-statement-on-ultrasound-exposure-in-the-first-trimester-and-autism-spectrum-disorders-pdf.html [Accessed 20/11/2019].

Schallware. 2019. *Ultrasound Simulator* [online]. Available at: https://www.schallware.de/ [Accessed 20/11/2019].

SCOR/BMUS. 2019. *Guidelines for Professional Ultrasound Practice* [online]. Available at: https://www.bmus.org/policies-statements-guidelines/professional-guidance/guidelines-for-professional-ultrasound-practice/ [Accessed 20/11/2019].

Sidhu, H. S., Olubaniyi, B. O., Bhatnagar, G., Shuen, V. & Dubbins, P. 2012. Role of simulation-based education in ultrasound practice training. *Journal of Ultrasound in Medicine*, 31, pp. 785–791.

Simpson, E. J. 1966. *The Classification of Educational Objectives, Psychomotor Domain* [online]. Available at: https://eric.ed.gov/?id=ED010368 [Accessed 15/11/2019].

Sonosim. 2019. *The SonoSim Ultrasound Training Solution* [online]. Available at: https://sonosim.com/ [Accessed 15/11/2019].

Spencer, J. 2003. Learning and teaching in the clinical environment. *British Medical Journal*, 326, pp. 591–594.

Sternberg, R. J., Roediger, H. L. III & Halpern, D. F. 2007. *Critical Thinking in Psychology*. Cambridge: Cambridge University Press.

Thomson, N. 2014. *Sonographer Workforce Survey Analysis* [online]. Available at: https://www.sor.org/learning/document-library/sonographer-workforce-survey-analysis [Accessed 7/11/2019].

Tolsgaard, M. G. et al. 2014. Which factors are associated with trainees' confidence in performing obstetric and gynecological ultrasound examinations? *Ultrasound in Obstetrics & Gynecology*, 43, pp. 444–451.

Tolsgaard, M. G. et al. 2017. The effects of simulation-based transvaginal ultrasound training on quality and efficiency of care: a multicenter single-blind randomized trial. *Annals of Surgery*, 265, pp. 630–637.

Turner, E. E. et al. 2015. Implementation and assessment of a curriculum for bedside ultrasound training. *Journal of Ultrasound in Medicine*, 34, pp. 823–828.

Tutschek, B. & Pilu, G. 2017. Pocket brain, an interactive, web-based ultrasound atlas of normal and abnormal fetal brain development. *Ultrasound in Obstetrics & Gynecology*, 49, pp. 431–432.

Williams, C. J., Edie, J. C., Mulloy, B., Flinton, D. M. & Harrison, G. 2013. Transvaginal ultrasound simulation and its effect on trainee confidence levels: a replacement for initial clinical training? *Ultrasound*, 21, pp. 50–56.

11 Health Profession Education and Training Using Virtual Reality

Pete Bridge
University of Liverpool
Liverpool, UK

CONTENTS

INTRODUCTION

Virtual reality (VR) has played a major role in education and training since the first flight simulators (Zazula et al. 2013) placed trainee pilots into a realistic environment. These early simulators provided users with full immersion through a combination of authentic hardware controls and visualisation equipment. Indeed, VR and simulation are still strongly associated with each other as a successful simulation must immerse the user in a realistic environment. This chapter, however, will focus on the use of VR simulation resources that make use of software and computer graphics to replicate reality. Most of the current VR training tools utilise immersive visualisation environments to enhance the realism and presence for the user.

It is convenient to classify VR training resources as 'open' or 'closed' (Bridge et al. 2007) depending on whether the environment is displayed on a screen for multiple users to engage with or delivered via a single-user interface. This distinction is important in health profession training depending on whether single-person or team-based procedures are being simulated. The choice of format is also to some extent driven by the scale of the equipment and environment being replicated. In general, closed systems offer users the most immersive experience but suffer from high cost per user. Open systems naturally facilitate interpersonal skills development by enabling users to train together and can be associated with reduced costs. This chapter examines the emerging role of VR in health profession training, illustrating key aspects with examples from the widening evidence base.

APPLICATIONS OF VR TO HEALTH PROFESSION EDUCATION AND TRAINING

Health professionals regularly call on a wide range of technical, interpersonal and clinical skills (Veale et al. 2014) in order to provide patients with competent and holistic care. Although training within the clinical environment provides a real-world grounding, there are some aspects of skills training that can be complemented or potentially replaced with simulation and in particular with high fidelity VR resources. Table 11.1 provides an overview of the recent publications related to VR use in training across a range of health professions.

The following sections of this chapter illustrate some established and emerging applications of VR to training within the technical, interpersonal and clinical domains of practice.

TECHNICAL SKILLS TRAINING

Most health professions require competence with technical skills including use of specialised equipment or techniques. VR applications have been used successfully to train learners with technical skills since the early days of flight simulators and enable learners to gain skills through repetition, coaching, use of extreme situations as well as through trial and error. Simulated scenarios within VR enable learners to develop hand-eye coordination, gain familiarity with controls and learn complex pathways and decision-making in a safe environment. As outlined previously, the VR resources currently used within healthcare training fall into two distinct categories; 'Open' and 'Closed'. Each of these formats offers completely different experiences to the user and have their own specific uses for different health professions.

TABLE 11.1
VR Training Publications by Profession

Profession	Publications	Indicative Uses
Radiotherapy	43	Treatment planning and delivery process and equipment simulation
Medical imaging	477	Imaging suite process and equipment simulation
		3D anatomical visualisation
Nursing	559	Decision making
		Process simulation
		Virtual patients
		VR cameras
Orthoptics	3	Anatomical visualisation
Occupational therapy	407	Wheelchair simulator
		Rehabilitation software
		Environmental simulation
Paramedics	319	Process and clinical simulation
		Triage training
		Resuscitation training
Physiotherapy	64	Game-based learning
		Virtual patients
		Anatomical visualisation
Surgery	1,866	Surgery process and technique simulation
		Teamwork in a surgical environment
		Performance assessment
General Medicine	504	Anatomical visualisation
		Virtual patients
		Process and technical simulations
		Teamwork applications

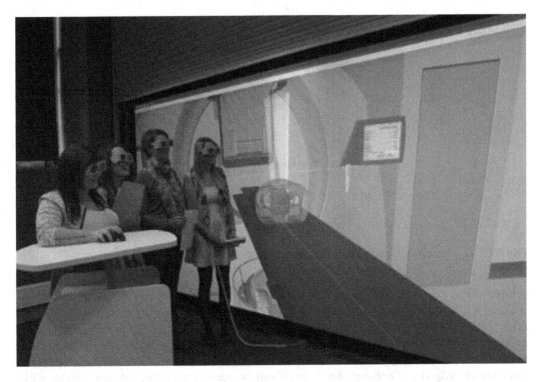

FIGURE 11.1   Collaborative learning in the VERT 'Open VR' radiotherapy training facility.

'Open' VR

Open VR systems employ screens to immerse the user in the virtual environment. Systems such as CAVES increase presence (Krijn et al. 2004) by surrounding the user with screens and use of shutter glasses or polarising systems provide stereoscopic visualisation to simulate large-scale environments. These systems are useful for those health professions requiring whole-room simulation where the learner must understand physical placement of equipment and patients. Good examples of these are found in radiotherapy and medical imaging training. One of the earliest (Bridge et al. 2007) and most widespread (Bridge et al. 2017) examples is the 'Virtual Environment for Radiotherapy Training'. Figure 11.1 shows users working together in the VERT. The shutter glasses provide the 3D visualisation necessary for practicing aligning the equipment and patient. The software also enables users to visualise structures inside the patient in 3D essential for learning relational anatomy and for checking radiation dose distribution. The large screen enables multiple users to interact while also enabling realistic depiction of the large-scale equipment and infrastructure.

Smaller scale desktop solutions in medical imaging such as Medspace.VR and Shaderware rely on standard PC workstation screens to simulate a range of radiography environments. Both enable users to practice patient setup and imaging. Medspace.VR provides increased user immersion through the use of stereoscopic glasses and highly realistic 3D environments as seen in Figure 11.2, while Shaderware retains conventional 2D imaging. These three solutions highlight one of the most useful aspects of VR for training by enabling learners to practice using radiation without endangering themselves or patients. The clinical equipment, particularly in radiotherapy, is extremely expensive and requires considerable supporting infrastructure. The rate of technological development also means that a VR solution that can easily be upgraded to simulate new equipment is more sustainable in the long term.

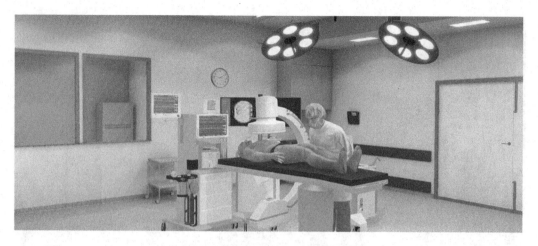

FIGURE 11.2 Typical medical imaging VR-simulated environment.

'Closed' VR

'Closed' VR systems provide users with heightened immersion and presence with evidence (Shu et al. 2018) suggesting an increase in engagement and learning. The rapidly evolving 3D headset market is now producing relatively inexpensive immersive visualisation equipment which can place individual users into highly realistic 3D environments surrounded by excellent 360 degree graphics and intuitive navigation. 'Grab and Move' controls allow users to interact directly with objects while laser pointers provide menu functionality. While external sensors can detect user position and map their movements within a large-scale virtual environment, these resources are perhaps best suited for small-scale simulations such as surgery.

The increasing of cameras in minimally invasive surgical procedures has provided an ideal opportunity for simulation of surgery using immersive headsets to deliver the camera view as seen in Figure 11.3. Increasingly advanced surgical simulators are widely used to provide trainee surgeons with safe opportunities to hone their skills across a diverse range of surgical techniques

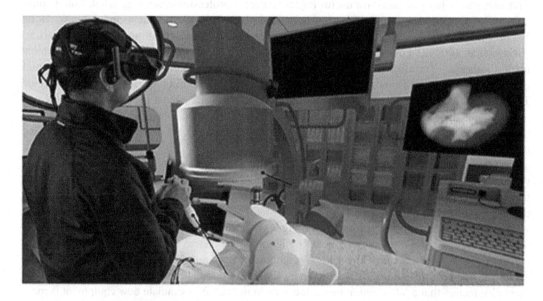

FIGURE 11.3 Typical VR surgical simulation.

(Jones et al. 2019, Sainsbury et al. 2020, Schmidt et al. 2019) in a safe environment without endangering the patient. Combined with haptic feedback controls (Jones et al. 2019, Sainsbury et al. 2020), these simulations enable surgeons to experience the realism of an operation within a 3D patient model with realistic clinical responses.

INTERPERSONAL SKILLS TRAINING

With increasing interest in simulation with allied health education, VR is providing key training in two key aspects. The most obvious class of VR educational applications encompasses skills and process training to hone individual learner's technical skills. These have already been discussed in the previous section. The other skills that can be provided by VR are those often referred to as 'soft skills' and include communication, leadership and team working.

A recent paper by Keith et al. (2018) concluded that team-based video gaming improved real-life team performance and a range of online games are available commercially that facilitate collaborative working. VR has also found use as an environment to facilitate communication. Early development of the vast 'Second Life' application allowed users to construct virtual replicas of real spaces and engage with other users via simplistic avatars. Research studies into educational use of Second Life have shown it to be a useful learning tool (Lorenzo-Alvarez et al. 2019), while Rahman et al. (2014) review highlighted a wealth of studies reporting its value for collaborative learning in a safe environment. More recently, emerging virtual worlds such as Sansar make use of headset technology to generate highly immersive and visually appealing environments and facilitate group interaction and online communication.

An interesting use of headset applications has been the commercial game 'Keep talking and nobody explodes'. In this application, a single user wears the headset while the rest of the group are 'outside' the VR environment. The headset user is the 'Defuser' who is faced with a ticking time-bomb and a complex series of controls, wires and buttons. The rest of the groups are the 'Experts' who cannot see the bomb but have access to a complex defusing manual. The game requires clear communication of complex concepts and visual descriptions and has been shown to enhance team-work and communication skills (Tidbury et al. 2019).

Closed headset systems have also found use within health profession training as resource for increasing dementia awareness. VR applications such as 'myshoes' (Basset et al. 2016) immerse the user in a world of confusion and miscommunication. The user is tasked with completing simple everyday tasks while the application moves objects to different locations, distorts the environment and confuses the user. Evaluation of these applications suggests that practical experience of a dementia simulation increases learner empathy and should improve patient care.

CLINICAL SKILLS TRAINING

Clinical competence represents the interface between technical competence, interpersonal skills and application of general healthcare principles to provide patients with the optimum service and experience. Many authors agree that clinical competence is multi-faceted with Church (2016) citing seven aspects (or clusters): understanding discipline knowledge; mastery of discipline-specific skills; ability to use sound judgement; adherence to professional standards; positive interpersonal relationships; situational application of skills and knowledge, and outcome evaluation by standards. Threaded through all of these is a firm focus on the patient and accordingly traditional healthcare training has emphasised patient contact as vital for gaining clinical skills. While this is still the case, anecdotal evidence from many healthcare students suggests that pressures of the workplace and the reluctance to place patients in the hands of inexperienced learners are combining to limit the effectiveness of this experience. Comments from students in some professions (Ketterer et al. 2019) indicate that much of their clinical training consists of observation and that opportunities for active engagement with tasks is sometimes limited. Simulation and use of VR can, however, provide learners with an opportunity to practice clinical skills within a virtual environment and research is currently ongoing into how this can translate into clinical skills development and ultimately feed into clinical competence.

One of the most exciting possibilities presented by VR-based training is that of partial replacement of clinical training time. Clinical departments are under increasing pressure not only from patient numbers but also from the increasing complexity of many professions, the introduction of new technology and the need to maintain staff training and updates. Unsurprisingly, providing learners with the opportunities for practice that they require can cause conflict and compromise their ability to gain important clinical skills. VR training could potentially ensure that learners gain a range of relevant skills before entering the clinical department and are then better placed to integrate with the teams and concentrate their efforts more on consolidating these skills with real patients. If this is achieved, then clinical training time will become more efficient and effective and ultimately could lead to a reduction in overall clinical hours required to gain competence.

This is a contentious issue, however, with many clinical staff uncomfortable with reducing patient contact time and citing concerns about training learners to be technically competent without a firm focus on the patients. Thoirs' et al. (2011) study highlighted how clinical medical imaging staff were resistant to replacement of clinical time by simulation-based training. Recent research, however, has demonstrated that clinical skills can be gained through use of integrated simulated placements with VR applications complementing lower fidelity simulation resources. Ketterer et al. (2019) randomised a cohort of radiotherapy learners into a simulated placement or a traditional clinical placement. After a fortnight of placement, all learners were subjected to a clinical assessment conducted by a blinded assessor in a novel environment. Equivalence of performance across a range of clinical skill domains was seen for most domains. Interestingly, results in the interpersonal domain demonstrated statistically significant improvement of communication skills in the simulation cohort. This had been influenced by the unique integration of simultaneous VR training and interactions with actors. Research from other professions also indicates the value of these extended simulation placements for training in occupational therapy (Imms et al. 2018), physiotherapy (Watson et al. 2012), osteopathy (Fitzgerald et al. 2017) and nursing (Larue et al. 2015).

In most cases, VR cannot provide all the aspects of a clinical placement and for partial replacement of clinical time, these resources must be combined with low-fidelity and human-based training and simulation. There is, however, a shift in current practice in many establishments from using VR as a complement to clinical training towards using it to front-load skills and therefore reduce clinical placement demands. Aside from gaining skills, there are two key aspects to VR-based training that are attractive to educators.

First, VR offers a safe learning environment (Norris et al. 2019) where learners can learn through experimentation and error. In many professions, learners need to practice complex or potentially dangerous tasks. Training in surgery and use of ionising radiation in particular present situations where trial and error are unacceptable in a clinical environment. It is important that, where possible, learners are able to learn from their mistakes and the use of VR training enables this without compromising patient safety.

The second aspect of VR-based training that is attractive to educators is the ability to present learners with consistent training experiences. Clinical training is opportunistic in nature (Veale et al. 2014) with learners' experience with particular procedures dependent on which patients happen to be present during their placement. While this reflects the varying nature of most healthcare departments, it can lead to inequity of experience, and in some cases, can compromise learners' abilities to meet assessment requirements. Training in VR means that each learner is presented with the same situations and that equity of training can be assured. Furthermore, by ensuring that all learners experience the same situations, peer support and joint reflection can be utilised to consolidate learning. Learners can discuss and compare their performance with the same situation and provide feedback to each other as demonstrated by Khan et al. (2018). It is challenging to support this in most clinical environments where each patient case is subtly or wildly different.

SKILLS ASSESSMENT

Beyond the use of VR resources as training tools, lies the potential for them to support assessment procedures. One of the cornerstones of assessment is the principle that candidates are provided with the same opportunities to gain marks and thus equity of examinations forms an important aspect of assessment. While standardised marking criteria and moderation can help to ensure that marking is equitable and fair, clinical assessment by its very nature presents candidates with different scenarios and opportunities. Many educators prefer to assess clinical skills with standardised cases. When assessing interpersonal and most clinical skills, these are excellent assessment tools, but in many cases are unsuitable for examining technical and interventional skills. Assessment of positioning in medical imaging (Gunn et al. 2018) and radiotherapy (Bridge et al. 2007) is increasingly being performed using VR environments. These offer many advantages including a safe environment, the ability to evaluate incorrect radiation exposure, equity of cases and, importantly, a quantitative measurement.

Surgical performance can also be measured using VR tools as evidenced by Schmidt et al. (2019) who developed a validated measure of laparoscopic surgery performance for trainees using VR simulation. This assessment tool combined different weightings of time, efficiency, safety, dexterity and outcome to enable learner performance to be compared quantitatively with rigour. Although many surgical techniques by their nature have reduced patient interaction, it is important that, where relevant, VR-based skills testing is complemented with interpersonal skills assessment. Recent work (Ketterer et al. 2019) has suggested that integration of VR resources with actors enables both skillsets to be demonstrated and assessed concurrently, as in real clinical practice.

DISCUSSION OF THE ROLE OF VR-BASED TRAINING

It is evident that VR resources offer some excellent benefits to clinical skills training and assessment but, as with any simulation equipment, care must be taken to ensure that they are used effectively and appropriately. In order to do this, it is important to consider the pedagogical context of the current evolution learning focus. Teaching practice is increasingly adopting a constructivist approach and shifting from a teacher-centred to a learner-centred model. More recent transitions are now transitioning towards learning centred pedagogy with a focus on supporting learning as opposed to learners. One of the challenges associated with exciting new developments in VR training is that frequently the technology drives the development of the application, leaving educators to try and embed it into their pedagogical framework. A more fruitful approach would be to integrate learning theory and learning outcomes into emerging applications at the design stage. This should be a key priority for development of emerging VR training solutions.

This is not an insurmountable obstacle, however, as VR and immersive technology in particular aligns well with three core elements of constructivist learning. These are 'Experiential' (active) learning, Situated Cognition (real world) learning and Collaborative learning. It is only the latter element that perhaps requires more development in relation to VR development as many of the existing applications offer a single user experience.

Most VR applications provide learners with a clinical problem or challenge where they must apply their technical skills to gain the appropriate solution. As such, these resources are well suited to problem-based learning strategies. An essential aspect of this is provision of context to the learners and this should be carefully planned. Enriching VR situations with realistic clinical data and notes including histories, diagnoses, investigations and relevant medical images is important to ensure that learners do not view their experience as a technical exercise but try to integrate it into a holistic approach to patient care. Provision of this context can also increase the realism of the experience and, some authors suggest, the rate of skills transfer.

Realism can also be increased through use of 'branding' to provide learners with the illusion that they are working within a particular department. Use of logos, consistently formatted patient notes and referrals or comments from virtual staff members can increase realism of the wider experience. Realism is also an important factor to consider when deciding on optimal format for a VR resource and there is some evidence to suggest that increased fidelity and presence is linked to increased skills transfer. Whatever level of presence is engendered by the VR hardware, it is essential that learners accept the situation as authentic and engage with the learning on that premise.

As with any simulation resource, debriefing is of paramount importance to consolidate learning and to ensure that learners have the opportunity to reflect on their performance and develop their own cognitive pathway or solution. As previously mentioned, peer reflection and feedback have enormous potential as part of a VR-based simulation strategy where learners can provide insight from their own experiences with the same situation.

Perhaps the most important aspect to consider when planning VR-based training is how to integrate this element with low-fidelity simulation and human resources in order to provide a more comprehensive and holistic training experience. Simple solutions such as integrating VR training with a live 'patient voice' can ensure that learners not only gain technical skills but also develop the ability to perform technical tasks while simultaneously providing support, explanation and reassurance to the patient.

At this point, the limitations of the evidence base relating to impact of VR resources must be acknowledged. Many papers (Cecil et al. 2018, Pulijala et al. 2018, Shao et al. 2020) report impact derived from evaluation of student self-perceptions and knowledge as opposed to testing clinical skills performance. More complex measurements involving time, pathway analysis (Sainsbury et al. 2020) or comparative outcome data (Gunn et al. 2018) have been used to provide useful evidence. It is recommended that more primary data relating to the specific impact of VR simulation on clinical skills be performed.

FUTURE TRENDS AND DEVELOPMENTS

Forecasting of technological developments is fraught with inaccuracy, but there are some currently developing trends in VR-based training that can be clearly identified.

FIDELITY

The first of these developments follows Moore's (1965) law regarding rates of increase in computing power and technological advancement. Development of VR delivery hardware is constantly evolving to increase user presence and realism of the virtual environment. While some of the current solutions such as Medspace.VR or VERT place the user in a realistic clinical environment, these are still a far cry from a perfect illusion, yet as modelling, animation and graphic capabilities increase this is likely to improve dramatically. Immersion and presence are strongly related to the impact of the application on the user in relation to their cognitive and emotional response. There is an interesting disconnect between medical applications of VR and non-medical in relation to this. Evidence from studies of non-medical VR applications suggest that presence does not impact on learning, whereas the medical context suggests a strong link. More work is needed to explain this phenomenon but perhaps the motivation of the trainee to gain skills prior to patient engagement is a key factor.

HAPTICS

Linked to fidelity is the use of haptic controls and feedback mechanisms. 'Haptic' relates to the sense of touch, and VR applications can simulate this through use of resistance or pressure in gloves or other worn items. This additional sense not only increases realism of the experience (Sainsbury et al. 2020) but can also help to ensure that learners apply the correct physical pressure in the real world. This technology is still developing, however, as identified in a recent systematic review (Rangarajan et al. 2020) and more research is needed to identify the specific training impact of haptic feedback.

AVATARS

Avatars are simulated humans that engage with users in the virtual environment. There are several potential uses of avatars in VR and development of these is an active area of research. The most obvious use of a responsive avatar is to provide learners with a more realistic patient. Simulation of patient response is commonplace in some of the high-fidelity simulation mannequins with breathing, heartbeat and verbal responses being programmable or controllable remotely (Erlinger et al. 2019). Erlinger's study compared student performance using these mannequins with a VR simulator and concluded that the VR technology was lagging. Indeed, transfer of this functionality into the VR domain is challenging as it not only requires real-time animation and control but also complex and immediate response to the actions of the learners. The element of realism possible with immersive VR, however, suggests that avatars have the ability to engender empathy in users as shown by Bouchard et al. (2013). Development of more complex and responsive avatars will ensure that VR training adopts a more holistic and empathic approach without resorting to integrated low fidelity resources.

The other use of avatars is as additional members of the team. Although most VR applications are designed for a single user, it is evident that many health professionals work alongside a colleague or in small teams. By including avatars in the simulated VR environment, learners will hopefully be able to gain a better understanding of team working skills. These avatars can be automatically controlled by the software or, for improved simulation of the real clinical workplace, by a fellow learner.

This is not a new concept, with many online games facilitating avatar interactions and complex team working between players. Games such as Call of Duty allow players from multiple locations to interact and collaborate via their onscreen avatars which players can sculpt or modify to better represent themselves. Each avatar responds to individual player controls and is capable of interacting fully with the environment in all players' viewpoints. This technology has the potential to help learners gain team working skills, particularly with complex technical tasks that require multiple staff input. Patient setup in radiotherapy, for example, demands that a member of staff is on each side of the patient to rotate and position the patient correctly. This ability for multiple users to input into the simulation is likely to increase the ability of VR solutions to nurture team working and interpersonal skills while simultaneously increasing realism and skill transfer.

AUGMENTED REALITY

Augmented reality is the blending of the virtual and the real world and this has tremendous potential for training. Technology such as Google glasses, or the Hololens, can overlay the real world with additional data. Applications such as 'HoloPatient' from GIGXR can place a 3D virtual patient into a specific space inside a classroom or conventional simulation suite and enable learners to inspect them from all angles as seen in Figure 11.4. Additional visualisation of vital signs can be overlaid to provide an augmented experience.

Although most existing Hololens or Google glasses applications in medical education are mostly restricted to delivering video resources or teleconferencing remotely to learners (McCoy et al. 2019), it is easy to imagine an explosion of future clinical training applications of this technology. Learners could utilise the skills of a virtual tutor or access prompts while undertaking procedures with patients. Perfect positioning for some technical tasks could be overlaid on patients to act as a template for learners to follow. Development of this technology with learning theory at its heart

FIGURE 11.4   Virtual 3D patient present in a real room.

could revolutionise how our learners develop high level clinical skills and shift the balance of simulation use back into the clinical workplace while maintaining a safe learning approach.

REFERENCES

Basset T, Adefila A, Graham S, Clouder L, Bluteau P, Ball S. (2016) myShoes – The future of experiential dementia training? *The Journal of Mental Health Training, Education and Practice*; 11(2): 91–101

Bouchard S, Bernier F, Boivin É, Dumoulin S, Laforest M, Guitard T, Robillard G, Monthuy-Blanc J, Renaud P. (2013) Empathy toward virtual humans depicting a known or unknown person expressing pain. *Cyberpsychology, Behavior and Social Networking*; 16(1): 61–71

Bridge P, Appleyard R, Ward J, Phillips R, Beavis A. (2007) The development and evaluation of a virtual radiotherapy treatment machine using an immersive visualisation environment. *Computers and Education*; 49(2): 481–494

Bridge P, Giles E, Williams A, Boejen A, Appleyard RM, Kirby M. (2017) International audit of VERT academic practice. *Journal of Radiotherapy in Practice*; 16(4): 375–382

Cecil J, Gupta A, Pirela-Cruz M. (2018) An advanced simulator for orthopedic surgical training. *International Journal of Computer Assisted Radiology and Surgery*; 13(2): 305–319

Church C. (2016) Defining competence in nursing and its relevance to quality care. *Journal for Nurses in Professional Development*; 32(5): E9–E14

Erlinger LR, Bartlett A, Perez A. (2019) High-fidelity mannequin simulation versus virtual simulation for recognition of critical events by student registered nurse anesthetists. *AANA Journal*; 87(2): 105–109

Fitzgerald K, Denning T, Vaughan B. (2017) Simulated learning activities as part replacement of clinical placements in osteopathy: A case study. *International Journal of Osteopathic Medicine*; 26: 44–48

Gunn T, Jones L, Bridge P, Rowntree P, Nissen L. (2018) The use of virtual reality simulation to improve technical skill in the undergraduate medical imaging student. *Interactive Learning Environments*; 26(5): 613–620

Imms C, Froude E, Chu EMY, Sheppard L, Darzins S, Guinea S, et al. (2018) Simulated versus traditional occupational therapy placements: A randomised controlled trial. *Australian Occupational Therapy Journal*; 65(6): 556–564

Jones B, Rohani SA, Ong N, Tayeh T, Chalabi A, Agrawal SK, Ladak HM. (2019) A virtual-reality training simulator for cochlear implant surgery. *Simulation & Gaming*; 50(2): 243–258

Keith MJ, Anderson G, Gaskin J, Dean D. (2018). Team video gaming for team building: Effects on team performance. *AIS Transactions on Human-Computer Interaction*; 10(4): 205–231.

Ketterer S, Callender JA, Warren M, Al-Samarraie F, Ball B, Calder K, Edgerley J, Kirby M, Pilkington PA, Porritt B, Orr M, Bridge P. (2019) Simulated versus traditional therapeutic radiography placements: A randomised controlled trial. *Radiography*; Online. https://doi.org/10.1016/j.radi.2019.10.005

Khan Z, Rojas D, Kapralos B, Grierson L, Dubrowski A. (2018) Using a social educational network to facilitate peer-feedback for a virtual simulation. *Computers in Entertainment*; 16(2): 1–15

Krijn M, Emmelkamp PMG, Biemond R, de Wilde de Ligny C, Schuemie MJ, van der Mast CAPG. (2004) Treatment of acrophobia in virtual reality: The role of immersion and presence. *Behaviour Research and Therapy*; 42: 229–239

Larue C, Pepin J, Allard E. (2015) Simulation in preparation or substitution for clinical placement: A systematic review of the literature. *Journal of Nursing Education and Practice*; 5(9): 132–140

Lorenzo-Alvarez R, Ruiz-Gomez MJ, Sendra-Portero F. (2019) Medical students' and family physicians' attitudes and perceptions toward radiology learning in the second life virtual world. *American Journal of Roentgenology*; 212(6): 1295–1302

McCoy CE, Alrabah R, Weichmann W, Langdorf MI, Ricks C, Chakravarthy B, Anderson C, Lotfipour S. (2019) Feasibility of telesimulation and google glass for mass casualty triage education and training. *Western Journal of Emergency Medicine*; 20(3): 512–519

Moore GE. (1965) Cramming more components onto integrated circuits. *Electronics*; 38(8): 114ff

Norris MW, Spicer K, Byrd T. (2019) Virtual reality: The new pathway for effective safety training. *Professional Safety*; 64(6): 36–39

Pulijala Y, Ma M, Pears M, Peebles D, Ayoub A. (2018) Effectiveness of immersive virtual reality in surgical training – A randomized control trial. *Journal of Oral & Maxillofacial Surgery*; 76(5): 1065–1072

Rahman MHA, Yahaya N, Halim NDA. (2014) Virtual world for collaborative learning: A review. 2014 *International Conference on Teaching and Learning in Computing and Engineering*, Kuching, pp. 52–57.

Rangarajan K, Davis H, Pucher PH. (2020) Systematic review of virtual haptics in surgical simulation: A valid educational tool? *Journal of Surgical Education*; 77(2): 337–347

Sainsbury B, Łącki M, Shahait M, Goldenberg M, Baghdadi A, Cavuoto L, Ren J, Green M, Lee J, Averch TD, Rossa C. (2020) Evaluation of a virtual reality percutaneous nephrolithotomy (PCNL) surgical simulator. *Frontiers in Robotics and AI*; 77(2): Online

Schmidt MW, Kowalewski K, Schmidt ML, Wennberg E, Garrow CR, Paik S, Benner L, Schijven MP, Müller-Stich BP, Nickel F. (2019) The Heidelberg VR score: Development and validation of a composite score for laparoscopic virtual reality training. *Surgical Endoscopy and Other Interventional Techniques*; 33(7): 2093–2103

Shao X, Yuan Q, Qian D, Ye Z, Chen G, Zhuang K, Jiang X, Jin Y, Qiang D. (2020) Virtual reality technology for teaching neurosurgery of skull base tumor. *BMC Medical Education*; 20(1): 1–7

Shu Y, Huang Y, Chang S, Chen M. (2018) Do virtual reality head-mounted displays make a difference? A comparison of presence and self-efficacy between head-mounted displays and desktop computer-facilitated virtual environments. *Virtual Reality*; 23(4): 437–446

Thoirs K, Giles E, Barber W. (2011) The use and perceptions of simulation in medical radiation science education. *Radiographer*; 58(3): 5–11

Tidbury L, Jarvis K, Bridge P. (2019) Initial evaluation of a virtual reality bomb-defusing simulator for development of undergraduate healthcare student communication and teamwork skills. *BMJ Simulation and Technology Enhanced Learning*; Online. doi: 10.1136/bmjstel-2019-000446

Veale P, Carson J, Coderre S. (2014) Filling in the gaps of clerkship with a comprehensive clinical skills curriculum. *Advances in Health Sciences Education*; 19(5): 699–707

Watson K, Wright A, Morris N, McMeeken J, Rivett D, Blackstock F, et al. (2012) Can simulation replace part of clinical time? Two parallel randomised controlled trials. *Medical Education*; 46(7): 657–667

Zazula A, Myszor D, Antemijczuk O, Cyran KA. (2013) Flight simulators – From electromechanical analogue computers to modern laboratory of flying. *Advances in Science and Technology Research Journal*; 7(17): 51–55

Section IV

Gamification and Virtual Reality in Contemporary Contexts

12 Exergaming in Multiple Sclerosis – Bridging the Evidence-Practice Gap

Sarah Thomas and Andy Pulman
Bournemouth University
Poole, UK

Jon Robinson
Teesside University
Middlesbrough, UK

CONTENTS

INTRODUCTION

In this chapter, we define and give a brief history of exergaming and consider the evidence base, focusing particularly on multiple sclerosis (MS). Next we describe two case studies based on MS research studies we have conducted. The first, a laboratory-based clinical trial comparing active video gaming (AVG) to traditional balance training, illustrates the concept of 'flow', arguably a key characteristic of AVG. The second, a pilot study, demonstrates the use of behaviour change techniques in a physiotherapist-supported home-based AVG programme. Having earlier highlighted the importance of understanding users' 'lifeworlds', we consider user perspectives of AVG (using MS as an example) as well as clinician perspectives. Finally, we reflect on why uptake of these technologies in clinical practice has been slow, suggest ways to improve the quality of exergaming research and reduce the evidence-practice gap and speculate on future developments in the field.

OBJECTIVES

The objectives of this chapter are to (i) provide a definition of exergaming (also known as AVG) and describe how it has been applied to health and rehabilitation contexts; (ii) highlight the importance of understanding users' 'lifeworlds' when designing AVG interventions; (iii) illustrate the concept of 'flow', arguably a key characteristic of AVG, via a case study; (iv) illustrate the potential role of behaviour change techniques in exergaming interventions via a case study; (v) consider user and therapist perspectives of AVGs; (vi) reflect on why the adoption of exergaming in clinical practice has been slower than anticipated; (vii) provide suggestions for improving the quality of exergaming research trials and (viii) speculate about technologies on the horizon.

BACKGROUND

Against a backdrop of an increasing emphasis on self-management and restrained health and social care systems, innovative interactive digital solutions can offer ways to provide tailored cost-effective solutions for improving aspects of health and social care (Deloitte Centre for Health Solutions, 2015; Marziniak et al., 2018). One innovation, known as AVG, active virtual reality (VR) or 'exergaming' combines gaming technology with physical activity. In 2010, Oh and Yang proposed a new research definition for the term 'Exergaming'. They defined it as "an experiential activity in which playing exergames or any videogames that requires [sic] physical exertion or movements that are more than sedentary activities and also include strength, balance, and flexibility activities" (pp. 10).

A BRIEF HISTORY OF EXERGAMING

A VG dates back to the video arcades of the 1970s (Bogost, 2005), where players stood at video cabinets putting immense effort into playing certain arcade games. In the second wave of gaming consoles (launched during the 1980s), explicit interest in alternatives to sedentary home gaming became more common (Bogost, 2005). In 1987, Exus released the Foot Craz pad controller for Atari 2600 (AtariAge, 2020), and shortly afterwards Nintendo and third-party developers released games such as Bandai's Dance Aerobics for the NES Power Pad (Wolf, 2012). Nintendo has continued to participate in this area – more so than other games manufacturers – with various strands of exergaming released through games for consoles subsequently introduced – such as the Nintendo 64, Gamecube and Wii (Nintendo, 2020a, 2020b, 2020c).

The Wii was introduced in 2006 and provided opportunities for increased physical activity at home via a variety of interactive games. It allowed users to perform activities using movements associated with activities in real life – for example, in Wii Sports, the remote could mimic the arc of a golf swing. The Wii Fit Plus (Nintendo, 2020d) – introduced in 2009 – consisted of activities which used the Wii balance board – a pressure-sensitive platform measuring a user's weight and centre of gravity via four sensors. Games required the user to shift their weight to control the action. Wii Fit Plus included a range of games encompassing strength training, yoga, stretches, aerobics and balance games. In October 2013, Nintendo confirmed it had discontinued production of the Wii in Japan and Europe, with its successor the Wii U (Nintendo, 2020e) – released in 2012 – ceasing production in January 2017.

Microsoft released their Kinect motion sensor for the Xbox in 2010. Used in combination with a range of exergames and other software, the Kinect sensor allowed the console to sense the depth of the players and the environment (Zhang, 2012). Kinect Sports Rivals used the Kinect motion-sensing camera to detect user movements and brought a lifelike experience through responsive tracking for games including bowling, rock climbing, target shooting, and tennis (Microsoft, 2020). Although manufacture of the Kinect sensor and adapter has been discontinued, Kinect technology continues to survive in products like the HoloLens, Cortana voice assistant, the Windows Hello biometric facial ID system, and a context-aware user interface (Microsoft Developer, 2020).

Sony also experimented with motion control technology on their PS3 console with the PlayStation Move (Sony, 2020) but when they transitioned over to the PS4 console in 2013 there were limited motion control options available for the PS4 as a standalone console, and a small pool of fitness video games released. Sony are, however, currently using the controllers to enhance PlayStation VR experiences with up to two PlayStation Move motion controllers, in compatible games (Sony, 2020).

LIFEWORLD AND PERSON-CENTRED DESIGN

A theoretical perspective which is helpful to consider when thinking about the development of technology to support people with long-term conditions (LTC) is the need to understand their everyday experiences and how they live with their condition – gaining an insight into their thoughts, feelings and beliefs and an understanding of what it is like living with an illness (Holloway, 2008). Todres, Galvin, and Dahlberg (2007) revisited the potential of Husserl's (1970) notion of the lifeworld and theorised how lifeworld-led care might provide important ideas and values central to the humanisation of healthcare practice (Dahlberg, Todres, and Galvin, 2009). They also noted ways of disseminating qualitative research findings to make them valuable to users and to help deepen professionals' lifeworld understanding (Ziebland, 2004).

Öhman, Söderberg, and Lundman (2003) suggested that it is important to gain a deeper understanding of how people with LTCs experience illness and life in order to understand the illness from the perspective of the individual and their relationships with family, friends, carers and health care professionals (HCPs). By considering the nature of people's needs who are living with an LTC and reflecting on how they cope and adapt to their situation, it enables both HCPs and technologists to obtain a better foundation for understanding systems which might have a basis for helping them. It has also been suggested that practitioners might like to consider how best to meet the needs of people with LTCs by seeking to understand their experiences and the social networks in which they are embedded, alongside how self-management might be supported by healthcare (Allen and Gregory, 2009; Robertson *et al.*, 2017; Balasubramanian *et al.*, 2018).

Socio-technical design theory was a response to the desire to create systems that were useful and apt (Faulkner, 2000). It began with the desire of a group of therapists, researchers and consultants to apply techniques they had developed for soldiers to help them regain psychological health (Mumford, 2006). Socio-technical system design (STSD) is still widely advocated in the field of health informatics for the development of health care applications (Ritter, Baxter and Churchill, 2014). The intellectual roots of User-Centered Design (UCD) lie in several areas of basic and applied research which includes that of STSD (Ritter, Baxter, and Churchill, 2014). As the recent MS Society report on MS and technology suggests (MS Society, 2018), rather than immediately reaching for technology or data solutions, the future needs to be built upon a culture and context enabling the potential of technology and data to give people affected by MS the best care possible. That means creating a truly collaborative and open culture between clinicians, commissioners, patients, the third sector and the technologists building solutions.

THE CONCEPT OF FLOW

A common question that many HCPs and instructors of physical activity ask is:

> Why do my patients stop, or don't even start, their prescribed exercises?

This is a difficult and complex question. A possible reason for this is because patients may find prescribed exercises boring, with little personal engagement or clear motivation, irrespective of factors such as pain, anxiety etc. In other words, the clear benefits of physical activity are often not enough to encourage and maintain adherence (Campbell *et al.*, 2001; Middleton, 2004; Jack *et al.*, 2010; Peek *et al.*, 2016).

AVG has the potential to offer a solution through what is called 'Flow Experience'. Flow is arguably one of the most important aspects of AVG technology. Although you may not know the specifics of the concept – yet – you will most certainly have experienced it. Take a moment to think of an activity you have greatly enjoyed despite little reward, where actions required little thinking and even became automatic, where time seemed to pass faster (or slower) than perceived – these are just three components of flow experience. Now imagine the exercise programme you provided possessed some of these same properties – enjoyment, automatic movements with little thought, and an altered perception of time – exercise prescription would be a much easier task. Flow experience, herein called 'flow', is synonymous with the popular sports phrase of "being in the zone". However, it is not only found in sports; it can occur in nearly any activity given the right environment and contributing factors. Here we will introduce this concept and its link to AVG.

Csikszentmihalyi, the founder of flow theory, refers to flow state as pure engagement in an activity, with high enjoyment (Csikszentmihalyi, 1990), and has investigated the occurrence of flow through a range of activities including rock climbing, dancing, chess, music and sporting activities (Weinberg and Gould, 2011). The concept of flow is related to a person's overall well-being, considered a state of optimal human experience (Seligman, 2011) and linked to high levels of performance (Jackson and Eklund, 2002). Increasing levels of attained flow equate to an intrinsically motivating optimal state, and a desire to repeat the activity (Csikszentmihalyi, 1990); put differently, a want or need to continue said activity.

Flow plays an important role in understanding engagement and positive experiences in the context of gaming (Nah *et al.*, 2014) as commercial games are often designed using these very concepts (Nacke and Lindley, 2008; Sharek and Wiebe, 2011; Kiili, Arnab, and Lainema, 2012; Huang *et al.*, 2018). Cowley *et al.* (2008) comment that "Flow is a well-established construct for examining experience in any setting and its application to game-play is intuitive" (p. 2).

Moreover, modern video games are able to provide and manipulate many of the components/ facilitators of flow through providing the user with constant feedback, clear goals, automatically adapting to the skill of the player and creating a balance between the challenge of the task and the ability of the player (Bressler and Bodzin, 2013; Fang, Zhang, and Chan, 2013). When flow is achieved the gamer or user will experience a high level of engagement and involvement in addition to an altered perception of time. However, it is important that there is a balance between the skill/ability of the individual and the challenge/difficulty of the situation. If a task is too easy the individual will lose interest and become bored; if it is too hard, the individual will present with increased stress and anxiety (Figure 12.1). As Figure 12.1 suggests, a balance is needed in order to attain a sense of flow.

A systematic review of the role of flow in exercise (Jackman *et al.*, 2019) identified that the most frequently used validated measure of flow was the Flow State Scale (FSS) created by Jackson and Marsh (1996). The FSS is based on the measurement of nine specific subscales:

1. **Autotelic Experience** – the activity is intrinsically rewarding and undertaken for its own sake
2. **Clear Goals** – clear idea of what needs to be accomplished
3. **Challenge-Skill Balance** – balance between the challenge of the activity and personal skills
4. **Concentration on the Task at Hand** – complete focus on the task
5. **Paradox of Control** – a belief of being in complete control of own actions, without any conscious or exertive effort at the task
6. **Unambiguous Feedback** – clear and immediate feedback

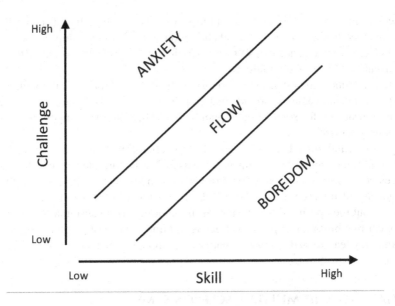

FIGURE 12.1   Flow theory representation. (Derived from Csikszentmihalyi, 1990.)

7. **Action-Awareness Merging** – involvement in the task; actions become automatic

8. **Transformation of Time** – altered perception of time; either speeding up or down

9. **Loss of Self-Consciousness** – no concerns with appearance; focused only on the activity

In addition to providing an alternative means of being physically active, there are also potential psychological benefits of exergaming technology, which may be related to flow. Studies have reported that exergaming is perceived as significantly less strenuous and more enjoyable than traditional therapy (Brumels *et al.*, 2008), as well as positively affecting self-esteem, mood and motivation (Staiano and Calvert, 2011). AVG may also be more acceptable and engaging than traditional means of exercise (Warburton *et al.*, 2007) – which may positively affect uptake and adherence.

ACTIVE GAMING AND REHABILITATION

Although originally designed for entertainment/fitness purposes AVG has been increasingly applied to health (Tripette *et al.*, 2017) and rehabilitation contexts (Bonnechère *et al.*, 2016; Coudeyre, 2017). It has been used for balance and gait training (Casuso-Holgado *et al.*, 2018; Porras *et al.*, 2018; Fang *et al.*, 2019; Lei et al., 2019), for upper limb training (Thomson *et al.*, 2014), to increase physical activity levels (Street, Lacey, and Langdon, 2017), quality of life (Cacciata *et al.*, 2019) and for cognitive rehabilitation (Stanmore *et al.*, 2017; Sala, Tatlidil, and Gobet, 2019). It has been applied to a wide range of clinical conditions; examples include, stroke (Laver et al., 2017), Parkinson's disease (Garcia-Agundez *et al.*, 2019), cerebral palsy (Lopes *et al.*, 2018), cancer (Tough *et al.*, 2018), MS (Taylor and Griffin, 2015).

AVG may help to overcome some of the barriers that people face when undertaking rehabilitation or trying to become more active (Mat Rosly *et al.*, 2017). Such barriers include the need to travel to a hospital or rehabilitation centre, limited therapist time (Perrochon *et al.*, 2018) and the sometimes boring and repetitive nature of prescribed home rehabilitation exercises (Kato, 2010).

Commercial systems such as the Wii and the Kinect are relatively low cost, can be used in the home, have broad appeal and incorporate variables known to enhance motivation, self-efficacy and adherence (such as self-monitoring, prompts, peer comparison and feedback about progress and performance) in the context of fun purposeful tasks (Thomas *et al.*, 2014). They enable the user to accumulate practice in short bouts in the rehabilitation setting or potentially in the convenience of the home. As there is a wide choice of software, activities can be selected to suit differing functional abilities, current symptoms and preferences.

However, as commercial systems were not originally designed for clinical populations they may require inappropriate movements, provide inappropriate feedback, or incorporate games or peripherals with poor usability for some clinical populations. There are also important issues related to the safety of unsupervised use.

Findings from randomised controlled trials of exergaming have shown promise in a range of clinical conditions including Parkinson's disease (Garcia-Agundez *et al.*, 2019) and cerebral palsy (Lopes *et al.*, 2018). However, conclusions are limited by the small size of many studies, the heterogeneity of interventions and the lack of long-term follow-up (Laver et al., 2017) with little known about the optimal dose of exergaming. Findings from systematic reviews in the area of neurological rehabilitation suggest that active gaming is at least as good as traditional rehabilitation and may lead to better engagement and adherence (Bonnechère *et al.*, 2016; Mat Rosly *et al.*, 2017).

ACTIVE GAMING AND MULTIPLE SCLEROSIS (MS)

In this chapter, we focus on MS as an example of the clinical application of AVG. Multiple sclerosis is a neurological condition that affects the central nervous system. Symptoms are varied and include problems with mobility, balance, fatigue, pain, tremor, stiffness or spasms, vision, bowel and bladder, mood and cognition. Over 2.5 million people have MS worldwide (Kanavos *et al.*, 2016). Despite the well-known benefits of physical activity and exercise for MS and their possible neuroprotective effects (Dalgas *et al.*, 2019), people with MS are much less active than the general population (Heesen *et al.*, 2006; Kinnett-Hopkins *et al.*, 2017). Like others with LTCs, people with MS face additional barriers to physical activity or participation in exercise programmes (Smith *et al.*, 2019). These include physical (e.g. pain, fatigue, mobility limitations, balance impairments, overheating), psychological (e.g. fear, embarrassment, lack of confidence) and environmental factors (transport, cost, lack of suitable facilities/trained staff) (Ploughman, 2017). AVG may offer ways of overcoming some of these barriers.

Research studies using off-the shelf AVG technologies in MS have tended to use the Nintendo Wii (Guidi, Giovannelli, and Paci, 2013; Brichetto *et al.*, 2013; Nilsagård, Forsberg, and von Koch, 2013; Prosperini *et al.*, 2013; Kramer, Dettmers and Gruber, 2014; Pau *et al.*, 2015; Robinson *et al.*, 2015; Thomas *et al.*, 2017) apart from two randomised controlled trials (RCTs) (Gutiérrez et al., 2013; Ozdogar *et al.*, 2020) that used the Xbox 360 and Xbox One Kinect, respectively. To date, most RCTs have been small and focused on balance training with large variations in protocols used (number of sessions, intervention duration, outcomes) (see reviews by Casuso-Holgado et al., 2018; Parra-Moreno, Rodríguez-Juan, and Ruiz-Cárdenas, 2019). Only four studies have involved home-based practice (Plow and Finlayson, 2011; Gutiérrez et al., 2013; Prosperini *et al.*, 2013; Thomas *et al.*, 2017). One study looked at the impact of a dual task (Kramer, Dettmers, and Gruber, 2014) and Yazgan et al. (2020) compared the Wii Fit with 'Balance Trainer' (THERA-Trainer balo).

Findings from these studies are mixed but suggest that AVG is acceptable and enjoyable, appears safe for home use for those with mild MS symptoms (Plow and Finlayson, 2011; Thomas *et al.*, 2017) and might be beneficial for balance (Brichetto *et al.*, 2013; Guidi, Giovannelli, and Paci, 2013; Robinson *et al.*, 2015; Yazgan *et al.*, 2020) and increasing physical activity levels (Plow and Finlayson, 2011; Thomas *et al.*, 2017). Interestingly, after 12 weeks of home-based use of the Wii Fit Plus by pwMS, Prosperini *et al.* (2014) found evidence of neuroplasticity with changes in the microarchitecture of white matter tracts (though these changes did not persist beyond the intervention period).

CASE STUDY 1: AN EXAMPLE OF A LABORATORY-BASED AVG PROGRAMME, COMPARED TO TRADITIONAL BALANCE TRAINING IN PEOPLE WITH MS

One of the authors (JR) conducted the first RCT to compare AVG (Nintendo Wii Fit) with traditional balance training and a control group (Robinson *et al.*, 2015). Fifty-six adults with a clinical diagnosis of MS were randomised into one of three groups. Group 1 received balance training using the Nintendo Wii Fit (AVG), group 2 completed traditional balance training (non-exergaming), and group 3 received no intervention (control group). The AVG games were first assessed for their gross movement patterns and a more traditional balance training programme was created to mirror these movements, based on common stressors of postural control (American College of Sports Medicine, 2010), in an attempt to isolate any effects due to intervention type. The AVG and traditional balance training group received four weeks of twice weekly balance-oriented exercise (see Table 12.1). Physical outcomes included postural sway (using a Kistler force platform) and spatiotemporal parameters of gait (using a GAITRite computerised walkway). The questionnaire outcomes included technology acceptance (the Unified Theory of Acceptance and Use of Technology questionnaire), and flow experience (the FSS questionnaire).

Considering the results of this study in terms of flow, the Wii Fit was found to be superior to traditional balance training on several dimensions of flow experience. High flow state scores were observed at baseline for both intervention groups, suggesting that flow state is achieved in new users with first use. However, the Wii Fit group had significantly higher post-intervention scores when compared to traditional balance training for the following subscales: Clear Goals, Concentration of Task, Unambiguous Feedback, Action-awareness Merging, and Transformation of Time. These last two subscales may contribute to an important psychological property of AVG through 'user-distraction' (Kato, 2010).

Although the positive influences of flow have been explored in a number of research areas, Jackman *et al.* (2019) observed that few have considered the 'how' and 'why' of flow. They noted, however, that Barry *et al.* (2016) hypothesised that AVG may possess immersive properties, and Robinson *et al.* (2015) proposed that flow may offer 'distractive properties'. In the above case study many participants reported feeling that only a 'few minutes had passed', having completed the 40–60 minute AVG intervention (an altered perception of time). This would suggest that the participants had attained a degree of flow, where the perceptions of physical activity were secondary to

---

TABLE 12.1

Exercises Completed in the AVG and Traditional Balance Training Groups

	Nintendo Wii Fit Group	Balance Training Group
	Balance Games	**Balance Game Equivalent**
1	Heading (soccer) (BB)	Wall Taps
2	Ski Slalom (BB)	Standing with feet together; resist perturbation
3	Table Tilt (BB)	Wobble board inflatable
4	Tightrope (BB)	Heel-to-toe straight line walking
	Aerobic Workouts	**Aerobic Workouts Equivalent**
1	Boxing (HC)	Show Boxing (gentle, non-impact)
2	Step up class (BB)	Step ups
3	Hula Hoop (BB)	Standing hip rotations
	Muscle Workouts	**Muscle Workouts Equivalent**
1	Torso Twists (BB)	Torso Twists
2	Rowing Squats (BB)	Mini squats

BB: balance board; HC: handheld controllers.

their engagement with the AVG. Put another way, the participants were distracted from the fact they were exercising. In fact, research has produced some very interesting findings as a result of distracting users from not only the duration of physical activity, but also the intensity.

Warburton *et al.* (2009), in a healthy, non-clinical group, demonstrated that not only did the addition of gaming technology to stationary cycling result in higher physiological metabolic requirements, but it did so without increasing perceived exertion suggesting a distraction effect. This would indicate an important advantage of AVG technology over that of traditional balance training – the facilitation of flow by design. Finding greater flow state properties in a therapeutic exergaming intervention is potentially of great importance. In fact, the success of gaming technology as an adjunct to physical therapy is the result of increased engagement and motivation compared to the typically mundane and repetitive tasks associated with traditional physical therapy (Kato, 2010; Glen *et al.*, 2017). To provide some perspective with regards to exercise in people with MS, the average life expectancy for a female in England and Wales is 82 years (Office of National Statistics 2013), and the average age of MS diagnosis is 30 years (Milo and Kahana 2010). Multiple sclerosis is reported to reduce life expectancy by approximately ten years (Brønnum-Hansen, Koch-Henriksen, and Stenager, 2004). Therefore, a female aged 30 years with MS would be encouraged to undertake MS-related exercise or rehabilitation over a period of approximately 42 years. The likelihood of exercise adherence through traditional modes, over this period of time, is questionable. However, AVG may offer the opportunity for continued engagement through the variety of games available and an engaging environment, where the game is the primary focus, and the physical activity a secondary component. However, more research is needed to establish if flow can be maintained in the longer term.

REHABILITATION IN THE HOME

Supporting people with MS to use AVG systems in the home rather than in rehabilitation or hospital contexts could increase the likelihood of sustaining use in the longer term (Forsberg, Nilsagård, and Boström, 2015; Robinson *et al.*, 2015). However, a systematic review of home-based AVG interventions for individuals with a neurological condition found a higher level of dropout in the AVG intervention groups than in control groups (Perrochon *et al.*, 2018). Reasons for discontinuation reported in studies included technological issues, lack of space, environmental distractions, lack of social support, symptoms, fear that use of AVG may increase risk factors, lack of customisation and negative feedback (Perrochon et al., 2018). In Plow and Finlayson's (2011) non-randomised home-based study of the Wii Fit with people with MS they found that use declined over time and noted:

> It may be that a more patient-centered approach will help foster the internalisation of goals related to playing Wii Fit. Strategies such as motivational interviewing and building skills related to problem solving and decision-making should be explored in future studies. (p. 28)

It has been suggested that future interventions should incorporate effective behaviour change techniques and strategies and encourage participants to have more realistic outcome expectancies. (Plow and Finlayson, 2011; Perrochon *et al.*, 2018).

CASE STUDY 2: MII-VITALISE – AN EXAMPLE OF A HOME-BASED AVG INTERVENTION INCORPORATING BEHAVIOUR CHANGE TECHNIQUES

'Mii-vitaliSe' is a home-based, physiotherapist-supported Wii intervention for people with multiple sclerosis (pwMS) developed and piloted by one of the authors (ST) and a multi-disciplinary team (Thomas *et al.*, 2014, 2015, 2017). The Mii-vitaliSe pilot study was the first RCT in the United Kingdom involving home-based use of the Wii for people with MS (Thomas *et al.*, 2014, 2017). In the RCT 30 people with a clinical diagnosis of MS who had fulfilled a risk assessment (balance and the suitability of their home environment – see description in Table 12.2) were randomised to receive Mii-vitaliSe either immediately (for 12 months) or after a six month delay (for six months).

TABLE 12.2
A Description of the Mii-vitaliSe Intervention

Risk assessment (prior to enrolment in the study)

Undertaken by a senior physiotherapist in participants' homes. Participants were considered eligible if:

- They could maintain independent standing balance with eyes open for 1 min.
- The physiotherapist clinically judged they demonstrated adequate balance reactions while on the Wii balance board and were able to step off safely forwards, backwards and sideways.
- The home environment was suitable (with minor modifications, if appropriate and possible).

Session 1: Orientation to the Wii in the hospital setting (week 1)

The physiotherapist went through the guidance book with the participant to cover: the benefits of physical activity; a practical demonstration of using the software (Wii Sports, Wii Sports Resort and Wii Fit Plus) safely; creating a 'Mii'; familiarisation with the controls.

Main behaviour change techniques (BCTs): Education about benefits of physical activity; decisional balance; pros and cons; consideration of outcome expectancies; demonstration of the behaviour

Session 2: Training and individualised assessment in the hospital setting (week 2)

Participants were guided through warm up/cool down exercises and given the opportunity to try out activities safely under supervision. The physiotherapist and participant discussed appropriate activities, taking into account physical capabilities and personal preferences.

Main BCTs: Introduction to/information about goal setting and action planning; behavioural practice/rehearsal

Session 3: Installation of Wii at home and start individual programme (week 3)

The physiotherapist installed the Wii equipment in participants' homes to interface with their television. Participants were provided with Wii Sports, Wii Sports Resort and Wii Fit Plus, asked to use their Mii when playing, and to log their use or non-use of the Wii each day using a 'play log'. The physiotherapist reiterated safety advice (e.g., the importance of the safety strap, taking a graded approach and frequent rests, etc) and agreed an individualised programme with the participant, indicated ways this could be progressed and provided guidance on keeping the daily play log and on setting/recording goals.

Main BCTs: Goal setting and action planning; self-monitoring; positive reward; feedback; restructuring the physical environment.

Ongoing programme with support and monitoring

This was followed by independent home use supported by a combination of telephone support (week 5—to identify and resolve any early difficulties encountered; week 12 to encourage progress, support people to overcome any barriers) and home review visits (weeks 7 and 16) at which the physiotherapist reassessed the individual and provided support and discussed possible progression/modification of the activities. Thereafter, ongoing monitoring, support and encouragement was provided via monthly contact (phone/email) from the physiotherapist.

Main BCTs: Monitoring; self-monitoring; feedback (physiotherapist and console); revisiting outcome expectancies; goal setting; action planning, if-then coping planning including a review of individual barriers and facilitators; collaborative problem-solving.

Mii-vitaliSe materials

Guidance book which comprised 30 A4 pages divided into 12 sections and was presented in a ring bound folder. Its contents included the background to the research, the benefits of exercise, setting up a Wii, a guide to starting to exercise, how to continue and record activity, some quotations from pwMS, advice on setting goals, top tips, general health and safety advice and useful websites/ bibliography/resources.

A **personal activity workbook** comprising 35 pages designed to help prepare individuals for the orientation sessions with the physiotherapist and to encourage and support their continued use of the Wii.

Games descriptors that provided information about the Wii Sports, Wii Sports Resort and Wii Fit Plus games including aims, possible benefits, safety tips, guidance on how to play, duration, information about scoring and progression, risks and, where possible, suggested adaptations. Space next to each descriptor enabled the physiotherapist and/or participant to add their own notes or comments.

A **daily play log** that had a dual purpose of being part of the intervention to enable participants to monitor their patterns of Wii use as well as enabling the research team to gain insights into levels and patterns of use (date, whether played, adverse events (e.g. pain, tenderness, soreness, fatigue, dizziness, headaches, aching, stiffness, falls), games played, duration, intensity, enjoyment, fatigue, reasons for non-use, free text comments).

BOX 12.1   KEY PRINCIPLES GUIDING THE DEVELOPMENT OF MII-VITALISE PILOT RCT

- Use off-the-shelf-technology
- Encourage short bouts of exercise that can be accumulated
  - Incorporate home-based element
  - Include behaviour change techniques
  - Include health professional support
- Monitor adherence
- Include long-term follow-up

An initial consultation workshop involving eight service users with experience of using the Wii was held prior to the pilot RCT to inform the development of the Mii-vitaliSe intervention package. The research team for the RCT included two service users. The intervention drew on relevant psychological frameworks and theories (motivational interviewing, social cognitive, cognitive behavioural and self-determination theory) and incorporated behaviour change techniques.

Mii-vitaliSe consisted of two familiarisation sessions with the Wii in the hospital supervised by a physiotherapist followed by home use (Wii Sports, Sports Resort and Fit Plus software provided) with physiotherapist support and personalised resources (see Table 12.1). The rationale of Mii-vitaliSe was to support people with MS, via a collaborative problem-solving approach, to increase activity levels in their own homes using the Nintendo Wii. Mii-vitaliSe encouraged the internalisation of goals and aimed to provide individuals with skills, strategies and support to identify solutions to overcome barriers they encountered. To enhance personalisation and engagement, there was a personal activity workbook that included written activities and reflections.

What set Mii-vitaliSe apart from many other AVG interventions in MS was its theoretical underpinning, the involvement of people with MS in its development, the inclusion of behaviour change techniques (e.g. motivational interviewing, decisional balance, information provision, goal setting, problem-solving, feedback and (self-) monitoring, identifying and overcoming barriers, action and coping planning, and reframing setbacks), and the inclusion of a longer-term intervention and follow-up period.

No serious adverse events occurred during the trial (all adverse events were reviewed by an independent safety monitoring committee) but there were isolated reports of pain, discomfort, aggravation of an existing injury, four near-falls and one fall. Once alerted to such issues the physiotherapists followed up and referred on, if necessary. If appropriate, the participant was given advice about how best to adjust their Wii programme to minimise the likelihood of a reoccurrence.

The findings suggested that Mii-vitaliSe was well received by the majority of participants and acceptable to the physiotherapists delivering it (see next section for further detail). Study recruitment was successful, retention was good and attrition low. Over six months participants used the Wii on average two days per week for around 30 minutes which approaches aerobic activity guideline levels for MS (Latimer-Cheung *et al.*, 2013). At six months follow-up the majority of outcomes in this feasibility study (including clinical measures of balance and gait and self-reported physical activity levels) were in the direction of benefit for the Mii-vitaliSe group (though confidence intervals were wide and included zero). Overall, findings from this study were promising and suggested a full scale trial of effectiveness and cost-effectiveness would be feasible and warranted.

ILLUSTRATION OF THE APPLICATION FROM THE USERS' PERSPECTIVES

Qualitative studies can help to provide important insights about the acceptability of interventions and their perceived value and benefits as well as identify barriers encountered (O'Cathain *et al.*, 2013). To date in the field of exergaming consideration of users' and clinicians'/therapists' perspectives has been relatively limited (Celinder and Peoples, 2012; Meldrum *et al.*, 2012; Levac *et al.*, 2013; Tatla *et al.*, 2015; Taylor and Griffin, 2015; Wingham *et al.*, 2015; Paquin, *et al.*, 2016; Klompstra *et al.*, 2017; Hamilton *et al.*, 2018; Levac, Miller and Colquhoun, 2018; Rand, Givon, and Avrech Bar, 2018; Tobaigy *et al.*, 2018).

In the field of MS four qualitative studies (Plow and Finlayson, 2011; Forsberg, Nilsagård and Boström, 2015; Palacios-Ceña *et al.*, 2016; Thomas *et al.*, 2017) have explored the perceptions of people with MS who have received an AVG rehabilitation intervention. Palacios-Ceña *et al.* (2016) explored perceptions of a home-based Kinect intervention while the other three studies explored perceptions of the Wii. Plow and Finlayson (2011) and Thomas *et al.*'s (2017) studies involved home-based use of the Wii. Forsberg, Nilsagård, and Boström (2015) and Thomas *et al.* (2017) also explored the perceptions of the physiotherapists who delivered the AVG intervention.

PLOW AND FINLAYSON (2014)

This qualitative study in the USA involved telephone interviews undertaken with 30 pwMS before and 22 pwMS after a 14-week home-based Wii Fit programme (Plow and Finlayson, 2011). Participants found the Wii Fit fun and engaging, reported that it increased their confidence and sense of control, and removed barriers associated with going to a gym to exercise, and increased leisure activity. Participants reported enjoying competing against their previous scores and found it normalising to compete against others. Many did not like the negative nature of some of the default feedback and found it demoralising. Participants also reported usability barriers due to limitations in the Wii to accommodate different fitness and functional levels (some found the games too fast while others felt they were not challenging enough and some had fears of falling due to the small size of the balance board). Those who used the Wii regularly described the gaming experience as effortless (suggesting flow) and as a natural fit into their lives that had become part of their exercise self-identity. Despite finding the Wii games fun, some participants reported encountering common exercise barriers such as lack of time, environmental factors (such as space issues or interruptions), low motivation, vacations and boredom.

FORSBERG, NILSAGÅRD, AND BOSTRÖM (2015)

This qualitative study was embedded within a multi-centre RCT ($n = 84$) in Sweden that tested an intervention consisting of 12 sessions of Wii Fit balance games supervised by a physiotherapist in the rehabilitation setting (Nilsagård, Forsberg, and von Koch, 2013). Interviews were undertaken with 15 pwMS following the intervention and a focus group was held with nine physiotherapists involved in the study (who had either delivered the intervention or administered outcomes).

Both pwMS and physiotherapists felt that using the Wii was fun and engaging, that it challenged physical and cognitive capacities, could be used by a range of ages and levels of experience and on 'not so good' days. The physiotherapists found starting up the Wii Fit game was time consuming and unnecessarily complex involving numerous button presses. Participants found the Wii easy to use (though their use was supervised). People with MS noted that time in the sessions went by quickly (suggesting 'flow' states) when they were using the Wii and the physiotherapists observed that participants were highly motived and pushed themselves. Both the pwMS and the physiotherapists liked the fact that the Wii programme offered a wide range of games with the ability to progress to more challenging levels, enabling a degree of tailoring to users' needs and preferences. However, several games requiring rapid responses (such as snowboarding and slalom) were considered less usable. Some of the participants needed one-to-one supervision from the physiotherapist or a sturdy chair or rollator nearby for support. One limitation of the Wii Fit noted by the physiotherapists was its focus on static balance. A physiotherapist remarked that receiving negative visual feedback (in the form of a sad Mii character) was unnecessary and demoralising for some.

One of the main motivators reported by participants was the immediate feedback provided by the software enabling them to self-monitor their performance and compete against previous scores. The centre of gravity feedback helped participants to become more aware of their stability limits and increased their confidence. Improvements in balance were seen during the sessions and pwMS described situations where these improvements had translated into everyday life (e.g., being able to stand for longer, feeling safer when walking and being able to walk faster, feeling less dizzy). People with MS found the technology easy to use (but this was supervised use). Participants felt that having a Wii available to use in their own home could overcome barriers to being active, such as bad weather and limited time. However, they noted that a Nintendo Wii would not be affordable for all. The physiotherapists deemed an assessment of balance necessary prior to any home-based use of the Wii to ensure appropriate and safe use and that follow-up physiotherapist support would be needed to maintain engagement and intensity and assist in appropriate game selection.

PALACIOS-CEÑA ET AL. (2016)

This qualitative study was embedded within an RCT in Spain that compared a ten-week home-based Kinect intervention (with physiotherapist supervision and monitoring via video conference) with twice-weekly physiotherapy sessions (Gutiérrez *et al.*, 2013). All 24 participants who were randomised to the Kinect treatment arm were interviewed. The Kinect programme (Kinect Joy Ride, Adventures & Sports) involved four sessions per week, with session duration progressively increased (up to a maximum of 20 minutes).

Participants reported feeling more independent and a greater sense of control. Many were surprised by their functional improvements which seemed to be achieved without a sense of effort (suggesting a flow state). Participants reported that improvements often translated into their everyday lives. The Kinect offered opportunities for participants to interact with other family members across the generations, helping to normalise their MS. Participants described being motivated by a desire to improve their scores and try new games and liked being able to compete with family members and friends. Some participants described becoming 'hooked' on the games and exceeded the prescribed dose. Several noted that a preparatory physical conditioning programme prior to starting the Kinect home-based intervention would have been helpful.

THOMAS ET AL. (2017)

This qualitative study was embedded within a pilot RCT of Mii-vitaliSe (see earlier description of intervention). Interviews ($n = 19$) were undertaken with participants after 6 and/or 12 months of the Mii-vitaliSe intervention as well as with the two physiotherapists who delivered Mii-vitaliSe.

As with the previous studies, the majority of participants found the AVG (in this case the Wii) fun and engaging and more interesting than traditional exercises. However, similar to previous findings, some participants found certain games unsuitable (because they required fast responses or challenging movements) and several participants did not feel safe and/or confident using the Wii balance board. Nevertheless, participants were able to overcome some usability barriers by making accommodations such as having a perching stool in combination with the balance board or a chair for support or refraining from stepping off the balance board sideways.

Participants reported a wide range of physical and psychological benefits including increased physical activity, improved mood, reduced anxiety and stress, increased confidence, improved sleep and general well-being, improved balance, posture and core strength, better hand-eye coordination and dexterity, increased exercise endurance and some relief of symptoms (pain, fatigue). Similar to Forsberg, Nilsagård, and Boström (2015) and Palacios-Ceña et al. (2016), participants described how improvements had translated into their everyday lives (dropping fewer pegs when hanging out the washing, finding it easier to climb out of a shower or climb a ladder or reach things from shelves or bend down to pick things up, being able to dance more in time to music at a social event, greater accuracy and dexterity when reaching for door handles, being able to walk further).

As in the other home-based AVG studies, participants highlighted the convenience of using the Wii at home. They liked the fact that it could be used for brief periods of time without the need to go to the gym or wear sports clothing. When using the balance board, participants found the feedback provided about centre of gravity helpful for self-monitoring and for working on balance and posture and the physiotherapists found the balance board a useful assessment and therapeutic tool. As in the other studies, the social and normalising aspects of the Wii were mentioned by participants who liked the fact that it is an off-the-shelf commercial platform that can be played with friends and family. Several participants and the physiotherapists suggested that 'buddying' up or meeting as a group or using social media groups might have helped with motivation.

A sense of achievement was important in encouraging people to continue using the Wii and this was very individual (could be linked to beating previous scores, competing with others, spending a certain amount of time using the Wii, using the Wii on certain days, improving technique, balance or endurance or the social aspects). Participants reported being motivated by perceived improvements in their well-being and in their daily lives and the fact that they noticed deterioration when they were not using the Wii. Sometimes others noticed improvements that the individual had not been aware of.

One participant emphasised the importance of having realistic expectations and setting flexible goals given the unpredictability of MS. Those who managed to establish a habit and routine and incorporate the Wii into their lifestyle – "I fit it in, it's part of my life" – seemed to be more likely to continue to use the Wii. For some the Wii acted as a gateway to other physical activities and to greater self-confidence.

In the Mii-vitaliSe study, the physiotherapists undertook an initial risk assessment, undertook orientation sessions in the hospital, set up the Wii in participants' homes and provided ongoing support via home visits and phone calls. This contact was valued highly by participants and considered essential to ensure people were safe and adequately supported in their homes. Many participants felt this gave them confidence and an impetus to use the Wii. The physiotherapists were able to correct inappropriate techniques or compensatory movements, collaboratively problem solve solutions for overcoming barriers, suggest adaptations and progressions, provide encouragement and a push – or reassurance – when needed.

Similar to Plow and Finlayson (2014), the main barriers to use that participants reported included time, vacations, distractions in the home, home constraints (such as the TV being in use or having to move furniture around to make space for the Wii) and fine weather (when it may be preferable to be outside). Participants with cognitive or memory issues sometimes found it difficult to navigate the Wii menus and control the cursor. The physiotherapists provided additional support in such instances. Wii-related barriers to use included issues with configuration and set-up, dexterity challenges and safety concerns.

For some people, the Wii provided an opportunity to undertake physical activity that they would not otherwise have taken and for others it provided a means of topping up activity. When the weather was fine, people did tend to use the Wii less; conversely, the Wii was seen to be particularly useful in the winter when it might be difficult to go out and be active. For most people (but not all), the hot weather did not seem to be a major barrier in terms of fatigue as people tended to use it at cooler times of the day and wear less clothing.

Participants sometimes learnt things about the games and came up with strategies and accommodations that they were able to share with the physiotherapists. Some participants progressed much further in specific games than the physiotherapists ever had – making it challenging for the physiotherapists to advise them on what to expect next.

One of the main motivators to continued use was having a sense of accountability. For the majority, this accountability served as a positive push to encourage them to use the Wii and, for some, this motivation became internalised during the course of the study. This links in with Mohr, Cuijpers, and Lehman's (2011) construct of 'supportive accountability' that has been considered in the context of e-health interventions. There was a fine line between accountability being a motivator and it becoming an unhelpful pressure – sometimes the physiotherapists needed to provide reassurance to participants if they felt a sense of guilt or pressure. Those participants with high levels of intrinsic motivation often needed very little support from the physiotherapists. Others, who were encountering barriers (such as time or family/work commitments), sometimes found it difficult to sustain use, even with the support provided. In such instances, the physiotherapists sometimes found the follow-up phone calls challenging and slightly uncomfortable. They described how they attempted to build a rapport, offered reassurance if participants were feeling guilty and tried to find ways to support participants to link Mii-vitaliSe goals with wider personally meaningful goals. They emphasised to participants that they were not 'checking up on them' or 'telling them what to do' but rather were encouraging the individual to come up with their own solutions. This concurs with Mohr, Cuijpers, and Lehman's (2011) recommendations that 'it should be made clear that the aim of performance monitoring is to provide feedback, that failure to meet goals provides opportunity for self-reflection and growth, and that there are no negative consequences' (p. 4).

Overall, the physiotherapists found delivering Mii-vitaliSe a valuable experience that involved 'thinking outside the box'. It was a different way of working and they liked the collaborative approach. They noted that the Wii is more interesting and enjoyable than the standard balance exercises that are often prescribed.

The physiotherapists found it helpful seeing participants using the Wii in their own homes where they could observe and assess their technique and posture and talk them through goals. They were struck by the fact that people were often able to do a lot more in their own homes than in the hospital setting. It provided an opportunity to try out things together that participants might not have had confidence to try out on their own. It gave insights into what goes on 'behind the scenes' in people's lives after prescribing an exercise programme. The follow-up telephone calls were sometimes logistically challenging as at times it could be difficult to contact participants.

The physiotherapists found facilitating use of the Wii to be a steep learning curve, particularly becoming familiar with the Mii-vitaliSe resources and setting up the equipment in participants' homes. The risk assessment protocol was modified partway through the study to include more detail about the participants' own televisions after a couple of occasions when a different cable was needed to connect the console to the television. With practice the physiotherapists became more adept and confident (sometimes family members helped with set up) but suggested that it would be helpful to have a technician undertake the set up. The physiotherapists found the programme materials well written but felt they could be simplified and made more interactive. They suggested combining the handbook and workbook into one resource.

The collaborative approach meant that the physiotherapists were able to learn from participants' experiences and share this learning with other participants – for example, aspects about advanced levels of games they themselves had not reached, strategies and accommodations that people had found helpful or games that were particularly challenging and best avoided.

OVERALL

Findings from these qualitative studies suggest that AVG is generally liked and acceptable to people with MS and to physiotherapists and that it can lead to flow experiences. People with MS described a wide range of benefits and reported that some transferred to their everyday lives. Users found the games fun and appreciated the convenience of being able to use the equipment in the home with professional support available. However, some of the games were found to lack suitability, and the Nintendo balance board was not appropriate for those with more severe balance impairments.

DISCUSSION

Clearly off-the-shelf AVG offers huge potential for rehabilitation in clinical populations. It appears to be at least as good as conventional rehabilitation providing opportunities for rehabilitation at a greater intensity or frequency of repetition, potentially leading to better adherence (Coudeyre, 2017). We note, however, that conclusions in relation to effectiveness are limited by the preliminary nature and/or poor methodological quality of many studies – with larger, more rigorous studies needed.

As AVG can be used in the home setting, it offers possibilities for self-management and reducing burden on therapists (Feys and Straudi, 2019). However, there are limitations with commercial AVG. Off-the-shelf systems are not designed for clinical populations and may be overly complex. As we have seen, equipment/peripherals may not be appropriate/usable; games may not be at the right level of difficulty and default feedback may be upsetting and without feedback and advice the user may make maladaptive compensatory movements without realising.

Custom-designed or bespoke systems may avoid some of these issues (Taylor and Griffin, 2015; Feys and Straudi, 2019) by allowing greater configuration and tailoring to a particular rehabilitation purpose (Bonnechère, 2018). However, developers of bespoke systems/games can often only aspire to the level of sophistication and user experience – and low retail cost – of those created by large commercial companies with huge budgets at their disposal (Colder Carras et al., 2018). Additionally, individuals with relatively low levels of disability may prefer off-the-shelf systems as they can be used with friends and relatives in the home and can be purchased at relatively low cost for personal use.

There are technological, clinical, and economic barriers to the uptake of technology-aided rehabilitation in practice (Feys and Straudi, 2019). For example, bespoke fully immersive technologies may be costly to develop meaning that only specialised centres can afford them (Ravenek, Wolfe, and Hitzig, 2016). HCPs/therapists do not routinely use technology (Levac et al., 2017; Langan et al., 2018; Musselman, Shah, and Zariffa, 2018) and may lack the prerequisite confidence, knowledge or skills to incorporate it into their practice and have little or no clinical time allocated to learn about such innovations (Thomson et al., 2016; Glegg and Levac, 2018; Langan et al., 2018). They may see it as disrupting patient–clinician relationships or too time consuming to set up and administer or their facilities may lack dedicated space to store/use technology (Langan et al., 2018). Thus, there is a need to provide clinicians/therapists with adequate training and support to enable them to have the knowledge, confidence and time to use technology in their clinical roles (Levac et al., 2013; Liu et al., 2014; Glegg and Levac, 2017). Deutsch and McCoy (2017) note that 'Practicing therapists should consider the VR [virtual reality] and SG [serious games] systems as one more addition to their intervention toolbox that should be systematically analysed and applied judiciously' (p. 8). Online resources/tools such as the 'Wii-habilitation' and 'Kinect-ing with Clinicians (KwiC)' websites have been created to support therapists to incorporate and make decisions using AVG in practice (Levac et al., 2018). Signal et al. (2018) argue that in order to increase the likelihood of a technology becoming part of clinical practice, there needs to be greater consideration of contextual factors (such as values, processes, beliefs about professional roles) that underpin clinical decision making and meaningful involvement of clinicians in the development process.

Home-based use of exergaming potentially enables exercise/rehabilitation without requiring a visit to the gym or therapist time. However, findings from qualitative studies suggest that some therapist support is needed (Hamilton et al., 2018; Thomas et al., in preparation) and desired by most patients/users to provide advice, guide progression and encourage continued use. As patient activation levels (Feys et al., 2016) vary considerably, a personalised medicine approach could be used with intensity and format (Marziniak et al., 2018; Schez-Sobrino et al., 2019) tailored depending on individual need and preference. Supervision could potentially be conducted remotely via telerehabilitation (depending on a satisfactory assessment of risk and individuals' preferences and levels of activation), and the use of motion capture and artificial intelligence could enable feedback to be provided by a virtual therapist in real time. There could also be options for users to link up with others remotely, facilitating elements of competition and/or peer support.

When designing an exergaming intervention, a person-centred, interdisciplinary and collaborative approach that involves end users and carers (Wiemeyer *et al.*, 2015) along with other stakeholders (such as clinicians, service providers and commissioners) will help to ensure it is fit-for-purpose and will increase the likelihood of enjoyment and flow experiences (Huang *et al.*, 2018). It is also important to consider issues related to implementation and scalability early on (O'Cathain *et al.*, 2019), the likely lifespan of the technology (given the swift pace of technology) and whether principles and learning can be generalised to other emerging technologies.

We know little about the long-term sustainability of benefits from AVGs because many RCTs in the field to date have only involved small samples and not included any longer-term follow-up. There has been poor reporting of intervention (and control condition) characteristics such as dose, frequency and intensity. Even within a single clinical condition, there is often much heterogeneity between studies in terms of intervention and participant characteristics, outcome measures used and length of follow-up, making it difficult to compare studies or combine their findings. There is a need to follow the relevant reporting guidance (Consolidated Standards of Reporting Trials – CONSORT, 2010) and describe interventions (and control/comparator interventions) in adequate detail (e.g. Template for Intervention Description and Replication [TIDieR] checklist – Hoffmann *et al.*, 2014). Threapleton *et al.* (2016) have developed a useful practical checklist for use when conducting home-based VR research including aspects related to trial set-up, screening, equipment set-up and storage and participant training and monitoring.

Given the rapid pace of technology and the potential benefits of personalising aspects of an intervention (such as dose and delivery format), there is a scope for the use of innovative trial designs such as Sequential Multiple Assignment Randomised Trials (SMART) that can adopt a stepped-care approach allowing tailoring of resources to certain participant characteristics (Plow and Finlayson, 2019). Pragmatic hybrid trial designs enable testing of an implementation strategy alongside a consideration of effectiveness and help to close the evidence-implementation gap (Curran *et al.*, 2012). Plow and Finlayson (2019) also highlight the potential use of ecological momentary assessment protocols to gather outcome data.

Ensuring exergaming interventions have a clear theoretical underpinning and incorporate appropriate behaviour change techniques (Michie *et al.*, 2013; Michie, Atkins, and West, 2014; Sangelaji *et al.*, 2016; Motl, Pekmezi, and Wingo, 2018; Plow and Finlayson, 2019) may increase the likelihood of adherence and sustained behaviour change over the longer term. Creating an intervention logic model (see Table 12.3 – Mii-vitaliSe example) helps to specify the proposed mechanisms of action of an intervention. We recommend considering the use of mixed methods RCT designs to enable the incorporation of nested qualitative components and process evaluations as well as parallel economic evaluations to consider cost-effectiveness.

FUTURE TRENDS/DEVELOPMENTS

Anticipating future trends is difficult, but it is sensible to consider the point we are at in the lifecycle of gaming systems (Kooiman and Sheehan, 2015) at the time of writing. The three major games console manufacturers – Sony (*current console – PlayStation 4*), Microsoft (*Xbox One*) and Nintendo (*Switch*) – although still producing occasional health-related games have gradually withdrawn from this area since the late 2000s. As tradition and logistics have dictated, legacy and software support is drastically reduced over time for earlier console modules. For example, as of 2014, the Nintendo Wi-Fi Connection service was discontinued and it is no longer possible to use online features of Nintendo Wii software such as online play, matchmaking, competitions and leaderboards.

Microsoft discontinued its last surviving Xbox version of Kinect in October 2017 and requires an adapter for the newer Xbox One consoles. The Kinect sensor is now only available as a second-hand device. Microsoft may be heading back into this area with the Azure Kinect which was announced

TABLE 12.3
Mii-vitaliSe Logic Model

Inputs	Activities	Outputs	Impacts	Outcomes (Proposed for Full Trial)
Resources: *Human*	*Intervention contact:*	*Personalised/tailored Workbook:*		
• Physiotherapists to deliver intervention	• Orientation sessions (weeks 1 and 2) in the hospital setting with physiotherapist × 2	• Decisional balance table	• Increased use of Nintendo Wii	• Increases in levels of physical activity (GLTEQ and activPAL)
Equipment		• Notes and considerations (including games to try, individual preferences for and reflections about games, adaptations)	• Support to use Mii-vitaliSe via home visit and telephone support from physiotherapists plus Mii-vitaliSe resources	• Increases in confidence to do physical activity (SCI-ESES) and MS related self-confidence (MSSE – control subscale)
• Nintendo Wii • Software (Wii Sports, Wii Sports Resort, Booklet to support practice of home-based Wii	• Wii set up in participants' homes and starting individual programme (week 3)	• Practical considerations including set up and safety aspects	• Support to do Mii-vitaliSe at home via goal setting, action and coping planning, guidance, feedback and monitoring from physiotherapists and self-monitoring via goal setting, Wii feedback and play log	• Improvements in mood (HADS) and fatigue (FSI) and quality of life (SF-36, MSIS-29)
• 2 × Wii remotes × 2 Nunchuks • Wii balance board and non-slip cover • Wii remote charger • Battery charger • Spare batteries	• Home review visits face-to-face by physiotherapist (weeks 7 and 16) • Physiotherapist telephone support (weeks 5 and 12) • Ongoing monthly telephone/email support by physiotherapist	• Warm up and cool down exercises • SMART goals for Wii use at home • Action and coping plans for Wii use at home		• Improvements in measures of walking endurance (2MTW), gait (Gait Stride Time Rhythmicity) static balance (Static Posturography) and dynamic balance (Instrumented TUG), upper extremity function (9HPT)
Materials • Guidance book • Personalised activity book • Games descriptors • Play log		*Play log:* (date/time, whether played, adverse events, activities undertaken, duration, intensity, enjoyment, fatigue, reasons for non-use)		

GLTEQ – Godin Leisure Time Exercise Questionnaire; SCI-ESES – Spinal Cord Injury – Exercise Self-Efficacy Questionnaire; MSSE – Multiple Sclerosis Self-Efficacy Scale; HADS – Hospital Anxiety and Depression Scale; FSI – Fatigue Symptom Inventory; SF-36 – Short Form Survey (Quality of Life) 36 item; MSIS-29 – Multiple Sclerosis Impact Scale; 2MTW – 2 Minute Timed Walk; TUG Timed Up and Go; 9HPT – 9 Hole Peg Test.

in 2019 and is only available presently in a developer's kit (Microsoft Azure, 2020). The next-generation console from Microsoft (Xbox Scarlett) is due late in 2020. Sony – as mentioned earlier – chose not to update motion control functionality for the PS4 (although move controllers designed for the PS3 were compatible and could be used with the PS4 Sony VR headset) (Sony, 2020). The next-generation console from Sony (PS5) is also due in 2020 and Sony possibly might include a wireless VR headset as a part of the system at launch. Nintendo has altered its focus from motion gaming to mobile gaming, with Wii U and the subsequent Switch (Nintendo, 2020f), which currently seems to be the most suitable console for AVG. There are not many games available at present, but this might change in the future. Ring Fit Adventure (Nintendo, 2020g) launched in October 2019 merges an adventure game concept with exercising via a circular 'Ring-Con' (a motion sensing resistance ring) and a Leg Strap (affixed to one leg) which both house parts of the Switch console. This could be viewed as an innovative step back into this area by the manufacturer and herald a new wave of exercise and health gaming software if successful in sales.

Despite a lack of hardware upgrades, Nintendo, Microsoft and Sony's motion controllers and fitness games remain on the market but stocks are limited to buy – only currently sold by third party providers rather than the manufacturers. Depending on the build quality of existing products, these might still be able to be utilised in future projects, but at some point might be considered high risk in terms of how long they will stay working. Gaming consoles like other electrical products can experience reliability problems over time (Ungar, 2008). This might also impact on the likelihood of utilising these solutions for a globally rolled out health intervention due to the time taken to trial, disseminate and roll-out the results to a wider audience. Nevertheless, the principles and underlying methodological models of implementing game-based technology into healthcare environments are still applicable even if the technology and support require changes. Existing products – like the Wii or Kinect – would ideally need retesting to confirm what online functionality is still available or has been withdrawn, in addition to checking that legacy cables supplied still work on newer Smart televisions.

If console manufacturers are less keen on the concept of AVG, then it seems natural that others will start to fill the gap in the market over time. Community fitness innovations and concepts aimed at high impact sportsmen and women is a growth area. Examples include Zwift (2020) which allows players to ride bicycles on stationary trainers while navigating through five virtual worlds and Peloton (2020) which combines sophisticated group exercise class concepts and technology with fitness equipment. Examples currently tend to be via a static exercise bike at home linked through WiFi to an app via a phone or tablet and are targeted at high impact users such as cyclists and triathletes rather than casual and more sedentary exercisers. These types of innovations might be able to be either reconfigured or the concepts adapted for lower impact use in healthcare and rehabilitation. Conceptual models would carry over quite well by utilising the virtual group methodology and similar technological device integration, but ensuring that the exercises were not as demanding.

VR solutions like Oculus Rift (Oculus, 2020) also offer a potential pathway, although like other previous technologies suggested as groundbreaking in their potential scope such as Second Life (2020), there remain risks and contradictory views about their ability to break through into the mainstream (Rodriguez, 2020). This clouds the issue of estimated longevity. Sony VR headset sales for the PS4 console now exceed 4.2 million although this only equates to 4.4% of all PS4 console owners (Orland, 2020). This is also set against a small 0.99 million rough Superdata estimate for Oculus Rift headset sales to the end of 2018 (Rodriguez, 2020). However, VR continues to grow as an alternative area for health and medical technology research and possible interventions (Maier *et al.*, 2019), although perhaps not as swiftly as some have predicted. For home use there will also continue to be safety issues around utilising fully immersive headsets – perhaps causing balance problems and disorientation (Massetti *et al.*, 2018) – and, like gaming consoles, the complexities of installation and support for a less technically competent user base could remain problematic.

Wearable technology (worn directly on the body or located within clothes and trainers) should also be flagged as a potential source of possibilities (Brichetto *et al.*, 2019; Rodgers et al., 2019), but again there is a wide range of issues to consider. These include questions around linking up and monitoring data collected based on the number of different devices and how they interact and integrate with owned computers, tablets and phones. This area should also be considered as having great potential looking towards the future.

While offering incredible potential and positive applications for exergaming innovations in health, barriers will continue to exist around the lifecycle of gaming consoles, and it is possible that research will continue to find longevity of projects post-trial is often cut short by the vagaries of the gaming market demand and supply cycle. Principles and methodologies will be generalisable even as the software and hardware driving innovations in this area continue to change over time. Awareness of these issues at the outset and more open dialogue between the health sector and console manufacturers might be one way to address this issue in the future.

REFERENCES

Allen, D., and Gregory, J. (2009). The transition from children's to adult diabetes services: Understanding the 'problem'. *Diabetic Medicine.* 26 (2), 162–166.

American College of Sports Medicine. (2010). *ACSM's Resource Manual for Guidelines for Exercise Testing and Prescription.* 6th Ed. Philadelphia, PA and London: Lippincott Williams & Wilkins.

AtariAge. (2020). Controllers – Foot Craz [online]. AtariAge. [Viewed 29 January 2020]. Available from: https://atariage.com/controller_page.php?SystemID=2600&ControllerID=17

Balasubramanian, C.L.K., Li, C.Y.C., Saracino, D., Freund, J., and Vallabhajosula, S. (2018). Item-level psychometric analysis of the community, balance & mobility scale in community dwelling older adults. *Archives of Physical Medicine and Rehabilitation.* 99 (10), e88.

Barry, G., Van Schaik, P., MacSween, A., Dixon, J., and Martin, D. (2016). Exergaming (XBOX Kinect™) versus traditional gym-based exercise for postural control, flow and technology acceptance in healthy adults: A randomised controlled trial. *BMC Sports Science, Medicine and Rehabilitation.* 8 (1), 25.

Bogost, I. (2005). The rhetoric of exergaming. Proceedings of the Digital Arts and Cultures (DAC).

Bonnechère, B., Jansen, B., Omelina, L., and Van Sint Jan S. (2016). The use of commercial video games in rehabilitation: A systematic review. *International Journal of Rehabilitation Research.* 3 (4), 277–290.

Bonnechère, B. (2018). Serious games in rehabilitation. From theory to practice. In Bonnechère B. (ed): *Serious Games in Physical Rehabilitation.* Cham, Switzerland: Springer International Publishing. pp. 41–109.

Bressler, D.M., and Bodzin, A.M. (2013). A mixed methods assessment of students' flow experiences during a mobile augmented reality science game. *Journal of Computer Assisted Learning.* 29 (6), 505–517.

Brichetto, G., Spallarossa, P., de Carvalho, M.L.L., and Battaglia, M.A. (2013). The effect of Nintendo Wii on balance in people with multiple sclerosis: A pilot randomized control study. *Multiple Sclerosis Journal.* 19 (9), 1219–1221.

Brichetto, G., Pedullà, L., Podda, J., and Tacchino, A. (2019). Beyond center-based testing: Understanding and improving functioning with wearable technology in MS. *Multiple Sclerosis Journal.* 25 (10), 1402–1411.

Brønnum-Hansen, H., Koch-Henriksen, N., and Stenager, E. (2004). Trends in survival and cause of death in Danish patients with multiple sclerosis. *Brain.* 127 (4), 844–850.

Brumels, K.K.A., et al. (2008). Comparison of efficacy between traditional and video game based balance programs. *Clinical Kinesiology,* 62 (4), 26–31.

Cacciata, M., Stromberg, A., Lee, J., Sorkin, D., Lombardo, D., Clancy, S., Nyamathi, A., and Evangelista, L.S. (2019). Effect of exergaming on health-related quality of life in older adults: A systematic review. *International Journal of Nursing Studies.* 93, 30–40.

Campbell, R., et al. (2001). Why don't patients do their exercises? Understanding non-compliance with physiotherapy in patients with osteoarthritis of the knee. *Journal of Epidemiology and Community Health.* 55 (2), 132–138.

Casuso-Holgado, M.J., Martin-Valero, R., Carazo, A.F., Medrano-Sanchez, E.M., Cortes-Vega, M.D., and Montero-Bancalero, F.J. (2018). Effectiveness of virtual reality training for balance and gait rehabilitation in people with multiple sclerosis: A systematic review and meta-analysis. *Clinical Rehabilitation.* 32 (9), 1220–1234.

Celinder D., and Peoples, H. (2012). Stroke patients' experiences with Wii Sports during inpatient rehabilitation. *Scandinavian Journal of Occupational Therapy.* 19 (5), 457–463.

Colder Carras, M., Rooij, V.J.A., Spruijt-Metz, D., Kvedar, J., Griffiths, M.D., Carabas, Y., and Labrique, A. (2018). Commercial video games as therapy: A new research agenda to unlock the potential of a global pastime. *Frontiers of Psychiatry,* 8, 300.

CONSORT. (2010). Consort Reporting Guidelines. Available from: http://www.consort-statement.org/ [Accessed 29 January 2020].

Coudeyre, E. (2017). High tech-low cost, the growing place of Wii in rehabilitation. *Annals of Physical and Rehabilitation Medicine.* 60, 361–362.

Cowley, B., et al. (2008). Toward an understanding of flow in video games. *Computers in Entertainment.* 6 (2), 1.

Csikszentmihalyi, M. (1990) *Flow: The Psychology of Optimal Experience,* Praha: Lidové Noviny. London: Harper Perennial.

Curran, G.M., Bauer, M., Mittman, B., Pyne, J.M., and Stetler, C. (2012). Effectiveness-implementation hybrid designs: Combining elements of clinical effectiveness and implementation research to enhance public health impact. *Medical Care.* 50 (3), 217–226.

Dahlberg, K., Todres, L., and Galvin, K. (2009). Lifeworld-led healthcare is more than patient-led care: An existential view of well-being. *Medicine, Health Care and Philosophy.* 12 (3), 265–271.

Dalgas, U., Langeskov-Christensen, M., Stenager, E., Riemenschneider, M., and Hvid, L.G. (2019). Exercise as medicine in multiple sclerosis-time for a paradigm shift: preventive, symptomatic, and disease-modifying aspects and perspectives. *Current Neurology and Neuroscience Reports.* 19 (11), 88.

Deloitte Centre for Health Solutions. (2015). *Connected Health: How Digital Health is Transforming Health and Social Care.* London: Deloitte Centre for Health Solutions.

Deutsch, J., and McCoy, S.W. (2017). Virtual reality and serious games in neurorehabilitation of children and adults: Prevention, plasticity and participation. Pediatric physical therapy: The official publication of the Section on Pediatrics of the American Physical Therapy Association. 29 (S3) IV Step 2016; Conference Proceedings, S23.

Fang, Q., Ghanouni, P., Anderson, S.E., Touchett, H., Shirley, R., Fang, F., and Fang, C. (2019). Effects of exergaming on balance of healthy older adults: a systematic review and meta-analysis of randomized controlled trials. *Games for Health Journal.* 9 (1), 11–23.

Fang, X., Zhang, J., and Chan, S.S. (2013). Development of an instrument for studying flow in computer game play. *International Journal of Human-Computer Interaction.* 29 (7), 456–470.

Faulkner, X. (2000). *Usability Engineering.* Basingstoke: Palgrave.

Feys, P., Giovannoni, G., Dijsselbloem, N., Centonze, D., Eelen, P., and Lykee Andersen, S. (2016). The importance of a multi-disciplinary perspective and patient activation programmes in MS management. *Multiple Sclerosis Journal.* 22 (2S), 34–46.

Feys, P., and Straudi, S. (2019). Beyond therapists: Technology-aided physical MS rehabilitation delivery. *Multiple Sclerosis Journal.* 25 (10), 1387–1393.

Forsberg, A., Nilsagård, Y., and Boström, K. (2015). Perceptions of using videogames in rehabilitation: A dual perspective of people with multiple sclerosis and physiotherapists. *Disability & Rehabilitation.* 37 (4), 338–344.

Garcia-Agundez, A., Folkerts, A.K., Konrad, R., Caserman, P., Tregel, T., Goosses, M., Göbel, S., and Kalbe, E. (2019). Recent advances in rehabilitation for Parkinson's disease with Exergames: A systematic review. *Journal of Neuroengineering & Rehabilitation.* 16, 17.

Glegg, S.M., and Levac, D.E. (2017). Enhancing clinical implementation of virtual reality. In 2017 International Conference on Virtual Rehabilitation (ICVR) (pp. 1–7). IEEE.

Glegg, S.M.N., and Levac, D.E. (2018). Barriers, facilitators and interventions to support virtual reality implementation in rehabilitation: A scoping review. *PM&R.* 10 (11), 1237–1251.

Glen, K., Eston, R., Loetscher, T., and Parfitt, G., (2017). Exergaming: Feels good despite working harder. *PloS One.* 12 (10), e0186526.

Guidi, I., Giovannelli, T., and Paci, M. (2013). Effects of Wii exercises on balance in people with multiple sclerosis. *Multiple Sclerosis.* 19 (7), 965.

Gutiérrez, R.O., Galán Del Río, F., De La Cuerda, R.C., Alguacil Diego, I.M., González, R.A., and Page, J.C.M. (2013). A telerehabilitation program by virtual reality-video games improves balance and postural control in multiple sclerosis patients. *NeuroRehabilitation.* 33 (4), 545–554.

Hamilton, C., McCluskey, A., Hassett, L., Killington, M., and Lovarini, M. (2018). Patient and therapist experiences of using affordable technology in rehabilitation: A qualitative study nested in a randomized controlled trial. *Clinical Rehabilitation.* 32 (9), 1258–1270.

Heesen, C., et al. (2006). Physical exercise in multiple sclerosis: Supportive care or a putative disease-modifying treatment. *Expert Review of Neurotherapeutics.* 6 (3), 347–355.

Hoffmann, T.C., et al. (2014). Better reporting of interventions: Template for intervention description and replication (TIDieR) checklist and guide. *BMJ.* 348, g1687.

Holloway, I. (2008). *A-Z of Qualitative Research in Healthcare.* 2nd Ed. Oxford: Blackwell Publishing.

Huang, H.-C., et al. (2018) How to create flow experience in exergames? Perspective of flow theory. *Telematics and Informatics,* 35 (5), 1288–1296.

Husserl, E. (1970). *The Crisis of European Science and Transcendental Phenomenology.* Evanston: Northwestern University Press (Trans. D. Carr).

Jack, K., McLean, S.M., Moffett, J.K., and Gardiner, E. (2010). Barriers to treatment adherence in physiotherapy outpatient clinics: A systematic review. *Manual Therapy.* 15 (3), 220–228.

Jackman, P.C., et al. (2019). Flow states in exercise: A systematic review. *Psychology of Sport & Exercise.* 45; 1–16.

Jackson, S., and Eklund, R. (2002). Assessing flow in physical activity: The flow state scale-2 and dispositional flow scale-2. *Journal of Sport & Exercise Psychology.* 24 (2), 133–150.

Jackson, S.A., and Marsh, H. (1996). Development and validation of a scale to measure optimal experience: The flow state scale. *Journal of Sport & Exercise Psychology.* 18 (1), 17–35.

Kato, P.M. (2010). Video games in health care: Closing the gap. *Review of General Psychology.* 14 (2), 113.

Kanavos, P., Tinelli, M., and Efthymiadou, O., Visintin, E., Grimaccia, F., and Mossman, J. (2016). *Towards Better Outcomes in Multiple Sclerosis by Addressing Policy Change: The International MultiPlE Sclerosis Study (IMPrESS).* London: The London School of Economics and Political Science.

Kiili, K., Arnab, S., and Lainema, T. (2012). The design principles for flow experience in educational games. *Procedia Computer Science.* 15, 78–91.

Kinect-ing with Clinicians (KwiC). (2016). Available from: http://kinectingwithclinicians.com [Accessed 31 January 2020].

Kinnett-Hopkins, D., Adamson, B., Rougeau, K., and Motl, R.W. (2017). People with MS are less physically active than healthy controls but as active as those with other chronic diseases: An updated meta-analysis. *Multiple Sclerosis & Related Disorders.* 13, 38–43.

Klompstra, L., Jaarsma, T., Mårtensson, J., and Strömberg, A. (2017). Exergaming through the eyes of patients with heart failure: A qualitative content analysis study. *Games for Health Journal.* 6 (3), 152–158.

Kooiman, B.J., and Sheehan, D.P. (2015). Interacting with the past, present, and future of exergames: At the beginning of a new life cycle of video games? *Society & Leisure.* 38 (1), 55–73.

Kramer, A., Dettmers, C., and Gruber, M. (2014). Exergaming with additional postural demands improves balance and gait in patients with multiple sclerosis as much as conventional balance training and leads to high adherence to home-based balance training. *Archives of Physical and Medical Rehabilitation.* 95 (10), 1803–1809.

Langan, J., Subryan, H., Nwogu, I., and Cavuoto, L. (2018). Reported use of technology in stroke rehabilitation by physical and occupational therapists. *Disability & Rehabilitation: Assistive Technologies.* 13 (7), 641–647.

Latimer-Cheung, A.E., et al. (2013). Development of evidence-informed physical activity guidelines for adults with multiple sclerosis. *Archives of Physical Medicine & Rehabilitation.* 94 (9), 1829–1836.

Laver, K.E., Lange, B., George, S., Deutsch, J.E., Saposnik, G., and Crotty, M. (2017). Virtual reality for stroke rehabilitation. *Cochrane Database of Systematic Reviews.* Issue 11. Art. No.: CD008349. doi: 10.1002/14651858.CD008349.pub4

Lei, C., Sunzi, K., Dai, F., Liu, X., Wang, Y., Zhang, B., He, L., and Ju, M. (2019). Effects of virtual reality rehabilitation training on gait and balance in patients with Parkinson's disease: A systematic review. *PloS One.* 14 (11), e0224819.

Levac, D.E., Miller, P.A., Glegg, S., and Colquhoun, H. (2013). Integrating virtual reality video games into practice: Clinicians' experiences. *Physiotherapy Theory Practice.* 29 (7), 504–512.

Levac, D., Glegg, S., Colquhoun, H., Miller, P., and Noubary, F. (2017). Virtual reality and active videogame-based practice, learning needs, and preferences: A cross-Canada survey of physical therapists and occupational therapists. *Games for Health Journal.* 6 (4), 217–228.

Levac, D.E., Miller, P., and Colquhoun, H. (2018). How do the perspectives of clinicians with and without virtual reality or active video game experience differ about its use in practice? *International Journal of Child Health and Human Development.* 11 (2), 249–254.

Levac, D.E., Pradhan, S., Espy, D., Fox, E., and Deutsch, J.E. (2018). Usability of the 'Kinect-ing' with clinicians website: A knowledge translation resource supporting decisions about active videogame use in rehabilitation. *Games for Health Journal.* 7 (6), 362–368.

Liu, L., Miguel Cruz, A., Rios Rincon, A., Buttar, V., Ranson, Q., and Goertzen, D. (2014). What factors determine therapists' acceptance of new technologies for rehabilitation - A study using the Unified Theory of Acceptance and Use of Technology (UTAUT). *Disability & Rehabilitation.* 37 (5), 447–455.

Lopes, S., et al. (2018). Games used with serious purposes: A systematic review of interventions in patients with cerebral palsy. *Frontiers in Psychology.* 9, 1712.

Maier, M., Ballester, B.R. Duff, D., Duarte Oller, E., and Verschure, P.F.M.J. (2019). Effect of specific over nonspecific VR-based rehabilitation on poststroke motor recovery: A systematic meta-analysis. *Neurorehabilitation and Neural Repair.* 33 (2), 112–129.

Marziniak, M., Brichetto, G., Feys, P., Meyding-Lamadé, U., Vernon, K., and Meuth, S.G. (2018). The use of digital and remote communication technologies as a tool for multiple sclerosis management: Narrative review. *JMIR Rehabilitation and Assistive Technologies*. 5 (1), e5.

Massetti, T., et al. (2018). The clinical utility of virtual reality in neurorehabilitation: A systematic review. *Journal of Central Nervous System Disease*. 10, 1–18.

Mat Rosly, M., Mat Rosly, H., Davis Oam, G.M., Husain, R., and Hasnan, N. (2017). Exergaming for individuals with neurological disability: A systematic review. *Disability & Rehabilitation*. 39 (8), 727–735.

Meldrum, D., Glennon, A., Herdman, S., Murray, D., and McConn-Walsh, R. (2012). Virtual reality rehabilitation of balance: assessment of the usability of the Nintendo Wii(®) Fit Plus. *Disability & Rehabilitation: Assistive Technology*. 7 (3), 205–210.

Michie, S., Richardson, M., Johnston, M., Abraham, C., Francis, J., Hardeman, W., Eccles, M.P., Cane, J., and Wood, C.E. (2013). The behavior change technique taxonomy (V1) of 93 hierarchically clustered techniques: Building an international consensus for the reporting of behavior change interventions. *Annals of Behavioral Medicine*. 46 (1), 81–95.

Michie S., Atkins L., and West R. (2014). The behaviour change wheel. *A Guide to Designing Interventions*. Sutton, Great Britain: Silverback Publishing.

Microsoft. (2020). Kinect Sports Rivals [online]. Microsoft. [Viewed 29 January 2020]. Available from: https://www.microsoft.com/en-us/p/kinect-sports-rivals/brhsml8030zn?activetab=pivot:overviewtab

Microsoft Azure. (2020). Azure Kinect DK [online]. Microsoft Azure. [Viewed 29 January 2020]. Available from: https://azure.microsoft.com/en-gb/services/kinect-dk/

Microsoft Developer. (2020). Kinect for Windows [online]. Microsoft Developer. [Viewed 29 January 2020]. Available from: https://developer.microsoft.com/en-us/windows/kinect

Middleton, A. (2004). Chronic low back pain: Patient compliance with physiotherapy advice and exercise, perceived barriers and motivation. *Physical Therapy Reviews*. 9 (3), 153–160.

Milo, R., and Kahana, E. (2010). Multiple sclerosis: Geoepidemiology, genetics and the environment. *Autoimmunity Reviews*. 9 (5), A387–A394.

Mohr, D.C., Cuijpers, P., and Lehman, K. (2011). Supportive accountability: A model for providing human support to enhance adherence to eHealth interventions. *Journal of Medical Internet Research*. 13 (1), e30.

Motl, R.W., Pekmezi, D., and Wingo, B.C. (2018). Promotion of physical activity and exercise in multiple sclerosis: Importance of behavioural science and theory. *Multiple Sclerosis Journal. Experimental, Translational and Clinical*. 4, 1–8.

MS Society. (2018). *Improving Care for People with MS: The Potential of Data and Technology*. London: MS Society.

Mumford, E., (2006). The story of socio-technical design: Reflections on its successes, failures and Potential. *Information Systems Journal*. 16, 317–342.

Musselman, K.E., Shah, M., and Zariffa, J. (2018). Rehabilitation technologies and interventions for individuals with spinal cord injury: Translational potential of current trends. *Journal of Neuroengineering and Rehabilitation*. 15 (1), 40.

Nacke, L., and Lindley, C.A. (2008). Flow and immersion in first-person shooters, in Proceedings of the 2008 Conference on Future Play Research, Play, Share - Future Play '08. New York, NY: ACM Press, p. 81. doi: 10.1145/1496984.1496998

Nah, F.F., et al. (2014). Flow in gaming: Literature synthesis and framework development. *International Journal of Information Systems and Management*. 1 (1), 83–124.

Nilsagård, Y.E., Forsberg, A.S., and von Koch, L. (2013). Balance exercise for persons with multiple sclerosis using Wii games: A randomised, controlled multi-centre study. *Multiple Sclerosis*. 19 (2), 209–216.

Nintendo. (2020a). Nintendo 64 [online]. Nintendo. [Viewed 29 January 2020]. Available from: https://www.nintendo.co.uk/Corporate/Nintendo-History/Nintendo-64/Nintendo-64-625959.html

Nintendo. (2020b). Nintendo GameCube [online]. Nintendo. [Viewed 29 January 2020]. Available from: https://www.nintendo.co.uk/Corporate/Nintendo-History/Nintendo-GameCube/Nintendo-GameCube-627129.html

Nintendo. (2020c). Nintendo Wii [online]. Nintendo. [Viewed 29 January 2020]. Available from: https://www.nintendo.co.uk/Wii/Wii-94559.html

Nintendo. (2020d). Nintendo Wii Fit [online]. Nintendo. [Viewed 29 January 2020]. Available from: https://www.nintendo.co.uk/Games/Wii/Wii-Fit-283894.html

Nintendo. (2020e). Nintendo Wii U [online]. Nintendo. [Viewed 29 January 2020]. Available from: https://www.nintendo.co.uk/Wii-U/Wii-U-344102.html

Nintendo. (2020f). Nintendo Switch [online]. Nintendo. [Viewed 29 January 2020]. Available from: https://www.nintendo.com/switch/

Nintendo. (2020g). Ring Fit Adventure [online]. Nintendo. [Viewed 29 January 2020]. Available from: https://www.nintendo.co.uk/Games/Nintendo-Switch/Ring-Fit-Adventure-1638708.html

O'Cathain, A., et al. (2013). What can qualitative research do for randomised controlled trials? A systematic mapping review. *BMJ Open.* 2013 (3), e002889.

O'Cathain, A. et al. (2019). Guidance on how to develop complex interventions to improve health and health-care. *BMJ Open.* 9, e029954.

Oculus. (2020). Oculus Homepage [online]. Oculus. [Viewed 29 January 2020]. Available from: https://www.oculus.com/

Oh, Y., and Yang, S., (2010). Defining exergames & exergaming. *Proceedings of Meaningful Play.* 1, 1–17.

Öhman, M., Söderberg, S., and Lundman, B. (2003). Hovering between suffering and enduring: The meaning of living with serious chronic illness. *Qualitative Health Research.* 13 (4), 528–542.

Office of National Statistics. (2013). Interim Life Tables, England and Wales, 2009-2011 [online]. Available from: http://www.ons.gov.uk/ons/rel/lifetables/interim-life-tables/2009-2011/stb-2009-2011.html [Accessed 3 October 2013].

Orland, K., (2020). Putting Sony's 4.2 Million PSVR Sales in Context [online]. Ars Technica. [Viewed 29 January 2020]. Available from: https://arstechnica.com/gaming/2019/03/putting-sonys-4-2-million-psvr-sales-in-context/

Ozdogar, A.T., Ertekin, O., Kahraman, T., Yigit, P., and Ozakbas, S. (2020). Effect of video-based exergaming on arm and cognitive function in persons with multiple sclerosis: A randomized controlled trial. *Multiple Sclerosis and Related Disorders.* 40, 101966.

Palacios-Ceña, D., Ortiz-Gutierrez, R.M., Buesa-Estellez, A., Galan-Del-Rio, F., Cachon Perez, J.M., Martinez-Piedrola, R., and Cano-De-La-Cuerda, R. (2016). Multiple sclerosis patients' experiences in relation to the impact of the Kinect virtual home-exercise programme: A qualitative study. *European Journal of Physical and Rehabilitation Medicine.* 52 (3), 347–355.

Pau, M., Coghe, G., Corona, F., Leban, B., Marrosu, M.G., and Cocco, E., (2015). Effectiveness and limitations of unsupervised home-based balance rehabilitation with Nintendo Wii in people with multiple sclerosis. *BioMed Research International.* 2015, 916478.

Paquin, K., et al. (2016). Survivors of chronic stroke – Participant evaluations of commercial gaming for rehabilitation. *Disability & Rehabilitation.* 38 (21), 1–9.

Parra-Moreno, M., Rodríguez-Juan, J.J., and Ruiz-Cárdenas, J.D. (2019). Use of commercial video games to improve postural balance in patients with multiple sclerosis: a systematic review and meta-analysis of randomised controlled clinical trials. *Neurología.* 2019. https://doi.org/10.1016/j.nrl.2017.12.001

Peek, K., Sanson-Fisher, R., Mackenzie, L., and Carey M. (2016). Interventions to aid patient adherence to physiotherapist prescribed self-management strategies: A systematic review. *Physiotherapy.* 102 (2), 127–135.

Peloton. (2020). Peloton Homepage [online]. Peloton. [Viewed 29 January 2020]. Available from: https://www.onepeloton.co.uk/

Perrochon, A., Borel, B., Istrate, D., Compagnat, M., and Daviet, J.C. (2018). Exercise-based games interventions at home in individuals with a neurological disease: A systematic review and meta-analysis. *Annals of Physical and Rehabilitation Medicine.* 62 (5), 366–378.

Ploughman, M. (2017). Breaking down the barriers to physical activity among people with multiple sclerosis – A narrative review. *Physical Therapy Reviews.* 22 (3–4), 124–132.

Plow, M., and Finlayson, M. (2011). Potential benefits of Nintendo Wii Fit among people with multiple sclerosis. *International Journal of MS Care.* 13 (1), 21–30.

Plow, M., and Finlayson, M. (2014). A qualitative study exploring the usability of Nintendo Wii Fit among persons with multiple sclerosis. *Occupational Therapy International.* 21(1), 21–32.

Plow, M., and Finlayson, M., (2019). Beyond supervised therapy: Promoting behavioral changes in people with MS. *Multiple Sclerosis Journal.* 25 (10), 1379–1386.

Porras, D.C., Siemonsma, P., Inzelberg, R., Zeilig, G., and Plotnik, M. (2018). Advantages of virtual reality in the rehabilitation of balance and gait: Systematic review. *Neurology.* 90 (22), 1017–1025.

Prosperini, L., Fanelli, F., Petsas, N., Sbardella, E., Tona, F., Raz, E., Fortuna, D., De Angelis, F., Pozzilli, C., and Pantano, P. (2014). Multiple sclerosis: Changes in microarchitecture of white matter tracts after training with a video game balance board. *Radiology.* 273 (2), 529–538.

Prosperini, L., Fortuna, D., Giannì, C., Leonardi, L., Marchetti, M.R., and Pozzilli, C. (2013) Home-based balance training using the Wii balance board a randomized, crossover pilot study in multiple sclerosis. *Neurorehabilitation and Neural Repair.* 27 (6), 516–525.

Rand, D., Givon, N., and Avrech Bar, M. (2018). A video-game group intervention: Experiences and perceptions of adults with chronic stroke and their therapists: Intervention de groupe à l'aide de jeux vidéo: Expériences et perceptions d'adultes en phase chronique d'un accident vasculaire cérébral et de leurs ergothérapeutes. *Canadian Journal of Occupational Therapy.* 85 (2), 158–168.

Ravenek, K.E., Wolfe, D.L., and Hitzig, S.L. (2016). A scoping review of video gaming in rehabilitation. *Disability & Rehabilitation: Assistive Technology.* 11 (6), 445–453.

Ritter, F.E., Baxter, G.D., and Churchill, E.F., (2014). *Foundations for Designing User-centered Systems.* London: Springer.

Robertson, M.C., Tsai, E., Lyons, EJ., Srinivasan, S., Swartz, M.C., Baum, M.L., and Basen-Engquist, K.M. (2017). Mobile health physical activity intervention preferences in cancer survivors: A qualitative study. *JMIR Mhealth Uhealth.* 24 (5), e3. doi: 10.2196/mhealth.6970

Robinson, J., Dixon, J., Macsween, A., Van Schaik, P., and Martin, D. (2015). The effects of exergaming on balance, gait, technology acceptance and flow experience in people with multiple sclerosis: A randomized controlled trial. *BMC Sports Science, Medicine and Rehabilitation.* 7 (1), 8.

Rodgers, M.M, Alon, G., Pai, V.M, and Conroy, R.S. (2019). Wearable technologies for active living and rehabilitation: Current research challenges and future opportunities. *Journal of Rehabilitation & Assistive Technologies Engineering.* 6, 1–9.

Rodriguez, S. (2020). Facebook Will Never Break through with Oculus, Says One of the VR Company's Co-founders [online]. CNBC Tech. [Viewed 29 January 2020]. Available from: https://www.cnbc.com/2019/07/12/facebook-oculus-will-never-break-through-co-founder-jack-mccauley.html

Sala, G., Tatlidil, S., and Gobet, F. (2019). Still no evidence that exergames improve cognitive ability: A commentary on Stanmore et al. (2017). Neuroscience and Biobehavioral Reviews. 21 Nov.

Sangelaji, B., et al. (2016). The effectiveness of behaviour change interventions to increase physical activity participation in people with multiple sclerosis: A systematic review and meta-analysis. *Clinical Rehabilitation.* 30 (6), 559–576.

Schez-Sobrino, S., Vallejo, D., Glez-Morcillo, C., Castro-Schez, J.J., and Albusac, J. (2019). Toward precision rehabilitation for neurological diseases: Data driven approach to exergame personalization. *Proceedings of the 13th International Conference on Ubiquitous Computing and Ambient*

Second Life. (2020). Second Life Homepage [online]. Second Life. [Viewed 29 January 2020]. Available from: https://secondlife.com/

Seligman, M. (2011) *Flourish: A Visionary New Understanding of Happiness and Well-Being.* New York: Simon and Schuster.

Sharek, D., and Wiebe, E. (2011). Using flow theory to design video games as experimental stimuli. *Proceedings of the Human Factors and Ergonomics Society Annual Meeting.* 55 (1), 1520–1524.

Signal, N.E., Scott, K., Taylor, D., and Kayes, N.M., (2018). What helps or hinders the uptake of new technologies into rehabilitation practice? In Masia, L., Micera, S., Akay, M., and Pons, J.L. (eds.) *International Conference on NeuroRehabilitation.* Cham: Springer, pp. 265–268.

Smith, M., Neibling, B., Williams, G., Birks, M., and Barker, R. (2019). A qualitative study of active participation in sport and exercise for individuals with multiple sclerosis. *Physiotherapy Research International.* 2019, e1776.

Sony. (2020). PlayStation Move Controller [online]. Sony. [Viewed 29 January 2020]. Available from: Sony. 2019. PlayStation Move Controller. https://www.playstation.com/en-gb/explore/accessories/playstation-move-motion-controller/

Staiano, A.E., and Calvert, S.L. (2011). The promise of exergames as tools to measure physical health. *Entertainment Computing.* 2 (1), 17–21.

Stanmore, E., Stubbs, B., Vancampfort, D., de Bruin, E.D., and Firth, J. (2017). The effect of active video games on cognitive functioning in clinical and non-clinical populations: A meta-analysis of randomized controlled trials. *Neuroscience & Biobehavioral Reviews.* 78, 34–43.

Street, T.D., Lacey, S.J., and Langdon, R.R. (2017). Gaming your way to health: A systematic review of exergaming programs to increase health and exercise behaviors in adults. *Games for Health Journal,* 6 (3), 136–146.

Tatla, S.K., et al. (2015). Therapists' perceptions of social media and video game technologies in upper limb rehabilitation. *JMIR Serious Games.* 10 (3), e2.

Taylor, M.J., and Griffin, M. (2015). The use of gaming technology for rehabilitation in people with multiple sclerosis. *Multiple Sclerosis.* 21 (4), 355–371.

Thomas, S., Fazakarley, L., Thomas, P.W., Brenton, S., Collyer, S., Perring, S., Scott, R., Galvin, K., and Hillier, C. (2014). Testing the feasibility and acceptability of using the Nintendo Wii in the home to increase activity levels, vitality and well-being in people with multiple sclerosis (Mii-vitaliSe): protocol for a pilot randomised controlled study. *BMJ Open.* 4 (5), e005172.

Thomas, S., et al. (2015). Mii-vitaliSe - Development of a physiotherapist-supported Nintendo Wii intervention to encourage people with multiple sclerosis to become more active in the home. *Multiple Sclerosis Journal*. 21, 775.

Thomas, S., Fazakarley, L., Thomas, P.W., Collyer, S., Brenton, S., Perring, S., Scott, R., Thomas, F., Thomas, C., Jones, K., and Hickson, J. (2017). Mii-vitaliSe: a pilot randomised controlled trial of a home gaming system (Nintendo Wii) to increase activity levels, vitality and well-being in people with multiple sclerosis. *BMJ Open*. 7 (9), e016966.

Thomson, K., Pollock, A., Bugge, C., and Brady, M. (2014). Commercial gaming devices for stroke upper limb rehabilitation: A systematic review. *International Journal of Stroke*. 9 (4), 479–488.

Thomson, K., Pollock, A., Bugge, C., and Brady, M.C. (2016). Commercial gaming devices for stroke upper limb rehabilitation: A survey of current practice. *Disability and Rehabilitation: Assistive Technology*. 11 (6), 454–461.

Threapleton, K., Drummond, A., and Standen, P. (2016). Virtual rehabilitation: What are the practical barriers for home-based research? *Digital Health*. 2, 1–11.

Tobaigy, A., Alshehri, M.A., Timmons, S., and Helal, O.F. (2018). The feasibility of using exergames as a rehabilitation tool: The attitudes, awareness, opinions and experiences of physiotherapists, and older people towards exergames. *Journal of Physical Therapy Science*. 30 (4), 555–562.

Todres, L., Galvin, K., and Dahlberg, K. (2007). Lifeworld-led healthcare: Revisiting a humanising philosophy that integrates emerging trends. *Medicine, Health Care and Philosophy*. 10 (1), 53–63.

Tough, D., Robinson, J., Gowling, S., Raby, P., Dixon, J., and Harrison, S.L. (2018). The feasibility, acceptability and outcomes of exergaming among individuals with cancer: A systematic review. *BMC Cancer*. 18 (1), 1151.

Tripette, J., Murakami, H., Ryan, K.R., Ohta, Y., and Miyachi, M. (2017). The contribution of Nintendo Wii Fit series in the field of health: A systematic review and meta-analysis. *Peer Journal*. 5, e3600.

Ungar, L.Y. (2008). The economics of harm prevention through Design for Testability. In 2008 IEEE International Test Conference (pp. 1–8). IEEE.

Warburton, D.E., Bredin, S.S., Horita L.T., Zbogar, D., Scott, J.M., Esch, B.T., and Rhodes R.E. (2007). The health benefits of interactive video game exercise. *Applied Physiology Nutrition and Metabolism*. 32, 655–663.

Warburton, D., et al. (2009). Metabolic requirements of interactive video game cycling. *Medicine and Science in Sports and Exercise*. 41 (4), 920.

Weinberg, R.S., and Gould, D. (2011). *Foundations of Sport and Exercise Psychology*. 7th Ed. Champaign, Illinois: Human Kinetics.

Wiemeyer, J., Deutsch, J., Malone, L.A., Rowland, J.L., Swartz, M.C., Xiong, J., and Zhang F.F. (2015). Recommendations for the optimal design of exergame interventions for persons with disabilities: Challenges, best practices, and future research. *Games for Health Journal*. 4 (1), 58–62.

Wii-habilitation. (n.d.). Available from: http://www.wiihabilitation.co.uk/?page_id=1382 [Accessed 31 January 2020].

Wingham, J. et al. (2015). Participant and caregiver experience of the Nintendo Wii Sport™ after stroke: Qualitative study of the trial of Wii™ in stroke (TWIST). *Clinical Rehabilitation*. 29 (3), 295–306.

Wolf, M.J. (2012). Encyclopedia of Video Games: AL (Vol. 1). ABC-CLIO.

Yazgan, Y.Z., Tarakci, E., Tarakci, D., Ozdincler, A.R., and Kurtuncu, M. (2020). Comparison of the effects of two different exergaming systems on balance, functionality, fatigue, and quality of life in people with multiple sclerosis: A randomized controlled trial. *Multiple Sclerosis & Related Disorders*. 39, 101902

Zhang, Z. (2012). Microsoft Kinect sensor and its effect. *IEEE Multimedia*. 19 (2), 4–10.

Ziebland, S. (2004). The importance of being expert: The quest for cancer information on the Internet. *Social Science & Medicine*. 59 (9), 1783–1793.

Zwift. (2020). Zwift Homepage [online]. Zwift. [Viewed 29 January 2020]. Available from: https://zwift.com/uk

13 Participant Adherence to a Video Game-Based Tele-rehabilitation Program
A Mixed-Methods Case Series

Gerard G Fluet, Qinyin Qiu, Amanda Cronce, Eduardo Sia, Kathryn Blessing, Jigna Patel, and Alma Merians
Rutgers The State University of New Jersey
Newark, New Jersey

Donghee Yvette Wohn and Sergei Adamovich
New Jersey Institute of Technology
Newark, New Jersey

CONTENTS

INTRODUCTION TO STROKE REHABILITATION

As a result of current service delivery models in the United States, recovery of hand function continues to be a source of frustration and enduring disability for patients recovering from stroke. Current models allow quick, mobility-focused in-patient rehabilitation stays and sporadic, short outpatient rehabilitation sessions, which together restrict the amount of hand-focused rehabilitation a patient receives, and may contribute to persistent hand related disability. This persistent disability dramatically increases the cost of care for older persons with strokes (Ma et al., 2014) and has a substantial negative effect on the productivity of younger persons post stroke (Kissela et al., 2012).

A large majority of the home-based, technology-supported upper extremity interventions for people with stroke target gross movements of the shoulder and elbow (Laver et al., 2012). This is due to the technological limitations of the off-the-shelf, camera-based interfaces designed for commercial gaming systems (e.g. Kinect™), which are often used. These systems are not able to differentiate the complex configurations of finger position necessary for hand rehabilitation. This said, a large majority of the rehabilitation priorities identified by persons with stroke involve hand function (Harris and Eng, 2004, Timmermans et al., 2009).

Direct, in-person interactions between rehabilitation therapists and persons with rehabilitation needs are travel, time and labor intensive, making rehabilitation services costly and limiting patient access. At the same time, private payer reimbursement for rehabilitation services in privatized health care systems, and rehabilitation budgets in nationalized healthcare systems are trending down. These impingements on in-person, facility-based interaction between patient and therapist all point to a need for rehabilitation that can be performed by patients independently, in the home (Kairy et al., 2009).

ACCESSIBILITY AND ADHERENCE

Persons with upper extremity paresis after stroke are one of the many patient groups that require extensive rehabilitation to maximize their return to pre stroke levels of function (Kwakkel, 2006). As stated above, this volume of therapist-supervised, facility-based rehabilitation is seldom available to the substantial number of persons that could benefit from it (Lang et al., 2009, Simpson et al., 2018) see below). Unfortunately, adherence to home-based rehabilitation in persons with stroke is quite low (Jurkiewicz et al., 2011, Kairy et al., 2009), and independent exercise in fitness facilities performed by persons with stroke is far lower than recommended levels as well (Rimmer et al., 2008). Barriers to adherence to home exercise and activity recommendations in persons with stroke include facility-based factors (cost, access) as well as personal factors (low motivation, low levels of self-efficacy (Rimmer et al., 2008). Several authors have suggested video game-based rehabilitation in the home as an approach to surmounting these barriers. While this approach is associated with higher levels of adherence than traditionally presented activities, participation still lags behind recommended levels (Standen et al., 2015, Mihelj et al., 2012, Popović et al., 2014). Barriers to participation for this approach overlap (low motivation, low self-efficacy) and diverge (computer literacy, competing interests in the home) (Lam et al., 2015, Emmerson et al., 2018).

The New Jersey Institute of Technology Home Virtual Rehabilitation System (NJIT-HoVRS) was developed to address these barriers to participation in home rehabilitation (Qiu et al., 2009) while simultaneously studying the impact of motivation on home exercise adherence in persons with stroke. NJIT-HoVRS consists of a Leap Motion controller, a mechanical arm support and five videogame-based upper extremity rehabilitation activities. The Leap Motion Controller is used to collect finger position data and acts as an interface between the subject and the simulated rehabilitation activities. A recently published pilot study of HoVRS by Fluet et al. describes higher adherence levels to a program of hand rehabilitation games that were designed utilizing motivation techniques developed by the video game industry when compared to a group playing unenhanced versions of the same games (Fluet et al., 2019).

This chapter examines the subjects with the highest and lowest levels of adherence from each of the two groups. Qualitative data, collected via semi-structured interview, is combined with psychological, demographic, motor function and adherence data to present a rich examination of factors associated with adherence to these two protocols. It is hypothesized that this examination will underscore the fact that adherence is influenced by highly varied and complex constellations of personal and environmental factors that make one size fits all interventions, or even simple sets of best practices, insufficient for achieving adherence targets in a broad population.

CASE STUDIES

The four cases discussed in this chapter were the subjects from each intervention group in a larger pilot study that demonstrated the highest adherence and the lowest adherence to the training protocol as measured by total minutes of training over the 12 weeks.

TREATMENTS

The subjects in this study performed one of two training protocols, the enhanced motivation protocol (EMP) or the unenhanced control protocol (UCP). All subjects used the NJIT-HoVRS system to play video games designed to practice movements of their shoulder, arm and fingers. A team consisting of a technologist and a physical therapist set up the system in each subject's home during the first visit and taught them how to use it. Subjects practiced in their homes independently with support (on-line or in-person) as needed. All subjects were told to use the simulations as much as possible, but no less than 20 minutes daily for 12 weeks.

The EMP group played three games that present users with 8–12 levels of increasing complexity and difficulty. As each level advanced, there was an announcement and the graphics changed substantially. Scoring scale and bonus scoring opportunities changed at each level as well. The UCP group subjects played the same three games, and difficulty was changed utilizing an algorithm that was based on the subject's performance. These were incremental changes that made them undetectable for most subjects, with no announcement or substantial change in graphics as subjects progressed. Scoring and graphics stayed the same as difficulty increased (see Figure 13.1).

The three games played were designed to train hand and arm motions that are difficult for persons with stroke. *Speed Bump* trains hand opening, pronation and supination. The lowest levels train these motions independently. Speed and movement amplitude requirements increase gradually. Higher levels require combinations of pronation and supination and hand opening, and course complexity increases as subject's progress toward the highest levels.

Urban Aviator trains wrist movement. Subjects collect targets by navigating a plane over and around buildings and control the pitch of the plane with wrist flexion and extension movements. Difficulty is increased with larger vertical distance variation between targets and faster obstacle/target flow.

Maze Runner trains horizontal plane shoulder and elbow movements combined with hand opening. Lower levels require progressively larger forward, back, left, and right reaching movements

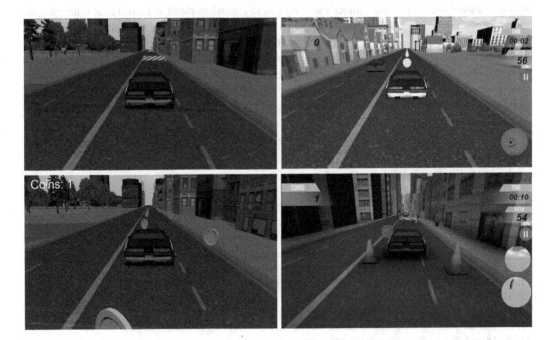

FIGURE 13.1   Enhanced motivation protocol and unenhanced control protocol presentations of speed bump simulation. (a) UCP easiest version of this simulation. (b) UCP difficulty has increased and a new skill has been added. Note that avatar does not change and graphics do not change despite the added challenge. (c) EMP easiest version of this simulation. Note the pace car in the left lane and speedometer and scoreboard, providing motivation to increase speed as this level is performed. (d) EMP difficulty has increased and a new skill has been added. Note the new avatar (car) and changing environment to enhance the sense that conditions have changed, adding a sense of achievement corresponding with the increased challenge.

to elicit avatar direction changes. Higher levels increase the size and number of direction changes, narrow paths, and obstacles that the avatar must jump over, which is elicited by the player opening and closing their hand.

Fruit Catching integrates the horizontal shoulder and elbow movements addressed by *Maze Runner* with pronation and supination movements. Participants control a basket that they use to catch oranges as they fall from trees. They score points by transporting the oranges to a bin and dumping them from the basket using a pronation movement. Lower levels emphasize the reaching movement, while higher levels emphasize pronation and supination. The highest levels require faster movements by making the oranges fall faster and more accurate movements by using a smaller basket.

MEASUREMENTS

For the larger pilot study, we chose three measures with well-established benchmarks for clinically meaningful change in persons with chronic stroke. Impairment was measured with the Upper Extremity Fugl – Meyer Assessment (UEFMA), which has substantial literature supporting its use as an outcome measure for upper extremity intervention trials, and also serves as a proxy measure of stroke severity. The UEFMA also has a fairly low floor, allowing it to capture subtle change in subjects with severe strokes (Deakin et al., 2003). We measured activity level improvement using the Box and Blocks Test (BBT) (Mathiowetz et al., 1985) and the Action Research Arm Test (ARAT) (Yozbatiran et al., 2008). We utilized this combination of tests to offset high floor effects we have observed when using the ARAT.

The impact of game mechanics on intrinsic motivation was measured with the Intrinsic Motivation Inventory (IMI) (McAuley et al., 1989) (see Table 13.1).

Adherence was measured by examining total training minutes, average minutes per session, which was used to evaluate the subject's willingness to choose to play the games over other participating in other activities, and average number of sessions per week, used to evaluate the subject's tendency to play the games for longer periods of time.

We also interviewed each of the four subjects after participation. We had a trained research assistant who did not participate in recruiting, data collection, or the provision of the intervention conduct the interviews. The interview questions fell under four broad themes: (1) previous use of technology, (2) system-based issues, (3) intrinsic motivation factors, and (4) extrinsic motivation

---

TABLE 13.1

Intrinsic Motivation Inventory Questions

1. IE – I enjoyed playing these games very much
2. EI – It was important to me to do well at these games
3. EI – I tried very hard on this activity
4. PC – I think I am pretty good at playing these games
5. VU – I would be willing to playing these games again because they have some value to me
6. EI – I put a lot of effort into playing these games
7. VU – I believe that playing these games could be of some value to me
8. PC – I am satisfied with my ability to play these games
9. PC* – This was an activity that I could not do very well
10. VU – I think this is important to do because it can help me control my fingers better
11. IE – I would describe these games as very interesting
12. IE – Playing these games was fun

Each question is answered on a scale from one (strongly disagree) to seven (strongly agree). Questions come from four of the intrinsic motivation scales; value-usefulness (VU), interest – enjoyment (IE), effort – importance (EI) and perceived competence (PC). * = reverse scaled question.

TABLE 13.2

Interview Questions – Technology Literacy Questions

First I want to discuss a little bit about your technology usage.

How tech-savvy do you think you are?

How long have you been using a computer?

Do you have a smart phone?

  How long have you been using it?

  What do you do on it?

Have you ever played computer games or videogames?

  Prior to stroke or current/continuing?

Do you play mobile games?

Extrinsic Motivation Questions

What made you participate in this home study?

Before you had the stroke, how physically active were you?

How do you feel about physical therapy in general?

Why do you want to improve your hand movement?

Out of the activities that you used to enjoy before you had the stroke but can no longer do easily, what do you miss most?

What is the most important part of your day?

Game System Performance Issues

How easy or difficult was the system to use?

What was the most difficult aspect?

What do you think would improve that?

Intrinsic Motivation Questions

How did you decide when to play the games?

How did this fit in with your daily routine?

What prompted you to play the games?

How you did you decide how long you wanted to play?

What did you like about the games?

What did you dislike about the games?

Did you pay attention to the points?

Did you keep track of your points?

Do you think the games were effective at getting you to exercise?

Did you exercise when you were not playing the games?

Did you enjoy playing the games?

Which was your favorite? Why?

What did you not enjoy? What could be improved?

How physically tired were you after a game session?

Were the game useful to you? How?

Would you be willing to continue playing the games even if you were no longer required to?

Do you feel like playing the games affected other aspects of your life? In what way?

factors. An interview script was used to ensure all topics of interest were covered, but subjects were encouraged to add their own comments (Table 13.2). Recorded data from the interviews was transcribed using Temi TEMI, a web-based voice-to-text transcription system. The research assistant and the principal investigator generated transcripts from the audio recordings and a third team member checked the transcripts for accuracy.

CASE 1: BJ

BJ is an 82-year-old man that started the training program seven years after his stroke. He lives with his wife in a private home, in a community with a median household income of $148,000. BJ's

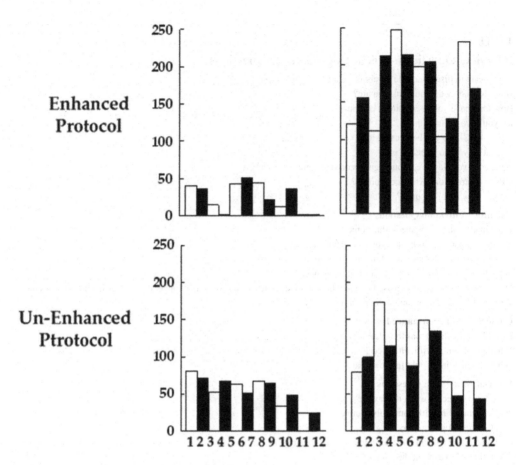

FIGURE 13.2 Weekly use of system by individual subjects in minutes. (a) WA participation times. (b) DM participation times. *Top* (c) SK participation times. (d) BJ participation times.

highest level of education achieved is an MBA, and prior to his retirement, he was the owner of a recruiting firm that services the financial industry. His technology use is limited to emailing and information searches, both on an iPad. He does not own a smart phone. BJ does not take medications for depression but his medical history during the trial was positive for hypertension. He was a regular exerciser and played golf often before his stroke and returned to his previous exercise habits with modification, but not golf, soon after his stroke, and he exercised throughout the training study period. BJ had not participated in physical or occupational therapy for several years prior to participating in the study and did not participate in therapy during the study; however, he has participated in multiple trials of rehabilitation research interventions over the years. BJ's impairments were relatively mild as he is an unlimited community ambulatory and he drives.

During the study BJ averaged 100 (43) minutes of training per week, making him the UCP group member with the highest adherence. He accomplished this averaging 4.6 (1.1) sessions per week. Average session length was 21.7 (6.5) minutes. Peak participation occurred during week seven with 150 training minutes. Lowest participation time was for week 12 with 44 minutes. BJ never trained less than three times in a week (see Figure 13.2). BJ's IMI composite scores were identical at pre- and post-test at 66 points. None of BJ's subscale scores changed over the course of the training period. Subsequent to training, BJ demonstrated a 5-point improvement in UEFMA and an 11-point increase in ARAT score (see Figure 13.3).

BJ did not play computer games prior to participating in the study. "That's for my grandkids," was his response to this question. When asked why he participated in this study BJ responded

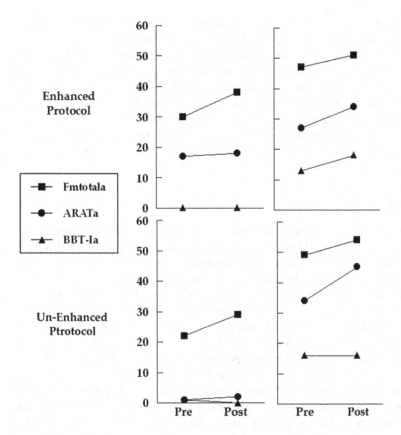

FIGURE 13.3   Clinical test scores measured at pre and posttest demonstrated by individual subjects. (a) WA clinical test changes. (b) DM clinical test changes. (c) SK clinical test changes. (d) BJ clinical test changes.

"Because I want to get better. Every opportunity I can take to get better and to improve I try to take advantage."

BJ cited the fact that having the system in his home made getting his training in easier. He described the system as "well-designed" and the team that taught him to use the system as "very thorough and very easy to work with". BJ also hypothesized that his mild impairment level might be the reason that he did not have trouble using the system. "Because my disability is not severe, most activities I have to do, is not as hard as it is for many other stroke survivors".

When asked how the games could be improved, BJ told researchers that they could be more challenging and also suggested multi-player games. When asked if boredom became an issue over the course of the training program he agreed that "… doing the same thing over and over.…. I guess you become a little more bored with it, it becomes a little more monotonous."

BJ played all four games an equal amount of time until the last month. In the last month, he played the Maze game as much as the other the games put together. When asked why, he stated, "That's because it was the toughest. More challenging. I had to make more adjustments and so forth; It was just more difficult and I wanted something more challenging. I feel challenging means I have a greater chance to improve."

As stated previously BJ's responses resulted in a high *effort/importance* subscale score on the IMI, which emphasizes the amount of effort that he exerted during the training activities. "I always look at every activity as a positive experience," BJ explained. "Because what I wanted to be is challenged, to improve, to get better, and the only way you can get better is by working, playing these games, whatever the activity is. I'm all in."

When asked which activities he was able to perform more effectively after training BJ cited carrying packages and that he was able to shake hands with somebody with his affected arm. During his post-testing session, he also relayed to the therapist assessing him that he could put the keys in the ignition of his car without leaning forward, which he was unable to do before training.

When asked how the games were useful to him BJ shared, "You're involved, doing something meaningful. Rather than sitting around and looking for something to do. For example, on a day like today, it's raining out, you will have activities which will get you involved. It is always a plus." This said, BJ rated the *value/usefulness* of the training program (at both collection points) relatively low compared to other subjects. When asked why, BJ stated "If a game is not useful it's because it is too easy, you know?"

Case 2: WA

WA is a 35-year-old man that started the training program six months after his stroke. He lives with his wife and three-year-old daughter in a private home, in a community with a median household income of $62,000. WA's highest level of education achieved is an MBA, and he was employed full time as a comptroller during the training program. WA uses computers extensively as part of his employment and during his leisure time. He regularly solves his information technology problems without professional support. He owns a smart phone and utilizes multiple mobile applications (apps) to manage his day. WA played computer games using a variety of gaming platforms, including mobile games, prior to his stroke. WA does not take medications for depression, and his medical history during the trial was positive for hypertension. He ran, lifted weights, rode a bicycle for exercise prior to his stroke, and rode a motorcycle for recreation. He is struggling to return to exercise since his stroke and expresses frustration at not being able to participate in all of the physical activities his daughter is capable of, particularly swimming. He exercised sporadically and participated in physical therapy during the training study period. This study was WA's first experience with rehabilitation research. He began participating in a different upper extremity training study prior to completing the home training study described in this chapter. WA's impairments are moderate. He is a limited community ambulator and he drives.

During the study WA averaged 25 (19) minutes of training per week making him the EMP group member with the lowest adherence. He accomplished this averaging 2.1 (1.3) sessions per week. Average session length was 9.9 (6.9) minutes. Peak participation occurred during week six with 51 training minutes. Lowest participation times were for weeks four and 12, with no training logged, and week 11 with a single 1.8-minute session (see Figure 13.2). WA's IMI composite score was 72 at pre-test and 64 at post-test. WA's subscale scores for *interest/enjoyment*, *effort/importance* and *value/usefulness* all decreased over the course of the training period. Subsequent to training, WA demonstrated an 8-point improvement in UEFMA score and a 1-point increase in ARAT score (see Figure 13.3).

WA expressed frustration with his formal therapy experience describing it as overly simple and that it "might have produced results in the 1980s." WA felt that he has accomplished his level of recovery to date by "thinking of ways to kind of think outside the box to challenge my body and my mind." When asked why he chose to participate in the study, WA expressed several ideas including that he felt that the study looked like an "out of the box" way to work toward getting better. He also thought the study might be a way to return to gaming, which was a source of pride for him before his stroke. WA considers himself a numbers guy and felt that game scores would allow him to see his "progress over a course of time" (and that this might) "be beneficial and motivate me to continue."

When asked if the system was easy or difficult to use he answered, "It's very self-explanatory, especially for me, given the fact that I'm very technical. So, you know, I learned rather easily." When asked about the most difficult aspect of using the system WA answered, "I would say user friendliness. The difficulty was sometimes the device would freeze. Sometimes it would freeze and then you'd have to restart it. It also didn't calibrate your hand. If you had less function in your hand at the beginning and then later on as you start to get a little bit more function it wouldn't recognize the hand. Then you had to have the team calibrate it. Sometimes it did not recognize that I was trying to move and there'd be a little bit of a delay. There were bugs, which I would expect as a result of the software being in its Beta form. So I wouldn't expect it to be like some of the other recovery games I am using."

WA expressed being particularly frustrated when trying to play the *Speed Bump* simulation. The game trains hand opening and closing in its lowest levels, and hand opening was particularly difficult for WA. He did not play *Speed Bump* more than once a week for more than six minutes after the third week of the study. He played the two other games more often and for longer periods of time. WA explained his reasons for leaving the *Speed Bump* behind stating: "It was very, very difficult and I wasn't really getting any feedback body-wise from it. I was trying to open and close [my hand] and it was just creating a lot of resistance. My body was resistant to what my mind was commanding it to do, so I didn't really want to waste my time."

When asked if he liked the training program WA responded affirmatively. "I love video games," he explained. "I have always played video games to reduce stress." According to WA, participating in the study "allowed me to not only enjoy playing video games, but it also allowed me to challenge my mobility. I challenged my body and my mind to work together. My mind was wanting to move a lot faster than my body. Through repetition, my body did its best to keep up. I would say this was very good for me. I can't say that it would be the same for everyone, but for me I think it was good."

When asked if he paid attention to game scores, WA stated "That's really what drove me. Especially for *Fruit Catch* and also the *Maze Runner* game. When I first played it, I was only able to get 200. Then after a while I was able to get 1000. If you're not really number-based you might not be motivated by that. I would say that that helped me at least to try. [I said to myself] Only a little, only 200 or 300 more. That helped a lot for me." When asked about emphasizing competition between players WA saw the potential for good and bad. "It's kind of like a double-edged sword, right? If you're really not motivated, if your recovery process is not going how it should, if you're looking at others and they are hitting it out of the park, you might say 'Wow, I should just give up.' But if you're motivated and you're seeing that overall everyone is getting about 1000 it would make me say 'Okay, I need to step my game up a little bit.' It really depends on your motivation." When asked with whom he would want his scores to be compared, WA chose "the person who is leading. I always try to finish first. I don't know if you've ever watched Talladega Nights, Will Ferrell said, 'The [person in] second place is the first loser,' so I always try to make sure I'm first."

WA's participation time decreased substantially over the last weeks of the trial. When asked why he stated, "One chapter was closing and another one was opening. I was gearing up to participate in a four-month study (at a large rehab hospital in the area). I also had a full-time job, acupuncture and therapy."

CASE 3: SK

SK is a 55-year-old man that started the training program 17 years after his stroke. He lives with his significant other, who is also a stroke survivor, in a townhome, in a community with a median household income of $91,000. SK's highest level of education achieved was received at a trade school. Prior to his stroke, he was the owner/operator of a tractor-trailer. He does not use a computer, does not have email or social media accounts and does not own a smart phone. When probed regarding his technology use he laughed and said, "I'm old-school." SK does not take medications for depression. His medical history during the trial was positive for hypertension and occasional pain medication use. He did not exercise prior to his stroke, but had a strenuous occupation, and has exercised and participated in rehabilitation intermittently since. He has participated in multiple trials of rehabilitation research interventions and did not participate in therapy during the study. SK's impairments were moderate as he is an unlimited community ambulatory and he drives.

During the study SK averaged 168 (52) minutes of training per week making him the EMP group member with the highest adherence. He accomplished this averaging 5.8 (1.2) sessions per week. Average session length was 29.4 (6.9) minutes. Peak participation occurred during week five with 246 training minutes. Lowest participation time was for week 11 with 96 minutes. SK never trained less than four times in a week and trained all seven days in four of the 12 weeks (see Figure 13.2). SK's IMI composite scores were 80 at pre-test and 81 at post-test. None of SK's subscale scores varied more than a point over the course of the training period. Subsequent to training, SK demonstrated a 4-point improvement in UEFMA score and a 7-point increase in ARAT score. SK was the only subject that demonstrated an improvement in the BBT, with an increase of 5 points (see Figure 13.3).

When asked about his experiences with therapy after his stroke SK expressed frustration and compared this experience to therapy that he had for a torn rotator cuff: "I'll tell you one thing, I had my shoulder redone and that was fun, I could see it change automatically. But as far as the stroke, it's frustrating and no matter how much I try, I don't, I didn't get the results and yeah, it's frustrating for me."

During the study period SK's significant other's mother was ill. He credited this situation as a factor in his extensive participation in training: "I had a lot of time. That is the reason. I had plenty of time on it. Every day she would have to go out and see her mom. She was taking care of her mother and I had free time and I was determined to do that."

The games that SK played the most were the *Speed Bump* and *Maze Runner* simulations, and he expressed enjoyment playing both. When asked why he liked the *Speed Bump*, he stated, "I guess I wanted to do better. To race. Yeah I liked to be challenged." When asked what he liked about *Maze Runner* he said, "It was a challenge. It was good. It was challenging because I had to move with my arm." This said, SK disliked the game he had the most difficulty, with the *urban aviator* simulation. When asked why he found this game difficult he stated, "I guess because I had to raise my arm, my elbow – that's always been problematic. Raising my arm is difficult for me. That was frustrating, and *Urban Aviator* asked me to do that a lot, raise my arm. And I couldn't do it as well as the other ones."

When asked if he would continue to use the games if he were no longer involved in the study SK was polite but less than enthusiastic. He said "It was intriguing; I have to say that. This would be great 20 years ago when I had the stroke." When asked why the frequency of training sessions decreased a bit (from six or seven sessions per week to four or five) over the last four weeks of the study, SK shared, "I guess, you know, I got a little frustrated because I wanted to see the results. I guess I burned out or something. I got discouraged…I have done this for 20 years. I get fucked up and after a while I get defeated – not defeated, but you know, it's like I need a break."

When the interviewer asked SK if he paid attention to the game scores he said, "Yes I did in the beginning, but after a while I wasn't concerned about the points as much. I guess I wanted to see how I could do better but then I really focused on doing it more correct."

When asked why his *effort and importance* scale scores were higher than the other subjects in the study SK stated, "I guess, I was trying to work with you. You are trying to work with me and I am trying to work with you. That's the best way I can put it." When asked why he felt he was better at playing the games by the end of the trial SK laughed and said repetition. When asked if playing the games made him feel better he stated that he felt the same.

CASE 4: DM

DM is a 56-year-old man that started the training program three years after his stroke. He lives with his significant other in a private home, in a community with a median household income of $98,000. DM's highest level of education achieved was received at a trade school. Prior to his stroke, he was an auto mechanic working for a large dealership and he used computers extensively at work. He owns a computer and uses it for searching, email and banking. DM does not take medications for depression. His medical history during the trial was positive for hypertension and diabetes. Before his stroke, he did not exercise, but had a strenuous occupation and did "all my own work on the house. Lawn work…. I mean you name it and I could do it." He has exercised regularly and participated in rehabilitation intermittently since his stroke, and participated in Physical Therapy during the study. DM's impairments are moderate to severe and is a limited community ambulator and he drives.

During the study DM averaged 55 (12) minutes of training per week making him the UCP group member with the lowest adherence. He accomplished this averaging 5.7 (1.2) sessions per week with the average session length at 9.9 (1.1) minutes. Peak participation occurred during week four and eight with 68 training minutes. Lowest participation time was for week 12 with 24 minutes. DM never trained less than four times in a week and trained all seven days in three of the 12 weeks (see Figure 13.2). DM's IMI composite scores were 70 at pre-test and 71 at post-test. Subsequent to training, DM demonstrated a 7-point improvement in UEFMA score and a 1-point increase in ARAT score (see Figure 13.3).

When asked why he participated in the study DM stated, "I'd do anything to get moving again." DM has had positive experiences with rehabilitation. He describes Physical Therapy as pretty good and states, "If it wasn't for insurance I would always go." DM had a regular early morning exercise routine which was prescribed by his physical therapist. After this, he did his home exercises from physical therapy, and finally the study simulations.

DM played each of the three simulations an average of three to five minutes per session. He performed the *Speed Bump* simulation more often than the other two but did not describe it as his favorite. He stated, "I mean it's like you do the car game (*Speed Bump*), and you just go down the same road, and you gotta do the same thing." He expressed that monotony was a significant issue multiple times by stating, "Well the one where you gotta climb the ladders...it was all black and white. And, uh, you sit there and do them but you would almost fall asleep. And they were...every day they were the same. So in other words, like it's not a program where, okay...you did this, you did [these] two levels, now you're up to your next level. You know, you do your levels, then you go right back to them again. The ladder game was boring and I mean, they're boring games. That was the major thing, why I stopped playing them. They're really boring. It was like playing that old pong system. Have you looked at them? Even when I was doing them I was falling asleep. A little more color. A little more action."

When asked why his *perceived competence* sub-scale scores increased, he responded, "Because it was easy. You're doing the same thing over and over and over again; it's not like you're doing anything different." When asked why his *value/usefulness* sub-scale scores were higher compared to other subjects he stated, "They're valuable in a way that they get you moving. Instead of sitting there watching the TV, you might be playing a game and moving your arm. That's what I meant by I value it more. I meant like I could watch TV or I could play the game. Well, the game helps me move my arm." When asked why his *effort/importance* sub-scale score decreased he answered, "When I got it I was 'Gung Ho,' then I realized that I could do it with no problem. Why am I doing it? I'd rather do something that was harder."

DM was frustrated by the results he experienced during the trial. "I was told this will make my arm better, and basically it freed it up a little bit, but it wasn't that great." He compared it to the excellent results he experienced getting acupuncture treatments for his stroke-affected arm: "I got acupuncture and my arm just dropped, it's back to almost normal position. I couldn't lift it up, but it wasn't against my chest."

DM suggested that a scaffold approach to difficulty might have made the games more interesting. "Every day they were the same. So in other words, it's not a program where...You did this, level, now you're up to your next level. [In these games] you do your levels, then you go right back to them again." DM also expressed that the system might have been more appropriate for someone early in their recovery after stroke stating, "When you first have a stroke, the game would be useful, but now it's not useful. You're in worse condition when it first happens. It wasn't early enough after my stroke. I was already too advanced." When asked if he would be willing to continue playing the games even if he wasn't participating in a study, he responded with an unqualified no.

DISCUSSION

An initial examination of these four cases should begin to suggest that maximizing adherence across a broad sample of subjects will not be achieved by designing an intervention that addresses a set of four or five diagnosis – or population-specific barriers to participation. None of the subjects cited that they did not understand how to use the system, and it was provided to them in their home, free of charge, which eliminated knowledge, access and cost barriers. Computer literacy did not seem to have the same impact on adherence as in other studies in the literature. The two highest adhering subjects reported limited computer literacy, while the two lower adhering subjects both described using computers professionally for many years. Age is frequently cited as a barrier to adherence in computer supported rehabilitation activities (Standen et al., 2015, Emmerson et al., 2018, Lam et al., 2015); however, the four subjects in this presentation do not conform to this pattern. BJ, the oldest subject in the study, used the system four times as much as WA, the youngest. The patterns identified in this series align more closely to those of a recent study of a system specifically designed to be user friendly for persons with stroke found which no correlation between compliance and age or computer literacy (Taub et al., 2013).

Motivation is another factor that impacts adherence (Standen et al., 2015, Rimmer et al., 2008). Methodologically, this is a complex construct to address from both an intervention and measurement standpoint. Our attempts to quantify the motivational impact of the two different training programs demonstrated questionable effectiveness. BJ, who demonstrated the second highest adherence scores and expressed almost 100% positive comments about the system, had the lowest pre-participation IMI score and the second lowest post-participation IMI. DM who had low adherence and almost 100% negative comments about the training program demonstrated the second highest IMI scores, and his post participation IMI did not go down. In addition to the poor agreement between adherence, IMI scores, and qualitative comments, three subjects demonstrated almost identical pre- and post-participation IMI scores. While stable IMI after a long training program could be interpreted positively, it also raises concern that there may be confounding factors regarding this measurement or questions regarding its validity. In addition, we feel that this underscores the need for open-ended qualitative data collection performed in a setting that increases the subjects' ability to share open and honest feedback about the intervention (i.e. a phone interview with an interviewer that did not participate in recruiting or delivery of the intervention).

Interestingly, all four subjects cited a strong drive to improve their function. Our two high-adhering subjects' prior participation in rehabilitation research might be a telling common factor. Extrinsic motivation (the drive to get "better") may have been stronger in these two subjects (Meldrum et al., 2012). Another interesting common factor shared by the two high-adhering subjects was their pre-stroke occupations. Both BJ and SK were self-employed prior to their strokes whereas WA and DM were employed by others. Current employment situation may have had an impact on adherence as well. WA, the lowest adhering subject, was the only subject in this series that was employed throughout the study. Employment is a clear example of the barrier "competition for time," and hours spent working outside the home will be a variable used to interpret our adherence rates in larger studies.

DM and WA were both more impaired than BJ and SK, suggesting that HoVRS might be better suited to less impaired subjects and result in better adherence in this group. This said, *Maze Runner* was designed based on programs of horizontal planar reaching tasks that have been associated with improvements in motor function by moderate to severely impaired persons with stroke. DM and WA's improvements in UEFMA score were consistent with these studies, and it suggests that at least this activity was "impairment appropriate" for them (Bosecker et al., 2010). WA shared that he stopped playing one of the simulations, *Speed Bump*, altogether because it was too difficult and frustrating for him. This may have had a negative impact on his overall adherence. Technology-supported rehabilitation interventions are often criticized for taking a "one intervention fits all" approach to rehabilitation (Dobkin, 2017). While this criticism might be true of technology-supported rehabilitation research (which often employs uniform interventions within a group to increase internal validity), the equipment and training program utilized in this study could be adapted to individualized needs in more clinical settings, outside of the confines of a controlled study (Fluet et al., 2012).

While all four subjects demonstrated "clinically significant improvements" in motor function, only BJ expressed satisfaction with the outcome and was able to point to functional activities that he performed better after participation. It is important to note that the HoVRS system is not well suited to the gross motor function goals described by WA, the dystonia reduction goals cited by DM, or the arm elevation goal cited by SK. This mismatch may have been a source of frustration for these subjects and negatively impacted the adherence of WA and DM. A patient's perception that an intervention is helping them and they are capable of improving their function meaningfully (self-efficacy) are both related to motivation and adherence in the rehabilitation literature (Dobkin and Carmichael, 2016). Several studies of technology-supported and traditional rehabilitation utilize a behavior modification approach in which a therapist helps the participant identify a *realistic* real-world goal to work toward during treatment. Factors that render this goal realistic include patient impairment level, age and acuity level, as well as the stimuli provided by the treatment matching the patient's goals (and vice versa). This approach to establishing a cognitive link between intervention and a patient-identified goal is associated with higher levels of perceived self-efficacy and higher levels of adherence (Borstad et al., 2018).

DM, who had the most negative comments about the system and the study, explained that the game did not change at all which made it "lame" and "boring." He cited changes in music or visual environments as a possible relief from this boredom. DM differs substantially from BJ, who used the system almost twice as much, and did not comment on the game's aesthetics at all. It is worth noting that tens of thousands of people play *Candy Crush*™, a game that does not vary. DM and BJ both cited the addition of competition as a change that might improve the impact of the system on motivation, these comments are in line with a newer study by Thielbar et al. that suggests that competition may have a positive impact on motivation and adherence (Thielbar et al., 2020). WA suggested that he would enjoy simulations that are more competitive, but backtracked by saying that other people might find competition demoralizing. The computer gaming industry has identified and exploited this individualized response to competition. *Call of Duty*™, a popular mass-marketed computer game, is presented in both a highly competitive (*Modern Warfare*™ developed by Infinity Ward) and a more aesthetically oriented format (*Black Ops*™ developed by Treyarch). Going forward, we are considering adding personality testing to our data collection plan in order to identify markers that will allow us to leverage the flexibility of computer games in order to configure personalized interventions in a manner that will optimize motivation and adherence in a larger number of persons with stroke.

The process of examining these cases has uncovered several aspects of the larger study that we will need to bolster for it to be effective. In addition, some themes have begun to emerge (the possible role of extrinsic motivation, a need to enhance the perception of self-efficacy, etc.) that we will attempt to confirm with a more formal qualitative study of the entire pilot cohort. We feel that the complex interplay of the interventions and individual subject characteristics might suggest that study of the enhancement of long-term behaviors in a heterogeneous clinical population like stroke may be too complex to meaningfully reduce to a well-controlled, a-priori designed, randomized clinical trial (or series of RCT), resulting in the need for mixed methods and/or more iterative approaches.

REFERENCES

Borstad, A. L., Crawfis, R., Phillips, K., Lowes, L. P., Maung, D., Mcpherson, R., Siles, A., Worthen-Chaudhari, L. & Gauthier, L. V. 2018. In-home delivery of constraint-induced movement therapy via virtual reality gaming. *Journal of Patient-Centered Research and Reviews*, 5, 6–17.

Bosecker, C., Dipietro, L., Volpe, B. & Krebs, H. I. 2010. Kinematic robot-based evaluation scales and clinical counterparts to measure upper limb motor performance in patients with chronic stroke. *Neurorehabilitation and Neural Repair*, 24, 62–69.

Deakin, A., Hill, H. & Pomeroy, V. M. 2003. Rough guide to the Fugl-meyer assessment: upper limb section. *Physiotherapy*, 89, 751–763.

Dobkin, B. H. 2017. A rehabilitation-internet-of-things in the home to augment motor skills and exercise training. *Neurorehabilitation and Neural Repair*, 31, 217–227.

Dobkin, B. H. & Carmichael, S. T. 2016. The specific requirements of neural repair trials for stroke. *Neurorehabilitation and Neural Repair*, 30, 470–478.

Emmerson, K. B., Harding, K. E., Lockwood, K. J. & Taylor, N. F. 2018. Home exercise programs supported by video and automated reminders for patients with stroke: a qualitative analysis. *Australian Occupational Therapy Journal*, 65, 187–197.

Fluet, G. G., Merians, A. S., Qiu, Q., Lafond, I., Saleh, S., Ruano, V., Delmonico, A. R. & Adamovich, S. V. 2012. Robots integrated with virtual reality simulations for customized motor training in a person with upper extremity hemiparesis: a case report. *Journal of Neurologic Physical Therapy*, 36, 79.

Fluet, G. G., Qiu, Q., Patel, J., Cronce, A., Merians, A. S. & Adamovich, S. V. 2019. Autonomous use of the home virtual rehabilitation system: a feasibility and pilot study. *Games for Health Journal*, 8, 432–438.

Harris, J. E. & Eng, J. J. 2004. Goal priorities identified through client-centred measurement in individuals with chronic stroke. *Physiotherapy Canada*, 56, 171.

Jurkiewicz, M. T., Marzolini, S. & Oh, P. 2011. Adherence to a home-based exercise program for individuals after stroke. *Topics in Stroke Rehabilitation*, 18, 277–284.

Kairy, D., Lehoux, P., Vincent, C. & Visintin, M. 2009. A systematic review of clinical outcomes, clinical process, healthcare utilization and costs associated with telerehabilitation. *Journal of Rehabilitation Research and Development*, 31, 427–447.

Kissela, B. M., Khoury, J. C., Alwell, K., Moomaw, C. J., Woo, D., Adeoye, O., Flaherty, M. L., Khatri, P., Ferioli, S. & La Rosa, F. D. L. R. 2012. Age at stroke temporal trends in stroke incidence in a large, biracial population. *Neurology*, 79, 1781–1787.

Kwakkel, G. 2006. Impact of intensity of practice after stroke: issues for consideration. *Disability and Rehabilitation*, 28, 823–830.

Lam, M. Y., Tatla, S. K., Lohse, K. R., Shirzad, N., Hoens, A. M., Miller, K. J., Holsti, L., Virji-Babul, N. & Van Der Loos, H. F. M. 2015. Perceptions of technology and its use for therapeutic application for individuals with hemiparesis: findings from adult and pediatric focus groups. *JMIR Rehabilitation and Assistive Technologies*, 2, e1.

Lang, C. E., Macdonald, J. R., Reisman, D. S., Boyd, L., Kimberley, T. J., Schindler-Ivens, S. M., Hornby, T. G., Ross, S. A. & Scheets, P. L. 2009. Observation of amounts of movement practice provided during stroke rehabilitation. *Archives of Physical Medicine and Rehabilitation*, 90, 1692–1698.

Laver, K., George, S., Thomas, S., Deutsch, J. E. & Crotty, M. 2012. Virtual reality for stroke rehabilitation. *Stroke*, 43, e20–e21.

Ma, V. Y., Chan, L. & Carruthers, K. J. 2014. Incidence, prevalence, costs, and impact on disability of common conditions requiring rehabilitation in the United States: stroke, spinal cord injury, traumatic brain injury, multiple sclerosis, osteoarthritis, rheumatoid arthritis, limb loss, and back pain. *Archives of Physical Medicine and Rehabilitation*, 95, 986-995.e1.

Mathiowetz, V., Volland, G., Kashman, N. & Weber, K. 1985. Adult norms for the box and block test of manual dexterity. *American Journal of Occupational Therapy*, 39, 386–391.

Mcauley, E., Duncan, T. & Tammen, V. V. 1989. Psychometric properties of the intrinsic motivation inventory in a competitive sport setting: a confirmatory factor analysis. *Research Quarterly for Exercise and Sport*, 60, 48–58.

Meldrum, D., Glennon, A., Herdman, S., Murray, D. & Mcconn-Walsh, R. 2012. Virtual reality rehabilitation of balance: assessment of the usability of the Nintendo Wii((R)) Fit Plus. *Disability and Rehabilitation: Assistive Technology*, 7, 205–210.

Mihelj, M., Novak, D., Milavec, M., Ziherl, J., Olenšek, A. & Munih, M. 2012. Virtual rehabilitation environment using principles of intrinsic motivation and game design. *Presence: Teleoperators and Virtual Environments*, 21, 1–15.

Popović, M. D., Kostić, M. D., Rodić, S. Z. & Konstantinović, L. M. 2014. Feedback-mediated upper extremities exercise: increasing patient motivation in poststroke rehabilitation. *BioMed Research International*, 2014, 520374.

Qiu, Q., Ramirez, D. A., Saleh, S., Fluet, G. G., Parikh, H. D., Kelly, D. & Adamovich, S. V. 2009. The New Jersey institute of technology robot-assisted virtual rehabilitation (NJIT-RAVR) system for children with cerebral palsy: a feasibility study. *Journal of Neuroengineering and Rehabilitation*, 6, 40.

Rimmer, J. H., Wang, E. & Smith, D. 2008. Barriers associated with exercise and community access for individuals with stroke. *Journal of Rehabilitation Research and Development*, 45, 315.

Simpson, D. B., Breslin, M., Cumming, T., De Zoete, S., Gall, S. L., Schmidt, M., English, C. & Callisaya, M. L. 2018. Go home, sit less: the impact of home versus hospital rehabilitation environment on activity levels of stroke survivors. *Archives of Physical Medicine and Rehabilitation*, 99, 2216–2221.e1.

Standen, P. J., Threapleton, K., Connell, L., Richardson, A., Brown, D. J., Battersby, S., Sutton, C. J. & Platts, F. 2015. Patients' use of a home-based virtual reality system to provide rehabilitation of the upper limb following stroke. *Physical Therapy*, 95, 350.

Taub, E., Uswatte, G., Mark, V. W., Morris, D. M., Barman, J., Bowman, M. H., Bryson, C., Delgado, A. & Bishop-Mckay, S. 2013. Method for enhancing real-world use of a more affected arm in chronic stroke: transfer package of constraint-induced movement therapy. *Stroke*, 44, 1383–1388.

Thielbar, K. O., Triandafilou, K. M., Barry, A. J., Yuan, N., Nishimoto, A., Johnson, J., Stoykov, M. E., Tsoupikova, D. & Kamper, D. G. 2020. Home-based upper extremity stroke therapy using a multiuser virtual reality environment: a randomized trial. *Archives of Physical Medicine and Rehabilitation*, 101, 196–203.

Timmermans, A. A., Seelen, H. A., Willmann, R. D. & Kingma, H. 2009. Technology-assisted training of arm-hand skills in stroke: concepts on reacquisition of motor control and therapist guidelines for rehabilitation technology design. *Journal of Neuroengineering and Rehabilitation*, 6, 1.

Yozbatiran, N., Der-Yeghiaian, L. & Cramer, S. C. 2008. A standardized approach to performing the action research arm test. *Neurorehabilitation and Neural Repair*, 22, 78–90.

14 Virtual Reality Game-Based Exercises with Lead Motion Applied to Developmental Disorders

Ana Grasielle Dionísio Corrêa, Natália Regina Kintschner and Silvana Maria Blascovi-Assis
Mackenzie Presbyterian University
São Paulo, Brazil

CONTENTS

INTRODUCTION

Developmental Disorders context covers the study of several disabilities, whether they are due to physical, intellectual, sensory or language, and behavioral characteristics. Among these most common deficiencies, we highlight [DSM-5]: Intellectual Disability, Communication Disorders (language and speech), Autism Spectrum Disorder (ASD), Attention Deficit Hyperactivity Disorder (ADHD), Specific Learning Disorders, Motor Disorders, and also emotional and behavioral problems.

New technologies to encourage positive changes in developmental disorders often appear in the literature (Pavão et al., 2014; Tang et al., 2015; Garcia-Zapirain et al., 2017; Capelo et al., 2018). Thus, health professionals should be alert to know how to take advantage of these technologies to meet the society demands that is increasingly immersed in information and communication networks, whether as a way of communicating with other health professionals and patients but also as an intervention and an assessment tool.

There is a demand for computational tools that can facilitate and contribute to the area of Developmental Disorders in the context of their assessments and interventions. Virtual Reality (VR) is one of the technologies that has been widely used, especially in pediatric populations (Pavão et al., 2014; Garcia-Zapirain et al., 2017), with emphasis on those with ADHD, Autism, and Cerebral Palsy (CP). This is because children with developmental disorders are diagnosed and treated based on behavioral or motor criteria, which are particularly appropriate for VR interventions.

VR technology has advanced rapidly in the last two decades (Sherman and Craig, 2018). Today's computers are much faster and the quality of display devices, such as Head-Mounted Displays (HMD), has improved a lot. Besides, the cost of technology has significantly reduced, and consequently, more applications are being developed and disseminated in various areas such as health, education, entertainment, advertising, etc. In this sense, as the technology becomes affordable and enhances in terms of interaction, the effectiveness of VR as a therapeutic tool also increases (Burdea et al., 2011; Carvalho et al., 2014; Pavão et al., 2014; Wiederhold and Riva, 2019).

In the field of neurological rehabilitation, VR and interactive video games are beginning to be accepted as adjuvants therapeutic tools in the treatment of neurological patients, through real-time simulation and multiple channels, providing the opportunity to perform repetitive and rewarding activities (Oña et al., 2018). Commercial video game consoles, such as Nintendo Wii, PlayStation, and Microsoft Kinect, for example, are being rapidly adapted in therapeutic areas such as Physiotherapy and Occupational Therapy (Burdea et al., 2011; Carvalho et al., 2014; Tang et al., 2015). These video games have sensors that virtualize the user's body movements and translate them into actions in games, stimulating more global movements such as head, torso, legs, and arms inclination (Burdea et al., 2011; Carvalho et al., 2014).

The effects of this therapy model provide an increase in coordination activities of the upper limbs, manual dexterity, balance, and postural control (Menezes et al., 2015; Chen et al., 2018). However, the games are not designed for the execution of specific movements for wrists and fingers. This therapeutic objective can be solved with other types of technologies, such as hand motion detection sensors. There are several types, such as data gloves, which have force and feedback sensors (Rose et al., 2018). Nevertheless, these gloves are still expensive solutions for personal use and require training and dedicated equipment to use. Besides, users need to wear these gloves, which may not be compatible with all shapes and sizes of hands and fingers, requiring time for calibration, in addition to weight and the need for hygiene.

A portable, low-cost solution and independent from manual contact for use came to market in 2013. It is the Leap Motion device, a hand movement sensor, which works without physical contact and can be connected to the computer via the USB (universal serial bus) port (Vasconcelos and Aguiar, 2017). The device's sensor is on the table surface (next to the keyboard) and, when activated, it captures the gestures of the user's hands and fingers very precisely and transmits them to the computer. It is possible to develop applications in which the user has to interact with virtual objects, for example, moving an object from one place to another. There is also the advantage of being able to add Leap Motion to a VR headset (VR Gear, for example), enabling developers to build games and immersive virtual environments. Because it is a low-cost and portable technology, it can be easily used for data collection outside traditional laboratory or clinic environments, such as in schools and even at the patient's home (Niechwiej-Szwedo et al., 2018).

Leap Motion, despite its recent insertion in the market, has already been investigated as a technological resource to support motor rehabilitation interventions of upper limbs, since it allows capturing finer movements, which are essential for the rehabilitation of manual dysfunctions found in several conditions (Zhu et al., 2015; Niechwiej-Szwedo et al., 2018; Oña et al., 2018).

However, the literature consulted is far from exhausting the possibilities of using Leap Motion with groups of people with special needs, such as Down Syndrome (DS), ADHD, ASD, CP, among others. The applications of this technological resource can be quite diverse and meet the needs of stimulation to the characteristics and peculiarities of each clinical situation. Thus, neuro psychomotor aspects, as well as those of communication, learning or social interaction may appear, alone or in combination, among the objectives outlined for the use of Leap Motion, which, in this way, constitutes a resource that can be used in the interdisciplinary context.

This chapter aims to present Leap Motion and the possibilities of using this device in developmental disorders. The following section presents the characterization of the equipment and its forms of use (immersive and non-immersive). Then, works that use the device as a tool in rehabilitation programs with populations with ADHD, ASD, DS, and CP are described and some works that are in progress by teachers and students of the undergraduate course in Computer Science and the Postgraduate Program in Developmental Disorders at Universidade Presbiteriana Mackenzie. Finally, the final considerations of the chapter are presented.

LEAP MOTION CONTROLLER

The Leap Motion Controller (LMC) is an optical sensor device that allows capturing hand and fingering movements with sufficient precision. The sensor is connected to the computer via USB and was designed to be placed on the table, next to the keyboard in order to capture the movements of both hands (Figure 14.1).

The Leap Motion is a compact device, 8 cm wide by 3 cm high. The upper part of the device is made of smoked glass in order to hide the two image sensors and infrared LEDs (light emitting diodes) that work together to track the movements of the user's hands, while the base is rubberized to prevent it from sliding over the table. Leap Motion works with precision up to 1/100 mm and no visible latency in its field of view. The viewing range of Leap Motion is 60 cm above and around the device. This limit is due to the spread of LED light through space since it is difficult to infer the position of the hand from a certain distance.

It is possible to use Leap Motion connected directly to the computer in non-immersive experiments (Figure 14.2a). In this case, the video monitor is used as a display device. In immersive experiments, Leap Motion is used coupled to a pair of VR glasses, such as a VR Gear, for example (Figure 14.2b).

FIGURE 14.1   Connecting the leap motion device to a laptop (Leap Motion, 2020).

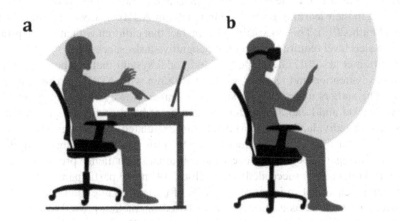

FIGURE 14.2   (a) Non-immersive leap motion; (b) immersive leap motion (Leap Motion, 2020).

Leap Motion allows measuring motor performance, such as reaction time, bimanual coordination, and the sequence of movements performed with the hands and fingers.

For this reason, this remote sensing technology has shown promise for the rehabilitation field, as it does not require the patient to use any motion detection device (for example, gloves with force and feedback sensors). Therefore, it provides a new way of interaction between the user and the computer, allowing a more natural and touchless interaction. Hand dexterity in patients with upper limb motor dysfunctions can be assessed using scheduled tasks with graphic objects added in the virtual world.

THE USE OF LEAP MOTION IN DEVELOPMENTAL DISORDERS

Studies on the use of Leap Motion can benefit children, young people, and adults with developmental disorders, associated or not, with cognitive or behavioral changes, with a wide range of activities in an educational or clinical environment (Ebner and Spot, 2015). The use of this tool can also be characterized by interdisciplinarity since the development of applications can occur in partnership among professionals in the areas of education, health, and computing. In the next subsections, we present some works involving VR and Leap Motion that have been conducted with people with ADHD, ASD, DS, and CP.

ATTENTION DEFICIT HYPERACTIVITY DISORDER (ADHD)

ADHD is one of the most common neurodevelopmental disorders of childhood (Goulardins et al., 2017; Neto et al., 2015). It is a brain disorder characterized by a continuous pattern of inattention and/or impulsivity due to hyperactivity that affects the development (Santos and Vasconcelos, 2010). In this syndrome, there are secondary symptoms associated with conduct disorders, emotional difficulties, and perceptual-visual-spatial skills (Trujillo-Orrego et al., 2012), which makes it difficult to perform fine hand movements (González et al., 2012), having an impact on learning and performance school and generally remain in adolescence and later in adulthood (Santos and Vasconcelos, 2010). Along with learning challenges, these children have problems when interacting with other peers due to a lack of attention to what is happening around them or what others are saying (Barkley, 2003). It is estimated that 5–15% of schoolchildren suffer from this deficiency (APA, 2003) and in adulthood, 60% of cases (Capelo et al., 2018).

In recent years, entirely new ways of interacting with information have become available due to a series of technological advances that make use of body gestures and eye movements. These new technologies are being increasingly used for the development of applications for teaching and support purposes. The use of these new learning tools and devices suggests that children are more actively involved in their learning tasks. Children without ADHD had worse results in the experiment (Yu and Smith, 2013). This is explained by the fact that children with attention problems need to maintain the alert level required to meet their cognitive tasks successfully.

Garcia-Zapirain et al. (2017) developed a multimodal system that promotes hand-eye coordination to improve attention and learning skills of children with ADHD during arithmetic learning activities. The authors united two physiological sensors: the eye-tracking device called Tobii X1 Light Eye Tracker and Leap Motion for recognition of hand gestures. Eye-tracking is a technique by which an individual's eye movements are calculated to find out where the person is looking at a particular time and the sequence in which his eyes are moving from place to place (Yamamoto and Nakagawa, 2002). The eye-hand coordination challenge proposed in the exercises (on mathematical operations) succeeded utterly (100% of the 19 participants). It proved that eye-hand coordination helps a lot to improve the skills of users with attention deficits. Although the tracking devices are still not as accurate, gesture-based interaction proved to be a great alternative compared to handwriting or keyboard and mouse, which were considered more boring and consequently distracting. The most important contribution of this work focuses on building a dual system to improve the knowledge and skills of children who have attention and learning difficulties, developing an application based on arithmetic games that uses an eye tracker and a manual tracker to interact with him.

Capelo et al. (2018) proposed a multisensory virtual game for the development of fine motor skills of children or adolescents aged 7–12 years with ADHD to attract their attention. The authors used Leap Motion as a means of interaction in the virtual environment to obtain higher concentration (attention) and motivation (differentiated interaction from the mouse and keyboard standard). The environment consists of three containers of different colors (red, green, and yellow), nine geometric figures (three spheres, three squares, and three rectangles), three for each color. Each container must have three distinct geometric figures, maintaining the respective color. The child must interact with the hands, using the Leap Motion device and concentrating on placing each object inside its respective containers. Filling each of the containers successfully, the child should list each of the colors as well as consider the corresponding figure for each of the containers, taking into account that each container receptacle has a hole of the size required to enter only the corresponding object, which is essential to improve concentration in children with ADHD. The study involved twenty children (ten with ADHD and ten without ADHD). Both groups maintained a high level of concentration and motivation during the interventions carried out over a week.

Cornejo and Martinez (2016) argue that technology itself is an attraction to keep children involved in the task, but only as long as it is new to them. Therefore, it is necessary to design diversified and good quality technological and instructional content to keep children involved in attention, cognition and memory skills. Based on this, the authors designed and implemented an educational software tool for children with ADHD to assess their attention and frustration levels while performing a short-term memory activity. It works like a memory game, where cards are presented to the player for a few seconds, and the child must select the cards shown previously immediately after the cards disappear. The digital mode (DF) was played with the Leap Motion device and the manual mode (FM) was played with tangible game cards. The authors used electroencephalography (EEG) to assess the emotional state of children as frustration, excitement and engagement. The results indicate notable differences in the medians of the frustration levels in both modalities. However, the Leap Motion modality presented a higher median of frustration than the manual activity (FM = 2.0 and DF = 4.0). The frustration, according to the authors, may be related to the fact that Leap Motion did not work as children expected, creating these significant levels of frustration. Even with these levels of frustration, the excitement levels were significantly higher with Leap Motion technology than with manual activity. This implies an interesting scenario in which children are frustrated with new digital learning activities, but they are also more excited and motivated to complete them.

In the three presented studies, there were reports of malfunctioning of Leap Motion. This is because the device is limited to specific movements and also depends on the position in which it is used (on the table or connected to a headset). However, we believe that children can adapt to the mode of technological interaction over time and with dynamic learning activities, it is possible to maintain levels of excitement and lower levels of frustration in a longitudinal study are expected. In short, a digital version with interactions of playful excitement may be the best option for educational and digital processes of children with ADHD.

Autism Spectrum Disorder (ASD)

ASD is a complex neurological development disorder that occurs in the first three years of life, characterized by central deficits in social communication, as well as repetitive patterns and restrictions in behavior, interests and activities (American Psychiatric Association, 2003). Many children with autism demonstrate severe impairments to reciprocal interaction, significantly affecting their joint attention (Kasari, Freeman & Paparella, 2006) and collaborative interaction and play (Machalicek et al., 2009), which reduces their ability to participate to meaningful learning, social interactions and leisure (Case-Smith & Arbesman, 2008).

Studies have shown that students with ASD prefer the use of electronic devices during the learning process due to the possibility of computer technology to provide students with a more predictable, natural, and less intrusive learning environment (Sansosti et al., 2015). The researchers suggested that students with ASD are more likely to interact with electronic devices than by personal contact since they have difficulties in understanding stimuli from the social context (Salter, Davey, & Michaud, 2014). In contrast, the general understanding of the effectiveness of using games activated by the Leap Motion device to train children with ASD is limited due to a small number of empirical studies.

Tang, Falzarano & Morreale (2018) developed a simple game where players can use hand or finger gestures to draw on the screen. The purpose was to examine whether the gesture-based manipulation with the fingers, using Leap Motion, would bring more fun and usability (in terms of ease of use) for children. The game supports just two simple actions: drawing with hand-finger aerial gestures and cleaning it with the gesture of a circle. Five children with Autism participated in the experiment, along with their parents. Both the children and their families showed high involvement in the drawing game and their level of attention was high throughout the game session. All children showed high levels of constant attention and, most often, while playing, they interacted with the help of their parents and their eyes remained focused on the visual stimulus.

A study by Zhu et al. (2015) was divided into two phases. In phase A, called the baseline, the child should place colored balls in their respective containers. In phase B, called the intervention phase, the children had daily training to place the balls in the corresponding containers, with the help of verbal instructions from the researcher. The results during phase A (baseline) showed that the children were not able to put the balls in the correct containers; however, after phase B (intervention), they performed the task with 100% accuracy. The results indicate that the game itself is not extremely difficult, but the hand skills for the 3D environment need training and encouragement.

Hu & Han (2019) investigated the efficacy of using gesture-based instructions to teach correspondence skills to schoolchildren with autism in China.

Syahputra et al. (2018) carried out a study based on a social history, which describes social interactions, situations, skills, concepts, and behaviors common to society in order to share accurate, convincing, and easy to understand social information for children with autism. This study was done through augmented reality and leap motion device in three autistic children. When using the system, they were accompanied and guided by a therapist who completed a questionnaire containing some questions and statements. Based on the results, all respondents chose to agree that if the social story app can be used as a learning means for children with autism, it helps mentors in the learning process, being an attractive, interactive, and easy to play game.

Down's Syndrome

Although children with DS follow the same stages of motor development as children without the syndrome, they have a delay in the acquisition of skills when compared to their neurotypical peers (Gutiérrez et al., 2019). Few studies were found involving the use of the Leap Motion device to people with DS. Among the studies that reported programs with the help of this device, it was observed that there are specific difficulties of this public concerning sensor usability. The difficulty in maintaining the proper distance of the hand to perform the proposed activities in the game seemed to be related to the fact that the player does not have an object in his hand, requiring constant mediation by the researchers (Pelosi et al., 2019).

Leap Motion was a tool in a study that analyzed some specific gestures performed by people with DS, comparing them to pairs without the syndrome. The authors made recommendations on how to develop inclusive 3D interfaces for individuals with DS. They alerted to the fact that usability studies are essential for checking the use of the equipment for different populations. They also observed that, in the case of DS, participants do not always follow the instructions provided. They conclude that individuals with DS can effectively use digital devices and that VR technology can be part of their daily interactions, considering their peculiarities (Gutiérrez et al., 2019).

CEREBRAL PALSY (CP)

CP, which is characterized by chronic non-progressive encephalopathy, whose motor sequel can impair various functions, may or may not be associated with other developmental changes, such as hearing or visual impairment, communication and language difficulties, and/or cognitive deficits (Rosenbaum et al., 2007). To the treatment adherence be maintained, motivating alternatives has long been sought, especially because CP commonly requires monitoring and prolonged physiotherapy. Leap Motion appears, then, as a possibility of a treatment program for this population, since the rehabilitation of the fine motor skills of the patient with CP is traditionally one of the main areas of concern for occupational and physical therapists (Batista et al., 2016).

The upper limb rehabilitation strategies aim at promoting motor recovery, being carried out through specific and repetitive exercises. The use of technology in rehabilitation is adequate to comply with these principles since it can provide exercises in a controlled, repetitive, intensive, interactive and motivating manner (Tarakci et al., 2019). The use of video games was considered more pleasant and preferable to conventional therapy because, together with the LMC, they can provide practice-oriented tasks as well as visual and auditory feedback on performance and gain, which motivates and involves even more to increase the intensity rehabilitation.

Alimanova's study (2017) presents a game proposal for manual function training, through the performance of different VR tasks using the LMC device, related to activities of daily living, such as picking up, moving and holding items present in a home. The main objective of the study was to introduce a simple and direct implementation of the LMC in the rehabilitation process. In conclusion, it is believed that the LMC technology proves to be beneficial in addition to being a tool capable of promoting the opportunity to see different applications of human–machine interaction in the field of activities of daily living due to its compact size, high performance, and accuracy.

In another study, carried out by Oliveira et al. (2016), individuals with CP and deficits in psychomotor development performed six activities present in conventional treatment in a virtual environment. The integration of the virtual environment with the LMC proved to be quite feasible when applied to the rehabilitation of patients with CP. Experts were highly optimistic about the inclusion of the use of the virtual environment, in addition to traditional treatments, as an alternative playful tool for cognitive and motor rehabilitation in this public.

Shah et al. (2019) presents a low-cost system that can provide fine motor rehabilitation through a rhythm-based game using the LMC. The device allowed the adaptability of the game so that it could be played by different movement ranges, showing promising preliminary results for the rehabilitation of various conditions, including the CP. As Postolache et al. (2017) designed and implemented a game for the VR scenario, in which an important work was done in relation to the interaction of the LMC, considering the customization of the game according to the needs of each patient. Future work involves additional testing with an extended number of users who are undergoing rehabilitation processes.

PPGDD PROPOSALS FOR THE USE OF LEAP MOTION FOR MOTOR REHABILITATION

To study the possibilities of applying virtual games using Leap Motion to stimulate manual functions, the research group linked to the Postgraduate Program in Developmental Disorders (Pós-Graduação em Distúrbios do Desenvolvimento – PPGDD) and the Computing and Informatics College (Faculdade de Computação e Informática – FCI), developed a project, with financial support from the Mackenzie Fund Research (MackPesquisa), with interdisciplinary aims to work with the population with CP. The PPGDD and FCI research groups are made up of professors and students, both in graduate and undergraduate courses, who may come from the areas of physiotherapy, physical education, occupational therapy, information technology, and engineering, which set goals for motor performance always aiming at the integration and inclusion of participants in social life.

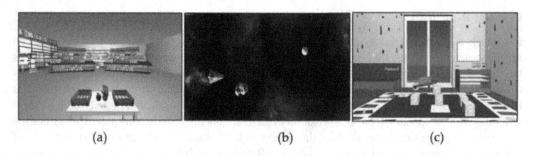

(a)   (b)   (c)

FIGURE 14.3   Screens of games created by PPGDD.

The group's objective was, initially, to create a motor self-rehabilitation system for upper limbs, based on VR games and controlled by wearable technologies (Leap Motion and integrated headset) for use in a therapeutic environment with CP clients. In a pilot study, the group developed three games that stimulate hand movements. In game 1 "Groceries" (Figure 14.3a), the player is encouraged to make moves to pick up and drop green and red apples into the fruit baskets. Red apples should be placed in the red basket, while green apples in the green basket. In game 2: "Space Invasion" (Figure 14.3b), the player must perform open and close hands movements to control a spaceship and avoid the meteors in space and collect energy. In game 3 "Stack Blocks" (Figure 14.3c), the player must stack a set of blocks with open and close hand movements. In all games, the therapist can configure and customize the game according to the specific needs of each participant, for example, selecting the order of hands to interact with the games, the time of each gameplay, and the mode of interaction (immersive or non-immersive). It is possible, for example, to start the game Groceries with the right hand. During three minutes of playing in immersive mode, passing to the left for another three minutes, and alternating hands for another three minutes.

The games are edited with verbal commands and with a brief written explanation on the screen to facilitate the understanding of the proposed tasks. The physiotherapists in the group defined the movements that can be worked on. The immediate feedback is due to the recording of the number of hits, errors and playing time in each level for each game, which is calculated by a mobile application in which it is possible to monitor the patient's progress. The choice of the complexity of the tasks is the therapist´s responsibility who conducts the session and it can be scheduled to be performed at home or in a therapeutic environment.

Games and support systems were designed and tested by the team, through necessary adjustments to meet the demands of physiotherapists to exercise hands. The movements of pronation, supination, flexion and extension of elbows, wrists, and fingers were prioritized.

The games were tested in a group of five adults with CP (aged among 26 and 43 years) at the Associação Nosso Sonho de Habilitação e Reabilitação de Pessoas com Deficiência, based in the city of São Paulo, to identify difficulties and interests. The intervention program occurred after the pilot study and the incorporation of some suggestions for technical adjustments to improve the usability of the games and the AppTherapy. The meetings took place twice a week, for two months, totaling 15 sessions. Each session lasted about 15–20 minutes and performed two games per day. In each session, two games were played, one day being immersive (headset version) and the other non-immersive (desktop version). At the end of each session, the participant gave a statement, which was recorded in a field diary by the researcher and considered for interface adjustments of the games that were necessary.

All participants had an important role in pointing out technical improvements in the games. Graphics color changes, auditory and visual feedback, difficulties with locking the games caused by the Leap Motion sensor, and layout changes were referred to as points to be improved for better execution of the proposed tasks.

All participants reported symptoms of tiredness or discomfort in any of the games and in most of them, there were moments of frustration, which occurred due to the specific difficulties of each of them. A usability study of Leap Motion carried out by King et al. (2018) showed that the device causes more significant fatigue in fingers, wrists, shoulders and neck. However, it was more preferred over other inputs, such as a mouse, due to the novelty of the technology. Only one of the participants experienced discomfort (vertigo and dizziness) (Venkatakrishnan et al., 2019) after using the VR headset. All the others played the games in both ways without complications. Preferences varied according to the adaptations of each one. All participants reported feeling an improvement in their performance and despite the frustrations encountered, they would like to continue playing in order to overcome them.

Overall, all participants were satisfied with the experience they had with Leap Motion. Some reports and adjustments are noteworthy since they were prominent in the intervention process both for the technical part and for the rehabilitation, as an advantage of using VR is the ease of adjusting the difficulty of the game.

FINAL CONSIDERATIONS

There is still a lot to research about the use of the Leap Motion device for rehabilitation, for example, its facilities and difficulties in use and application. The studies presented in this chapter show how the LMC has been a promising device in helping to improve the functional and cognitive performance of children with disorders related to developmental disorders. It may be a result of this technology that offers a more stimulating means of rehabilitation, since it is possible to integrate it with games, thus making the therapeutic interventions less tedious and more practical than conventional treatment for this public.

It is worth mentioning the limitations of the tool in terms of calibrating the speed and range of arm movements. As it deals with commercial software, many software components that make up its interface are closed like a black box; therefore, it is not possible to access them for changes and/ or adjustments. However, the functional tests performed so far with the use of the three presented games, have shown promise for use in the rehabilitation of upper limbs.

It is important to highlight that the device needs constant calibration for gesture detection, getting a challenge for future research to simplify the calibration processes to improve or recognize hand movements (Marin, Dominio & Zanuttigh, 2016).

ACKNOWLEDGMENTS

Thanks to the research sponsor Mackpesquisa, Project N. 181033, cojur-fmp-0286/2018 and PROEX N° 1133/2019. Thanks to the patients attending the Associação Nosso Sonho de Habilitação e Reabilitação de Pessoas com Deficiência who freely consented to participate in the tests with the games and their respective responsible ones.

REFERENCES

Alimanova, M., et al. (2017). 'Gamification of hand rehabilitation process using virtual reality tools: using leap motion for hand rehabilitation', In: *1th International Conference on Robotic Computing (IRC)*, pp. 336–339. doi:10.1109/IRC.2017.76.

American Psychiatric Association (2013). 'Diagnostic and statistical manual of mental disorders (DSM-5®)', In: *DSM–IV. Revised Text*. Masson, Barcelona.

Barkley, R. A. (2003). 'Issues in the diagnosis of attention-deficit/hyperactivity disorder in children', In: *Brain and Development*, 25(2), pp. 77–83. doi: 10.1016/S0387-7604(02)00152-3.

Batista, T. V., et al. (2016). 'Evaluating user gestures in rehabilitation from electromyographic signals', *IEEE Latin America Transactions*, 14(3), pp. 1387–1392. doi: 10.1109/TLA.2016.7459625.

Burdea, G., et al. (2011). 'Long-term hand tele-rehabilitation on the playstation 3: benefits and challenges', *IEEE Engineering in Medicine and Biology Congress (EMBC)*, Boston, MA, pp. 1835–1838. doi: 10.1109/IEMBS.2011.6090522.

Capelo, D. C., et al. (2018). 'Multisensory virtual game with use of the device leap motion to improve the lack of attention in children of 7–12 years with ADHD', In: *International Conference on Information Theoretic Security*, Springer, Cham, pp. 897–906.

Carvalho, B. A., Carrogi-Vianna, D., & Blascovi-Assis, S. M. (2014). 'Influência do uso do Nintendo® Wii- na destreza e na força de preensão manuais: estudo de caso na distrofia muscular de Becker', *ConScientia e Saúde*, 13(1), pp. 141–146. doi: https://doi.org/10.5585/conssaude.v13n1.4438.

Case-Smith, J., & Arbesman, M. (2008). 'Evidence-based review of interventions for autism used in or of relevance to occupational therapy', *American Journal of Occupational Therapy*, 62, 416–429. doi: 10.5014/ajot.62.4.416.

Chen, Y. P., Fanchiang, H. D., & Howard, A. (2018). 'Effectiveness of virtual reality in children with cerebral palsy: a systematic review and meta-analysis of randomized controlled trials', *Physical Therapy*, 98(1), pp. 63–77. doi: https://doi.org/10.1093/ptj/pzx107.

Cornejo, R., & Martinez, F. (2016). 'Exploring digital and manual modalities in educational activities for children with ADHD', *Research in Computing Science*, 129, pp. 37–44.

Ebner, M., & Spot, N. (2015). 'Game-based learning with the leap motion controller', In: Russell, D., & Laffey, J. M. (Eds.), *Handbook of Research on Gaming Trends in P-12 Education*, IGI Global: USA, pp. 555–565. doi: 10.4018/978-1-4666-9629-7.ch026.

Garcia-Zapirain, B., de la Torre Díez, I., & López-Coronado, M. (2017). 'Dual system for enhancing cognitive abilities of children with ADHD using leap motion and eye-tracking technologies', *Journal of Medical Systems*, 41(7), p. 111. doi: 10.1007/s10916-017-0757-9.

González, G. L., et al. (2012). 'Fine motor skills in attention deficit disorder with hyperactivity', *Cuban Journal of Neurology, Neurosurgery*, 3(1), pp. 13–17. doi: 10.1017/S0012162206000375.

Goulardins J. B., Marques J. C., & De Oliveira J. A. (2017). 'Attention deficit hyperactivity disorder and motor impairment: a critical review', *Percept Mot Skills*, 124(2), pp. 425–440. doi: 10.1177/0031512517690607.

Gutiérrez, M. S. D. R., Martin-Gutierrez, J., Acevedo, R., & Salinas, S. (2019). 'Hand Gestures in Virtual and Augmented 3D Environments for Down Syndrome Users' *Applied Sciences*, 9(13), 2641, pp. 1–16. doi: 10.3390/app9132641.

Hu, X., & Han, Z. R. (2019). `Effects of gesture-based match-to-sample instruction via virtual reality technology for Chinese students with autism spectrum disorders', *International Journal of Developmental Disabilities*, 65(5), 327–336. doi: 10.1080/20473869.2019.1602350.

Kasari, C., Freeman, S., & Paparella, T. (2006). 'Joint attention and symbolic play in young children with autism: a randomized controlled intervention study'. *J. Child Psychol. Psychiatry*, 47(6), pp. 611–620. doi: 10.1111/j.1469-7610.2005.01567.x

King, D., Horton, M., & Lamichhane, D. R. (2018). 'Evaluating the usability of the leap motion controller', In *Proceedings of the 32nd International BCS Human Computer Interaction Conference*, 32, pp. 1–5, July, 2018. doi: 10.14236/ewic/HCI2018.183.

Leap Motion. (2020). Disponível em < https://www.ultraleap.com/> (Acesso em 23 de janeiro de 2020).

Machalicek, W., Shogren, K., Lang, R., Rispoli, M., O'Reilly, M.F., Franco, J.H., & Sigafoos, J. (2009). `Increasing play and decreasing the challenging behavior of children with autism during recess with activity schedules and task correspondence training', *Res. Autism Spectr. Disord* 3(2), pp. 547–555. doi: 10.1016/j.rasd.2008.11.003.

Marin, G., Dominio, F., & Zanuttigh, P. (2016). 'Hand gesture recognition with jointly calibrated leap motion and depth sensor', *Multimedia Tools and Applications*, 75(22), 14991–15015. doi: 10.1007/s11042-015-2451-6.

Menezes L. D. C., et al. (2015). 'Motor learning and virtual reality in down syndrome; a literature review', *International Archives of Medicine*, 8, p. 119.

Neto F.R., et al. (2015).. 'Motor development of children with attention deficit hyperactivity disorder', *Revista Brasileira de Psiquiatria*, 37(3), pp. 228–234. doi: https://doi.org/10.1590/1516-4446-2014-1533.

Niechwiej-Szwedo E., et al. (2018). 'Evaluation of the leap motion controller during the performance of visually-guided upper limb movements', *PLoS One*, 13(3). doi: https://doi.org/10.1371/journal.pone.0193639.

Oliveira, J. M., et al. (2016). 'Novel virtual environment for alternative treatment of children with cerebral palsy', *Computational Intelligence and Neuroscience*, Article ID 8984379.

Oña E. D., et al. (2018). 'Effectiveness of serious games for leap motion on the functionality of the upper limb in parkinson's disease: a feasibility study', *Computational Intelligence and Neuroscience*, Article ID 7148427, p. 17.

Pavão S. L., et al. (2014). 'Impact of a virtual reality-based intervention on motor performance and balance of a child with cerebral palsy: a case study', *Revista Paulista de Pediatria*, 32(4), pp. 389–394. doi: https://doi.org/10.1590/S0103-05822014000400016.

Pelosi, M. B., Teixeira, P. O., & Nascimento, J. S. (2019). 'O uso de jogos interativos por crianças com síndrome de Down', *Cadernos Brasileiros de Terapia Ocupacional*. doi: https://doi.org/10.4322/2526-8910.ctoao1869.

Postolache, O., et al. (2017). 'Serious game for physical rehabilitation: measuring the effectiveness of virtual and real training environments', In: *IEEE International Instrumentation and Measurement Technology Conference (I2MTC)*, pp. 1–6. doi: 10.1109/I2MTC.2017.7969978.

Rose, T., Nam, C. S., & Chen, K. B. (2018). 'Immersion of virtual reality for rehabilitation', *Review. Applied Ergonomics*, 69, pp. 153–161. doi: https://doi.org/10.1016/j.apergo.2018.01.009.

Rosenbaum P., et al. (2007). 'A report: the definition and classification of cerebral palsy', *Developmental Medicine & Child Neurology*, 49, pp. 9–14. doi: 10.1111/j.1469-8749.2007.tb12610.x.

Shah, V., et al. (2019). 'A rhythm-based serious game for fine motor rehabilitation using leap motion', In: *58th Annual Conference of the Society of Instrument and Control Engineers of Japan (SICE)*, Hiroshima, Japan, pp. 737–742. doi: 10.23919/SICE.2019.8859927.

Salter, T., Davey, N., & Michaud, F. (2014). `Designing & developing QueBall, a robotic device for autism therapy', In *The 23rd IEEE International Symposium on Robot and Human Interactive Communication* (pp. 574–579). IEEE. doi: 10.1109/ROMAN.2014.6926314.

Sansosti, F. J., Doolan, M. L., Remaklus, B., Krupko, A., & Sansosti, J. M. (2015). `Computer-assisted interventions for students with autism spectrum disorders within school-based contexts: a quantitative meta-analysis of single-subject research', *Review Journal of Autism and Developmental Disorders*, 2(2), 128–140. doi: 10.1007/s40489-014-0042-5.

Santos, L. F., & Vasconcelos, L. A. (2010). 'Transtorno do déficit de atenção e hiperatividade em crianças: uma revisão interdisciplinar', *Psic.: Teor. e Pesq., Brasília*, 26(4), pp. 717–724. doi: https://doi.org/10.1590/S0102-37722010000400015.

Sherman, W. R., & Craig, A. B. (2018). *Understanding Virtual Reality: Interface, Application, and Design*. Elsevier: Morgan Kaufmann Publishers, 2nd edition, Cambridge, United State, pp. 5–6.

Syahputra, M. F., et al. (2018). 'Augmented reality social story for autism spectrum disorder', *Journal of Physics: Conference Series*, 2018. doi: 10.1088/1742-6596/978/1/012040.

Tang, T., Winoto, P., & Wang, R. (2015). 'Having fun over a distance: supporting multiplayer online ball passing using multiple sets of Kinect', In: CHI '15 *Extended Abstracts*, pp. 1187–1192. ACM Press, New York. doi: https://doi.org/10.1145/2702613.2732848.

Tang, T. Y., Falzarano, M., & Morreale, P. A. (2018). 'Assessment of the utility of gesture-based applications for the engagement of Chinese children with autism', *Universal Access in the Information Society*, 17(2), 275–290. doi: 10.1007/s10209-017-0562-8.

Tarakci, E., et al. (2019). 'Leap motion controller–based training for upper extremity rehabilitation in children and adolescents with physical disabilities: a randomized controlled trial', *Journal of Hand Therapy*, pp.1–8. doi: 10.1016/j.jht.2019.03.012.

Trujillo-Orrego, N., Ibáñez, A., & Pineda, D. A. (2012). 'Validity of the diagnosis of attention deficit disorder/hyperactivity disorder: what phenomenological to neurobiological (II)', *Reviews Neurology*, 54, pp. 367–379.

Vasconcelos, T. G., & Aguiar, Y. P. C. (2017). 'Leap Motion como tecnologia assistiva para pessoas com deficiência motora nos membros superiores', Paraíba: Universidade Federal da Paraíba: 2017. Disponível em: https://repositorio.ufpb.br/jspui/handle/123456789/4445.

Venkatakrishnan, R., et al. (2019). 'Towards an immersive driving simulator to study factors related to cybersickness', In *IEEE Conference on Virtual Reality and 3D User Interfaces (VR)*, pp. 1201–1202, March, 2019. doi: 10.1109/VR.2019.8797728.

Wiederhold, B. K., & Riva, G. (2019). 'Virtual reality therapy: emerging topics and future challenges', *Cyberpsychology, Behavior, and Social Networking*, 22(1), pp. 3–6. doi: 10.1089/cyber.2018.29136.bkw.

Yamamoto, K., & Nakagawa, S. (2002). 'Speech recognition under noisy environments using segmental unit input HMM', *Systems and Computers in Japan*, 33(8), pp. 111–112. doi: 10.1109/ICCCE.2010.5556819.

Yu, C., & Smith, L. (2013). 'Joint attention without gaze following: Human infants and their parents coordinate visual attention to objects through eye-hand coordination', *PLoS One*, 8(11), pp. 79659. doi: 10.1371/journal.pone.0079659.

Zhu G., et al. (2015). 'A series of leap motion-based matching games for enhancing the fine motor skills of children with autismo', In: *IEEE 15th International Conference on Advanced Learning Technologies*, pp. 430–431. doi: 10.1109/ICALT.2015.86.

Section V

Emerging Perspectives and Applications
of Virtual Reality in Practice

15 Applications of Virtual Reality in Aphasia Therapy
Findings from Research with EVA Park

Jane Marshall, Niamh Devane, Richard Talbot,
and Stephanie Wilson
University of London
London, UK

CONTENTS

INTRODUCTION

Imagine waking up in hospital with the realisation that you can no longer talk. When offered a pen and paper you discover that writing is also impossible. You are struggling to understand what others say and find written information mystifying. This is the situation faced by many stroke survivors with aphasia.

Aphasia is a communication disability caused by damage to the areas of the brain that control language. Its most common origin is stroke, although it can arise from any brain injury, particularly to the left hemisphere. About 45% of strokes cause aphasia, and in nearly a quarter of strokes the problems are persistent (Ali et al., 2015). It is estimated that 350,000 people are living with aphasia in the UK (see The Stroke Association, 2018).

Aphasia is typically accompanied by other stroke related impairments. Hemiplegia, or paralysis affecting one side of the body, occurs in 77% of strokes (Lawrence et al., 2001). There may also be sensory problems, such as loss of vision to one side of space, or cognitive difficulties affecting memory or executive function.

Aphasia varies across individuals. We have described a very severe case above. When problems are less severe, some speech and or writing may be retained, albeit with errors and/or hesitancies;

and understanding may only break down when complex language is used. Regardless of this variation, the consequences of aphasia are profound. Those affected often lose friends (Northcott & Hilari, 2011) and take part in fewer social activities than their age matched peers (Cruice et al., 2006). Feelings of isolation (Parr, 2007), depression (Kauhanen et al., 2000) and distress (Thomas & Lincoln, 2008) are common. Overall, people with aphasia experience worse quality of life than other stroke survivors with unaffected language (Hilari et al., 2012).

The problems of aphasia can be mitigated by rehabilitation (Brady et al., 2016). Therapy can help to recover language abilities, e.g. by supporting word retrieval (Wisenburn & Mahoney, 2009; Efstratiadou et al., 2018) and sentence production (Thompson, 2019). It can help family members and others to adapt their communication so that people with aphasia have better access to conversation (Simmons-Mackie et al., 2016). The psychological consequences of stroke and aphasia have also been tackled, with some evidence of success (Thomas et al., 2012).

In recent years, aphasia rehabilitation has made increasing use of digital technology (Zheng et al., 2015). Computer delivered therapies allow for self-directed practice of language exercises (Palmer et al., 2019), and can address a variety of modalities, including speech (Linebarger et al., 2001; Lavoie et al., 2017), comprehension (Fleming et al., 2017) and even gesture (Roper et al., 2016). Mainstream technologies, such as text-to-speech conversion and word prediction software, offer strategic compensations for some aphasic difficulties, particularly in relation to reading and writing (Marshall et al., 2018; Caute et al., 2019) and computers can help individuals to access therapy through remote treatment delivery (Woolf et al., 2016; Pitt et al., 2019).

Applications of virtual reality (VR) have been less extensively explored. Here there are a number of potential opportunities. Online VR treatment could reach isolated individuals and those who struggle to travel. The immersive experience of VR may increase engagement with therapy, and so encourage intensive language practice, with likely benefits for outcomes. Multi-user platforms allow for social interaction, so may help to reduce the isolation experienced by many people with aphasia and address some of their psychological needs. Perhaps most intriguingly, VR may help to tackle the problem of generalisation. We would all agree that aphasia therapy should aim to improve functional, everyday uses of language, rather than just performance on clinical tests (e.g. see Brady et al., 2016). Yet achieving (and demonstrating) change at this level is difficult. By using VR, therapists can embed language practice in simulated everyday contexts. For example, an exchange with a hairdresser can be conducted in a virtual salon. This holds out the possibility that treatment effects may generalise to real-life equivalents of those contexts.

Aspects of the virtual experience may promote such generalisation. One is immersion, where the user feels fully engaged in the virtual world and experiences it as real. This, in turn, can lead to presence, or the sense of being in a place. Such blurring of the distinction between real and virtual experiences may help to generalise skills that are developed in VR to everyday life. Konnerup (2019) describes how people with aphasia, interacting in the virtual environment of Second Life, frequently demonstrated both immersion and presence. For example, 'Michael' expressed concern to his wife about the possibility of being hit by a virtual tram, and another user described getting butterflies in her stomach when her avatar flew.

A further agent for generalisation may be the Proteus Effect (Yee & Bailenson, 2007). Users of a virtual environment have the opportunity to remodel themselves, for example by creating an avatar that is taller, younger and thinner than themselves. Such transformations not only influence other interactants in the virtual world, but, according to the Proteus Effect, can also affect the user's own behaviour. To illustrate, in one experiment, users with tall avatars were shown to behave more assertively in a money negotiating experiment than their shorter counterparts (Yee & Bailenson, 2007). In a follow up experiment, the same team demonstrated carry over to an offline version of the same task; i.e. users with tall avatars continued to display the more dominant characteristics in a subsequent face to face negotiation (Yee et al., 2009). To our knowledge, the Proteus Effect has not been explored with people with aphasia. However, benefits for rehabilitation might be hypothesised. For example, projecting a positive image of oneself in VR may subtly transform a user's self-perception, with possible benefits for their sense of efficacy and confidence.

To date, there have been few uses of VR in aphasia therapy (see review in Bryant et al., 2019). Cherney and colleagues adapted two treatment approaches for VR delivery (Cherney, 2010; Cherney et al., 2008). In both, treatment is administered by a virtual speech and language therapist, who leads the user through exercises, and models treatment responses. Another example is the Everyday Life Activities Virtual House (Stark et al., 2013). This is a virtual house through which the user travels while practicing language, for example by naming objects that are encountered. The VR Rehabilitation System provides rehabilitation exercises for a range of stroke impairments, including linguistic modules (Maresca et al., 2019). Finally, a team in Barcelona have delivered Intensive Language Action Therapy on a VR platform, and shown benefits across a number of individuals with aphasia (Grechuta et al., 2016, 2017).

While pioneering, these applications of VR are mainly for sole use, or enable interaction only with one other user. They are also vehicles for delivering specific language exercises. In contrast, EVA Park provides an imaginative, virtual space in which multiple users can interact. In this chapter, we describe EVA Park and its development. We outline the intervention research conducted with EVA Park and the findings from that research. We explore users' responses to EVA Park, through our qualitative and observational data. Finally, we discuss our findings and consider future directions.

EVA PARK

EVA Park is a multi-user online virtual island, built in on the Open Sim platform. It can be accessed from any typical home computer with processor and graphics capability to run a virtual world. Users are represented by personalised avatars, which are controlled via a standard computer keyboard, or a simplified key pad (see Figure 15.1). Communication takes place in real time mainly through speech, for which users wear headphones and a microphone. Users can also type messages which appear in a textbox on the screen. Avatars can walk, run or (if suitably shod) can roller skate round the island. They can also fly or teleport instantly to a new location. Five on screen icons enable avatars to execute a small repertoire of pre-programmed gestures, such as a wave, a dance

FIGURE 15.1 Simplified key pad for navigating EVA Park.

and a belly laugh. EVA Park contains a number of functional and fantastical locations, including houses, a restaurant, a café, a hair salon, a tiki bar, a narrowboat containing a planetarium, and a tardis, which houses a gaming hall. There are many attractive settings where avatars can meet, such as a roof top terrace with a sea view. The island is colourful with many visual jokes and surprises. For example, those who dive into the EVA Park lake will find a mermaid and a giant turtle, and if they click on the turtle, they will be taken for a ride. The EVA Park disco contains a glitter ball which, when clicked, makes the user's avatar dance.

EVA Park was designed as a platform for aphasia interventions. It does not deliver automated communication exercises. Rather, it is a virtual space in which people with aphasia can meet and receive support from a therapist or rehabilitation worker. Intervention is delivered in real time by the therapist, who is also represented as an avatar, and works remotely with his or her client(s).

EVA Park was created using a process of co-design (Wilson et al., 2015). The team consisted of speech and language therapy researchers, human-computer interaction researchers, a software developer and five consultants with aphasia, who were reimbursed for their time. By so involving people with aphasia in the design of EVA Park, we hoped to achieve an environment that was accessible and acceptable to the target user group, and relevant to their needs.

Ten co-design workshops were held over a one year period. These informed every aspect of EVA Park, such as what it should contain, the appearance of the avatars, and how the island should be navigated. Novel techniques were employed, which aimed to tap into the experiences of the consultants and enable them to contribute ideas despite their communication difficulties. For example, when considering the content of EVA Park, we asked the consultants to create photo diaries of situations in their daily lives that they found challenging. The resultant photos were mapped onto story grids, based on their attributes (e.g. 'noisy environment') and whether or not they were enjoyable and/or challenging. This stimulated a lot of discussion about the daily communicative experiences of people with aphasia and created a clear framework for selecting settings to include (and not include) in EVA Park.

Later workshops gave consultants experience with prototypes of EVA Park and collected their feedback. This took many forms. Firstly, they were observed using the prototype, for example to identify navigation barriers and assess their immediate response to the environment. Their verbal opinion was sought, and, if relevant, a rating response for a particular feature. They were also asked to use a 'Someone who isn't me' technique. This invited them to think about another person with aphasia, known to them, and imagine how they would respond to the feature. This technique enabled us to consider a range of responses to EVA Park, beyond those directly provided by the consultants.

The use of co-design shaped every aspect of EVA Park, with two key themes. Firstly, it provided crucial insights into the technological barriers that are faced by stroke survivors with aphasia. We therefore took all possible steps to simplify access to EVA Park and ease navigation. To give just one example, we incorporated increased space between objects and locations in EVA Park in order to minimise crowding. Secondly, the consultants made it clear that EVA Park must be convivial and fun to visit. As a result, the settings in EVA Park are mainly associated with socialisation and leisure, and opportunities for quirky, even bonkers content have been embraced.

RESEARCH WITH EVA PARK

Studies to date have investigated the feasibility of delivering different interventions via EVA Park and users' responses to those interventions. We have also explored benefits on a range of outcome measures, although these data are underpowered and therefore purely indicative.

STUDY 1

Our first study with EVA Park developed the platform (Wilson et al., 2015) and explored its capacity to host a virtual communication intervention (Marshall et al., 2016).

TABLE 15.1

Examples of Participants' Communication Goals in Study 1 and Associated Activities

Communication Goal	Activities
Improving word retrieval	Finding, photographing and naming all the animals in EVA Park
	Naming plants in the EVA Park greenhouse
	Naming kitchen items in the EVA Park house
Making functional requests	Asking for a haircut in the EVA Park Salon
	Ordering a drink in the EVA Park Café
	Ordering a pizza in the EVA Park restaurant
	(In all cases support workers played relevant roles, e.g. as a hairdresser or barista)
Giving explanations	Reporting a suspected crime at the EVA Park counter; with the support worker playing the role of a police officer
Making a speech	Giving a 'thank you' speech to fellow participants at the EVA Park lake-side
Talking in groups	Telling a personal story to fellow participants in the EVA Park treehouse
	Taking part in a meeting to discuss the creation of an EVA Park sports centre
	Performing the group facilitator role

Twenty people with aphasia were recruited to the intervention strand. All had mild or moderate aphasia following a stroke that occurred at least six months prior to the study (mean 62.1 months). We excluded people with severe aphasia as we were concerned that they might not be able to access the platform, e.g. because using EVA Park depends on some ability to produce and understand speech. We also excluded people with severe cognitive or sensory difficulties, for a similar reason. Participants had a wide age range (36–81, mean 57.8) and most, but not all, were prior computer users.

We ran four intervention periods, each involving five participants. Each person with aphasia was paired with a support worker. Support workers were all speech and language therapists or volunteers/assistants with extensive experience in aphasia. The intervention involved daily meetings in EVA Park with the support workers, over a period of five weeks (25 sessions in all). Meetings initially identified personally relevant communication goals that the person with aphasia wanted to address. Remaining meetings then targeted those goals, using conversation, role play and other language activities in EVA Park (see Table 15.1 for examples of participants' goals and the associated activities). Most sessions were 1:1 with support workers; but once a week, participants met in a group (five participants and five support workers). This was an opportunity for group discussions, e.g. about current affairs, music, and personal stories, and for group activities, such as a diving competition in the EVA Park lake. To further stimulate conversation, an election narrative was run during each intervention period. This presented five virtual candidates who were seeking election as the mayor of EVA Park. Billboards in EVA Park gave information about the candidates' policies and released gossip about them. For example, one candidate admitted to smoking cannabis at university. Participants had the opportunity to interact with this narrative, e.g. by discussing candidates' manifestos with their support workers. One participant regretted the fact that EVA Park did not contain a sports centre, so called a meeting to lobby the mayoral candidates on this issue.

Study 1 employed a randomised waitlist control design. Half the participants received intervention immediately after recruitment and assessment, in weeks 2–6, and half received intervention after a delay, in weeks 8–12. Thus, at week 7, we were able to compare outcomes across a treated and as yet untreated (control) group. The primary outcome measure was a standard assessment of functional communication in aphasia (Holland et al., 1999). Secondary measures explored aspects of language production (verbal fluency, word retrieval in conversation, and narrative), communicative confidence and feelings of social isolation. All measures were administered face to face, i.e. not in VR. We also interviewed participants and collected usability and user experience data.

Findings from the study were mixed. We showed that it was feasible to conduct this intervention in EVA Park and compliance was excellent. There was no attrition and 19 (/20) participants attended at least 21 sessions (the remaining participant attended 17 sessions). The responses of participants to the intervention were extremely positive (see Interview Data below). Results on the primary measure also indicated that the intervention was effective in improving functional communication skills. However, there were no changes on the other measures of language production, communicative confidence and feelings of social isolation. This suggested that more generalised benefits had not occurred, although responses to the interviews (see below) painted a more optimistic picture.

Interview Data

All participants were interviewed before and immediately after the intervention. Five randomly selected participants were also interviewed 12 months later. While covering a number of themes, the interviews aimed to determine participants' views about EVA Park and the perceived impact of the intervention received in EVA Park (Amaya et al., 2018).

Perceptions of EVA Park were overwhelmingly positive. Participants valued the different and attractive locations on the island, with some expressing feelings of presence, e.g.:

- "I'm sitting right on the ...... on a decking up the top ... of the houses... and I'm thinking oh God I'm on holiday here"

The opportunities for quirky activities, and to embed language practice in relevant contexts were also appreciated:

- Interviewer: "You enjoyed the dancing?" Respondent: "yeah, really enjoy it. Boom [mimes singing and dancing] really enjoyed it. It really was exciting"
- "We worked on various things well as in the, in the project like er ... doctor's appointments and ordering food or drinks in a bar and gen-generally they were quite good [laughs]"

No negative views were expressed about being represented as an avatar. One participant enjoyed the fact that his virtual self had no hemiplegia:

- "it's pretty good because you get to be a character with two arms [raises own arms and laughs]"

Another commented on the sense of release that she gained from being represented virtually:

- "It's nice to think that you can pretend to be you and, er, maybe even something silly and get away with it"

All participants commented positively on the relationship with their support worker, making it clear that a therapeutic bond could be formed in the virtual environment:

- "She said 'Don't worry ... carry on' and she wonderful"
- "help me with the ... the words ... couldn't say and she found/f//lɛf//lɛf/me feel free" Interviewer: "she let you feel free?" Respondent: "yeah, to talk"
- "My helper was very good... So helped me talk better and I got my confidence back again [smiles]"

Interactions with other participants were also valued:

- "and we could ... have um ... take the mickey out of one another ... and it's wonderful"

Negative responses were very few. One respondent found EVA Park difficult to navigate, feeling that it was too complicated: "there's too much to it". Other limitations were problems with sound and the possibility of making navigation errors. Conversely all participants said that they would recommend EVA Park to other people with aphasia.

In terms of impacts, all bar two participants felt that the intervention had improved their communication. For example, one said that he was talking more to family and friends and two reported on positive feedback from family members, e.g.:

- "My wife's and daughter said I hadn't spoken so much for ages [laughs]"

Eighteen participants also commented on positive changes in their activity levels following intervention, citing examples like attending a pottery class and using the telephone. Increased computer use was also identified:

- "I'm doing things I never thought I'd do again" Interviewer: "Such as?" Respondent: "Going to the computer and ordering my freedom pass"

A final strong theme was increased feelings of confidence, with 12 respondents reporting benefits in this area:

- "I'm not afraid now of saying what I want to say, so the words come out. So the confidence has come back now. Cos before I was full of it but now, after the stroke, it all disappears"

The evidence of perceived benefit that came from the interviews augments the outcome data. The fact that communication was seen to change corroborates the results on our primary measure. Encouragingly, three of those who were interviewed 12 months later still detected a change in their communication, suggesting that, at least for them, changes were long lasting. On the other hand, participants' observed changes in confidence were not in line with our quantitative findings. It is possible that the measure we used (Communication Confidence Rating Scale in Aphasia, Babbit et al), was not sensitive to change or was assessing different dimensions from those commented on by the participants.

Human Computer Interaction Observations

To provide insights into participants' use of EVA Park we aimed to record 40 individual sessions for analysis, with two from each participant drawn from the second week (early) and fifth week (late) of their intervention. As three recordings were lost to technical difficulties, the final sample was 37 sessions. We filmed both events in world (via screen capture software) and the person using EVA Park (via a camcorder).

The session recordings were coded and subject to thematic analysis (see Galliers et al., 2017). Coding was influenced by the gaming literature (Poels & Wijnand, 2007) and included markers of positive and negative affect, conversation types, immersion and social presence.

A key finding related to affect, where positive instances outweighed negative ones by a factor of over 4 to 1. Positive examples included laughing, joking, positive surprise and pride. To illustrate, one participant 'stroked' a cat in EVA Park by clicking on it, and laughed when it emitted a 'meow' sound. Another participant was invited to do sit ups and push ups in the EVA Park health centre by her support worker, and responded jokingly: 'Oo. Too much ups'. Some jokes played on the virtual nature of the world, for example when a participant declined to jump in the EVA Park lake because she had just washed her hair. The most common negative categories were 'displeasure' and 'frustration'. Several instances related to aphasic symptoms, for example when participants expressed frustration over word finding difficulties. Some arose from frustrations with the platform, for example when a participant was temporarily unable to stop her avatar from flying.

The coding data showed that a very wide range of conversations took place in EVA Park. The three main conversation categories were: conversations grounded in/stimulated by events in the virtual world; conversations grounded in/stimulated by events in the real world and conversations about using EVA Park. It was interesting that conversations about the real and virtual world were roughly equal in number in both the early and late recordings. The data indicate that participants had no difficulty segueing between virtual and real world topics. For example, in one instance an encounter with chickens in EVA Park led to a conversation about whether a participant had ever kept animals.

Instances of immersion (as marked by evidence of absorption) were few, which is perhaps surprising. There were, however, many other markers of engagement. Participants took part in a range of EVA Park activities, such as dancing in the disco, drinking at the bar, riding an elephant and diving in the lake. They also embraced opportunities for fantasy and transformation. For example, several elected to give their avatar wings and one woman in her 60s represented herself as a 1970s punk. Across the whole sample there were only eight instances of dis-engagement/detachment, and seven of these came from one participant.

Coding of social presence recorded instances in which participants sought or avoided the company of others, by, for example, greeting other avatars or moving away from them. Here, seeking the company of others was the dominant behaviour and significantly so in the late observations.

Summary from Study 1

This study showed that it was feasible to deliver daily language stimulation in EVA Park, using 1:1 interactions between support workers and people with aphasia. The intervention brought about improvements on a measure of functional communication, but not on other outcomes. Interviews and observations with participants indicated that the intervention was highly acceptable and strongly associated with markers of positive affect. These data also showed that EVA Park fostered social interaction and that therapeutic relationships could be formed with support workers. Participants perceived benefits from the intervention that both matched and exceeded those detected by the outcome measures.

STUDY 2

This study aimed to explore the range of interventions that can be delivered in EVA Park, and particularly whether EVA Park can host therapies that target specific language impairments. Using single case/case series designs, five different therapies were administered to individuals in EVA Park and outcomes were assessed. The participants with aphasia were also interviewed, to gain their views about receiving therapy in EVA Park. The study questions were: can these different therapies be conducted in EVA Park, what do participants feel about them, and do the outcomes mirror those that are achieved from face to face delivery?

The therapies that we tested aimed to improve noun and verb production, story telling, discourse and sentence production. We used therapies that have been described and evaluated in the aphasia therapy literature; and in all cases there was evidence that the therapies can be beneficial when administered face to face.

Here is a case example. 'Blake' (Marshall et al., 2018) was 60 years old and had longstanding aphasia, following a stroke five years prior to the study. He had very limited spontaneous speech and severe word finding difficulties. Before his stroke, he worked as a chemist. Blake received 20, one hour therapy sessions delivered over five weeks. Three sessions a week were led by a qualified speech and language therapist and one was led by a student of speech and language therapy. The treatment aimed to improve spoken noun production, using well evidenced cuing techniques. Blake was first shown a picture of an object and asked to say its name. If he could not, he was given graded cues about the meaning of the word and its first sound (e.g. for lemon: 'it's sour'; 'it starts with/l/'). We then used a technique called Semantic Feature Analysis (Boyle, 2004). Here the therapist asks a series of questions about an item, in order to stimulate the word (such as 'where would you find

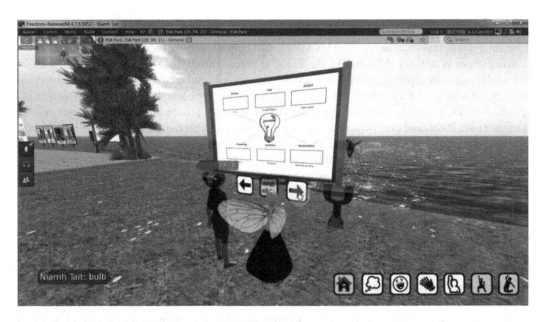

FIGURE 15.2 Projection of speech and language therapy materials in EVA Park.

this?', 'what would you do with it?'). This therapy was delivered in EVA Park via 1:1 sessions with the treating therapist. She used a billboard on which pictures and other therapy materials were projected (see Figure 15.2). The formal language exercises were supplemented with at least 10 minutes per session of informal tasks, in which Blake and his therapist visited different EVA Park locations to practice word production. For example, he requested items at the EVA Park fruit and flower stall and named objects in the EVA Park houses.

As in the first study, the feasibility results were good. Blake missed none of his treatment sessions and was positive about the experience of receiving therapy in VR. He described EVA Park as 'amusing' and rated his relationship with the treating therapist very highly, despite the fact that he only met her as an avatar. The fidelity of treatment delivery was checked, in order to determine whether the therapy was provided as intended. This was done by an independent therapist who rated recordings of 5 (randomly selected) treatment sessions against a checklist of required therapy components. Results were very positive with a 92% level of compliance. This showed that administering therapy in VR did not induce drift from the required procedures.

Were there measurable benefits from the therapy? When therapies of this type are administered face to face, typically practised words improve, but unpractised ones do not (Efstratiadou et al., 2018). We were therefore expecting similar results here. Fifty nouns were targeted in Blake's word finding therapy. His ability to name these nouns from pictures was tested twice before and twice after therapy. Before therapy, Blake could name half of the pictures (25/50). After therapy he named almost all of them (44/50). Fifty unpractised nouns were similarly tested. Blake's naming of these words did not change. He also showed no change on a standard measure of functional communication. We were interested in whether Blake himself detected any therapy benefits. When interviewed afterwards, he felt that therapy had improved his speech. He gave the example of now being able to ask for 'crabs' from a local fish stall (he had practised the word 'crab' in the Semantic Feature Analysis).

Summary from Study 2

A total of eight persons with aphasia received therapy in study 2. Of these, seven showed significant benefits on at least one of the outcome measures (see Carragher et al., 2018 for another example). In all cases the therapy proved feasible, and all participants reacted positively to the experience of receiving therapy in EVA Park.

Study 3

Our final study investigated whether EVA Park can help to address the social and psychological consequences of aphasia (Marshall et al., 2020). It is recognised that support groups can counter the social isolation experienced by many people with aphasia and reconnect individuals with community activities (Vickers, 2010). Benefits for quality of life have also been reported (Attard et al., 2018; van der Gaag et al., 2005; Corsten et al., 2014). However, groups are not universally available (Verna et al., 2009) and even when they are, they may be difficult to reach by isolated individuals. Remote group provision, using telehealth, has therefore been tested, and shown to be feasible and effective (Pitt et al., 2019). In this study, we investigated whether remote group support can be hosted by EVA Park and whether this support brought about gains, e.g. on measures of wellbeing and communication. We interviewed both the providers and recipients of the intervention to gauge their views.

Four groups of participants were recruited to the study. Each group comprised a co-ordinator, at least 2 volunteers and 8 or 9 people with aphasia. Co-ordinators were speech and language therapists or community stroke group leaders with extensive experience of aphasia. Volunteers also had prior experience of group support, and some were qualified therapists. The people with aphasia were aged between 40 and 77 (mean 58.09) and had mild or moderate language difficulties. All were at least eight months post stroke (mean 61 months).

Each group took part in a six month programme of intervention, consisting of 14 sessions delivered every two weeks. The content of the sessions was outlined in a manual, and was developed with reference to published group therapy techniques for people with aphasia (Attard et al., 2015; Elman, 2007; Holland, 2007). Sessions aimed to promote wellbeing and give participants the experience of communicative success. We aimed to foster social connections between group members and develop their strategies for coping with aphasia. Several sessions addressed personal issues, such as 'Aphasia' and 'Resilience'. Others focused on general topics, such as 'Comedy' and 'Music'. All sessions involved sharing views, stories and conversation. Following the principles of positive psychology (Seligman et al., 2005), participants were frequently invited to reflect on what gave meaning and value to their life, and on the personal strengths that enabled them to cope with aphasia. Two sessions were devoted to a group project. This enabled group members to use their skills in creating a collective, meaningful output. For example, one group produced a witty film about their experiences in EVA Park which was edited by a member who had worked in film before his stroke. Participants had unlimited access to EVA Park outside of scheduled sessions. To promote independent access, weekly cocktail hours and coffee mornings were set up. Group members were also encouraged to carry out assignments in EVA Park between the group meetings. For example, in one assignment they were invited to meet another participant for exercises in the EVA Park Health Centre.

Sessions were led in EVA Park by the group co-ordinators, with the support of the volunteers. Before intervention began, each team received eight hours of training from our Project Managers. This covered technical aspects of EVA Park, such as how to set up and support users, and therapeutic aspects, such as how to lead groups in the virtual environment. Regular supervision was provided throughout the intervention period. This involved monthly meetings with the co-ordinators and volunteers, which took place in a private area of EVA Park. Meetings discussed the progress of the groups, brain-stormed problems and looked ahead to future sessions. The research team also provided technical and other forms of support on an ad hoc basis, as requested by co-ordinators and volunteers.

Our first research question was whether or not this form of intervention was feasible in EVA Park. We were also interested in any indicative benefits. As in study 1, we used a waitlist controlled design. Two groups were randomised to receive intervention in months 6–11 of the study. The other two groups received intervention in months 13–17. Thus, at month 12, we could compare participants who had received intervention with a control group who had yet to be treated. The primary

TABLE 15.2

Illustrative Quotes from Co-ordinator Interviews in Study 3

Comments on group cohesion and communication	• 'It was wonderful we really saw the group forming and gelling as we went through.'
	• 'That was really genuine opening up and, and communication and debate and, erm, and agreement and disagreement as well … people were listening to different points of view and considering them and reacting erm, which I thought was fantastic in a virtual world with people that hadn't met, most of whom hadn't met each other in the real world.'
Comments on EVA Park limitations	• 'I would say if I'm honest at that, that was a, a frustration throughout… the technical side just seemed like a real barrier every week… And that felt as a coordinator [taps chest] that felt frustrating because you then couldn't really support them directly in that.'
	• 'There's no facial expression, there's no gesture, and all of these things that you very much rely on with somebody that's perhaps struggling with the spoken act, erm, so that was all stripped back.'

outcomes measured wellbeing and functional communication. Secondary outcomes assessed social connectedness, language and quality of life.

Feasibility findings were good. Thirty-four participants with aphasia were recruited, with seven lost to attrition. Reasons for drop out were mainly unconnected with the project, such as family problems and ill-health, although one person withdrew because he disliked groups. All four groups ran the intervention as planned and sessions were well attended. Those who began intervention attended a mean of 11.41 sessions (over 80% of the total programme). Almost all participants made some use of EVA Park outside the scheduled sessions. The mean number of unscheduled logins was 12.87 (range 0–81).

Findings from the outcome measures were less positive. There were no significant differences on any of our measures when the treated and control group were compared. Even when we combined data across all participants there were no pre to post intervention changes. We hope that these null findings will be illuminated by the results from the interviews with person with aphasia, which are still being analysed.

We interviewed all the group co-ordinators and held consensus discussions with them and their volunteers. These discussions explored the perceived benefits of EVA Park, and aspects that could be improved (see illustrative quotes in Table 15.2). Teams valued the safe space of EVA Park as a venue for sharing and trying new things. They appreciated the sense of community and group cohesion that developed during intervention and felt that EVA Park offered a bridge between therapy and the real world. Limitations included technological glitches, which were difficult to manage in the group context, and the fact that participants' facial expression and body language were not visible to group leaders.

Summary from Study 3

Study 3 showed that it is feasible to deliver group social support to people with aphasia in EVA Park, using a model in which delivery was delegated to community co-ordinators and volunteers. There were no changes on our outcome measures. Feedback from those who delivered the intervention identified many strengths of the platform, but also highlighted limitations.

DISCUSSION

Our research found that it was possible to employ co-design in order to build a therapeutic virtual world for people with aphasia. In so doing, we developed novel techniques that enabled people with language impairments to express their views about every aspect of the platform (Wilson et al., 2015 and see Galliers et al., 2012 for a previous application of co-design in aphasia). Such techniques are potentially informative for other teams wanting to involve marginalised groups in co-design.

Subsequent studies showed that EVA Park could host a range of interventions, targeting communication, language and psychological wellbeing. Different models of care were trialled, including one to one and group intervention. We showed that intervention in EVA Park can be led by qualified therapists, or delegated to support workers, students, co-ordinators and volunteers.

A key finding is that people with aphasia embrace this form of intervention. EVA Park proved accessible to our user community and was highly valued. Time and time again we saw people with aphasia entering into the joke of EVA Park. They decked their avatars with wings and roller skates and, in some cases, created a persona that was markedly younger and trendier than their real world selves. They were happy to get down in the EVA Park disco, or sing Beatles songs at the Tiki bar. One group rounded off their intervention with a sports day, which included a race round the EVA Park lake after patting the EVA Park donkey. Yet, at the same time, our participants used EVA Park to address communication goals, confront their linguistic impairments, and discuss the real world impacts of aphasia on their lives.

Feedback from users threw up two repeated themes. One is the enjoyment and fun that derived from EVA Park. The risk of depression following stroke is high (Paolucci, 2008), and that risk is magnified when aphasia is thrown into the mix (Kauhanen et al., 2000). In this context, the potential of EVA Park to elevate mood is not trivial. A second theme was the importance of the relationships formed in EVA Park. In our first study, participants spoke very warmly about the rapport with their support worker and the benefits that they derived from that rapport. In most cases, this was someone they had never met in the real world. In study 3, group co-ordinators spoke movingly about the sense of group cohesion that developed during the six months in EVA Park and the support that participants derived from one another. The fact that genuine social connection could form between individuals who met as avatars in a virtual space is an important finding.

Why has EVA Park worked? We would argue that a key factor was the use of co-design, which helped to ensure that the platform was accessible to and in tune with the target user community. While aphasia is often accompanied by stroke related cognitive difficulties it is not, in itself, a cognitive impairment. As a result, our participants could clearly cope with the virtual experience, and distinguish conversation about the real and virtual world. Neurological conditions that *do* impair cognition, such as dementia, might throw up more barriers to this form of intervention. It could be argued that we have not tested EVA Park with a typical population. In all studies, our participants were younger than most stroke survivors, where the average age is over 70 (Lee et al., 2011). Many were from professional backgrounds and most were prior computer users. It is possible that a more typical sample would be less amenable to receiving therapy in a virtual world. However, we did include people in their 80s, and people who had not used computers previously, with no difference in uptake. Assumptions about who is and is not amenable to technological interventions also need to be challenged, particularly as generational differences in technology use become less and less marked.

The introduction to this chapter highlighted factors like immersion and the Proteus Effect that could enhance benefits from VR treatments. Our outcome data have only partially fulfilled this promise. Some, but not all of our interventions, brought about change on at least one outcome measure. For example, in study 1 participants made significant improvements on an assessment of functional communication, which explores a person's capacity to communicate in everyday contexts. However, wider benefits have been difficult to demonstrate, for example in communicative confidence (study 1) or wellbeing (study 3). In part this may reflect problems of measurement, a view that is encouraged when we compare scores on our measures with the often more optimistic interview data. Lack of change may also be attributable to the content of intervention, rather than the virtual delivery. It should also be stressed that none of our studies is definitive. In all cases numbers are low, meaning that the power to detect change is lacking.

Before concluding, it is important to reflect on the limitations of EVA Park. In all studies, some technological problems occurred, particularly with sound. In study 3 these problems proved most obstructive, mainly because managing effects across a group was difficult. While most participants enjoyed, or even relished EVA Park, there were individuals who had a less positive experience. To date, numbers are too low to explore factors that might predict positive uptake, but this clearly merits further scrutiny. Finally, the lack of face to face information was regretted by a number of users. For example, in study 3 this made it difficult for group co-ordinators the gauge the emotional responses of group members during discussion. The fact that facial expression and natural gesture cannot be used in EVA Park also makes it unsuitable for many people with severe aphasia, who are highly dependent on these non-verbal modes of communication.

In conclusion, our findings suggest that employment of VR in aphasia therapy merits further research. In addition to larger scale studies, such research could explore candidacy for VR therapy, and further uses, such as peer befriending and conversation support. As a prototype, our platform has demonstrated the potential of VR for clinical practice. Clearly, further technological improvements are needed before general release, particularly if EVA Park is to be used with large numbers. With technological upgrade, EVA Park, or its like, could become a routine aphasia therapy resource.

ACKNOWLEDGEMENTS

The work reported in this chapter was supported by funding from The Stroke Association, The Tavistock Trust for Aphasia and City, University of London. We thank our many colleagues who have worked with us on the EVA Park Project: Tracey Booth, Anna Caute, Madeline Cruice, Julia Galliers, Helen Greenwood, Katerina Hilari, Anita Patel, Abi Roper, Celia Woolf, Nick Zwart. The work has also benefited from the support of numerous volunteers and student supporters. Finally, we thank the many people with aphasia and their family members who have made this work possible.

REFERENCES

Ali, M., Lyden, P. & Brady, M. (2015) Aphasia and dysarthria in acute stroke: recovery and functional outcome. *International Journal of Stroke*, 10, 400–406.

Amaya, A., Woolf, C., Devane, N., Galliers, J., Talbot, R., Wilson, S. & Marshall J. (2018) Receiving aphasia intervention in a virtual environment: the participants' perspective. *Aphasiology*. doi: 10.1080/02687038.2018.1431831.

Attard, M., Lanyon, L., Togher, L. & Rose, M. (2015) Consumer perspectives on community aphasia groups: a narrative literature review in the context of psychological well-being. *Aphasiology*, 29, 8, 983–1019. doi: 10.1080/02687038.2015.1016888.

Attard, M., Loupis, Y., Togher, L. & Rose, M. (2018) The efficacy of an inter-disciplinary community aphasia group for living well with aphasia. *Aphasiology*, 32, 2, 105–138. doi: 10.1080/02687038.2017.1381877.

Boyle, M. (2004). Semantic feature analysis treatment for anomia in two fluent aphasia syndromes. *American Journal of Speech-Language Pathology*, 13, 236–249.

Brady, M., Kelly, H., Godwin, J., Enderby, P. & Campbell, P. (2016) Speech and language therapy for aphasia following stroke. *Cochrane Database of Systematic Reviews*, 6. Art. No.: CD000425. doi: 10.1002/14651858.CD000425.pub4.

Bryant, L., Bruner, M. & Hemsley, B. (2019) A Review of virtual reality technologies in the field of communication disability: implications for practice and research. *Disability and Rehabilitation: Assistive Technology*. doi: 10.1080/17483107.2018.1549276.

Carragher, M., Talbot, R., Devane, N., Rose, M. & Marshall, J. (2018) Delivering storytelling intervention in the virtual world of EVA park. *Aphasiology*, 32, supl, 37–39. doi: 10.1080/02687038.2018.1484880.

Caute, A., Woolf, C., Wilson, S., Stokes, C., Monnelly, K., Cruice, M., Bacon, K. & Marshall, J. (2019) Technology-enhanced reading therapy for people with aphasia: findings from a quasirandomized waitlist controlled study. *Journal of Speech Language and Hearing Research*, 62, 12, 4382–4416. doi: 10.1044/2019_JSLHR-L-18-0484.

Cherney, L.R. (2010) Oral reading for language in aphasia (orla): evaluating the efficacy of computer-delivered therapy in chronic nonfluent aphasia. *Topics in Stroke Rehabilitation*, 17, 6, 423–431. doi: 10.1310/tsr1706-423.

Cherney, L.R., Halper, A.S., Holland, A.L. & Cole, R. (2008) Computerized script training for aphasia: preliminary results. *American Journal of Speech-Language Pathology*, 17, 1, 19–34. doi: 10.1044/1058-0360(2008/003).

Corsten, S., Konradi, J., Schimpf, E., Hardering, F. & Keilmann, A. (2014) Improving quality of life in aphasia—Evidence for the effectiveness of the biographic-narrative approach, *Aphasiology*, 28, 4, 440–452. doi: 10.1080/02687038.2013.843154.

Cruice, M., Worrall, L. & Hickson, L. (2006) Quantifying aphasic people's social lives in the context of non-aphasic peers. *Aphasiology*, 20, 12, 1210–1225. doi: 10.1080/02687030600790136.

Elman, R.J. (2007). *Group treatment of neurogenic communication disorders: The expert clinician's approach* (2nd ed.). Oxford, San Diego: Plural.

Efstratiadou, E.A., Papathanasiou, I., Holland, R., Archonti, A. & Hilari, K. (2018) A systematic review of semantic feature analysis therapy studies for aphasia. *Journal of Speech, Language & Hearing Research*, 61, 5, 1261–1278. doi: 10.1044/2018_JSLHR-L-16-0330.

Fleming, V., Krason, A., Leach, R., Leff, A. & Brownsett, S. (2017) Listen-in: high-dose home-based auditory comprehension therapy is achievable and effective. *International Journal of Stroke,* 12, 24.

Galliers, J., Wilson, S., Roper, A., Cocks, N., Marshall, J., Muscroft, S. & Pring, T. (2012) Words are not enough: empowering people with aphasia in the design process. *Proceedings PDC 2012*, Roskilde, Denmark.

Galliers, J., Wilson, S., Marshall, J., Talbot, R., Devane, N., Booth, T., Woolf, C. & Greenwood, H. (2017) Experiencing EVA park, a multi-user virtual world for people with aphasia. *ACM Transactions in Accessible Computing*, 10, 4, Article 15 (October 2017), 24 pages. https://doi.org/10.1145/3134227.

Grechuta, K., Rubio, B., Duff, A., Oller, E., Pulvermuller, F. & Verschure, P. (2016) Intensive language-action therapy in virtual reality for a rehabilitation gaming system. *Journal of Pain Management*, 9, 3, 243–254.

Grechuta, K., Bellaster, B., Munne, R., Bernal, T., Hervas, B., Segundo, R. & Verschure, P. (2017) The effects of silent visuomotor cuing on word retrieval in Broca's aphasics: a pilot study. *International Conference on Rehabilitation Robotics (ICORR)*, London, UK.

Hilari, K., Needle, J.J. & Harrison, K.L. (2012) What are the important factors in health-related quality of life for people with aphasia? A systematic review. *Archives of Physical Medicine and Rehabilitation*, 93, 1 SUPP, S86–S95.

Holland, A., Frattali, C. & Fromm, D. (1999) *Communication activities of daily living-2*. Austin, TX: Pro-Ed.

Holland, A. (2007) *Counseling in communication disorders: a wellness perspective*. San Diego, CA: Plural.

Kauhanen, M.L., Korpelainen, J.T., Hiltunen, P., et al. (2000) Aphasia, depression, and non-verbal cognitive impairment in ischaemic stroke. *Cerebrovascular Diseases*, 10, 455–461.

Konnerup, U. (2019) The emerging landscape of virtual environments. *Tidsskriftet Læring og Medier (LOM)*, 21, 1–10.

Lavoie, M., Macoir, J. & Bier, N. (2017) Effectiveness of technologies in the treatment of post-stroke anomia: a systematic review. *Journal of communication disorders*, 65, 43–53. doi: 10.1016/j.jcomdis.2017.01.001.

Lawrence, E.S., Coshall, C., Dundas, R., Stewart, J., Rudd, A.G., Howard, R. & Wolfe, C.D. (2001) Estimates of the prevalence of acute stroke impairments and disability in a multiethnic population'. *Stroke*, 32, 6, 1279–1284. doi: 10.1161/01.STR.32.6.1279.

Lee, S., Shafe, A. & Cowie, M. (2011) UK stroke incidence, mortality and cardiovascular risk management 1999–2008: time-trend analysis from the General Practice Research Database. *British Medical Journal Open*, 1, e000269.

Linebarger, M.C., Schwartz, M.F. & Kohn, S.E. (2001) Computer-based training of language production: an exploratory study. *Neuropsychological Rehabilitation*, 11, 1, 57–96.

Maresca, G., Grazia Maggio, M., Latella, D., Cannavo, A., De Cola, M., Portaro, S., Stagnitti, M., Silvestri, G., Torrisi, M., Bramanti, A., De Luca, R. & Calabro, R. (2019) Toward improving poststroke aphasia: a pilot study on the growing use of telerehabilitation for the continuity of care. *Journal of Stroke and Cerebrovascular Diseases*, 28, 10, 104303.

Marshall, J., Booth, T., Devane, N., Galliers, J., Greenwood, H., Hilari, K., et al. (2016) Evaluating the benefits of aphasia intervention delivered in virtual reality: results of a quasi-randomised study. *PLoS One*, 11, 8, e0160381. doi: 10.1371/journal.pone.0160381.

Marshall, J., Devane, N., Edmonds, L., Talbot, R., Wilson, S., Woolf, C. & Zwart, N. (2018) Delivering word retrieval therapies for people with aphasia in a virtual communication environment. *Aphasiology*, 32, 9, 1054–1074. doi: 10.1080/02687038.2018.1488237.

Marshall, J., Devane, N., Talbot, R., Caute, A., Cruice, M., Hilari, K., MacKenzie, G., Maguire, K., Patel, A., Roper, A. & Wilson, S. (2020) A randomised trial of social support group intervention for people with Aphasia: a novel application of virtual reality. *PLoS One*. https://doi.org/10.1371/journal.pone.0239715

Northcott, S. & Hilari, K. (2011) Why do people lose their friends after a stroke?: Friendship loss post stroke'. *International Journal of Language & Communication Disorders*, 46, 5, 524–534. doi: 10.1111/j.1460-6984.2011.00079.x.

Palmer, R., Dimairo, M., Cooper, C., Enderby, P., Brady, M., Bowen, A., Latimer, N., Julious, S., Cross, E., Alshreef, A., Harrison, M., Bradley, E., Witts, H. & Chater, T. (2019) Self-managed, computerised speech and language therapy for patients with chronic aphasia post-stroke compared with usual care or attention control (Big CACTUS): a multicentre, single-blinded, randomised controlled trial. *The Lancet Neurology*, 18, 9, 821–833. doi: 10.1016/S1474-4422(19)30192-9.

Paolucci S. (2008) Epidemiology and treatment of post-stroke depression. *Neuropsychiatric Disease and Treatment*, 4, 1, 145–154. doi: 10.2147/ndt.s2017.

Parr, S. (2007) Living with severe aphasia: tracking social exclusion. *Aphasiology*, 21, 1, 98–123. doi: 10.1080/02687030600798337.

Poels, Y. de K. & Wijnand, I. (2007). It is always a lot of fun! Exploring dimensions of digital game experience using focus group methodology. *Proceedings of the 2007 Conference on Future Play*.

Pitt, R., Theodoros, D., Hil., A. & Russell, T. (2019) The impact of the telerehabilitation group aphasia intervention and networking programme on communication, participation, and quality of life in people with aphasia. *International Journal of Speech-Language Pathology*, 21, 5, 513–523. doi: 10.1080/17549507.2018.1488990.

Roper A., Marshall, J. & Wilson, S. (2016) Benefits and limitations of computer gesture therapy for the rehabilitation of severe aphasia. *Frontiers in Human Neuroscience*, 10, 595. doi: 10.3389/fnhum.2016.00595.

Seligman, M.E.P., Steen, T.A., Park, N. & Peterson C. (2005) Positive psychology in progress. Empirical validation of interventions. *American Psychologist*, 60, 410–421.

Simmons-Mackie, N., Raymer, A., & Cherney, L.R. (2016) Communication partner training in aphasia: an updated systematic review, *Archives of Physical Medicine and Rehabilitation*, 97, 12, 2202–2221.e8. doi: 10.1016/j.apmr.2016.03.023.

Stark, J., Pons, C. & Daniel, C. (2013) Integrating face to face language therapy with virtual reality applications for persons with aphasia. *International Conference on Virtual Reality*, Philadelphia.

The Stroke Association. (2018) The state of the nation stroke statistics, February, 2018. https://www.stroke.org.uk/sites/default/files/state_of_the_nation_2018.pdf. (Accessed 15 January 2020).

Thompson, C.K. (2019) Neurocognitive recovery of sentence processing in aphasia. *Journal of Speech, Language, and Hearing Research*, 62, 11, 3947–3972. doi: 10.1044/2019_JSLHR-L-RSNP-19-0219.

Thomas S.A. & Lincoln N.B. (2008) Predictors of emotional distress after stroke. *Stroke*, 39, 1240–1245.

Thomas, S.A., Walker, M.F., Macniven, J.A., Haworth, H. & Lincoln, N.B. (2012) Communication and low mood (CALM): a randomized controlled trial of behavioural therapy for stroke patients with aphasia. *Clinical Rehabilitation*, 27, 5, 398–408. doi: 10.1177/0269215512462227.

van der Gaag, A., Smith, L., Davis, S., Moss, B., Cornelius, V., Laing, S. & Mowles, C. (2005) Therapy and support services for people with long-term stroke and aphasia and their relatives: a six-month follow-up study. *Clinical Rehabilitation*, 19, 4, 372–380. doi: 10.1191/0269215505cr785oa.

Verna, A., Davidson, B. & Rose, T. (2009) Speech-language pathology services for people with aphasia: a survey of current practice in Australia. *International Journal of Speech-Language Pathology*, 11, 3, 191–205. doi: 10.1080/17549500902726059.

Vickers, C. (2010) Social networks after the onset of aphasia: the impact of aphasia group attendance. *Aphasiology*, 24, 902–913.

Wilson, S., Roper, A., Marshall, J., Galliers, J., Devane, N., Booth, T. & Woolf, C. (2015) Codesign for people with aphasia through tangible design languages. *CoDesign*, 11, 1, 21–34.

Wisenburn, B. & Mahoney, K. (2009) A meta-analysis of word-finding treatments for aphasia. *Aphasiology*, 23, 11, 1338–1352. doi: 10.1080/02687030902732745.

Woolf, C., Caute, A., Haigh, Z., Galliers, J., Wilson, S., Kessie, A., Hirani, S., Hegarty, B. & Marshall, J. (2016) A comparison of remote therapy, face to face therapy and an attention control intervention for people with aphasia: a quasi-randomised controlled feasibility study. *Clinical Rehabilitation*, 30, 4, 359–373.

Yee, N. & Bailenson, J. (2007) The Proteus effect: the effect of transformed self-representation on behaviour. *Human Communication Research*, 33, 271–290.

Yee, N., Bailenson, J. & Ducheneaut, N. (2009) The Proteus effect: implications of transformed digital self-representation on online and offline behaviour. *Communication Research*, 36, 2, 285–312.

Zheng, C., Lynch, L. & Taylor, N. (2015) Effect of computer therapy in aphasia: a systematic review. *Aphasiology*, 30, 211–244. doi: 10.1080/02687038.2014.996521.

16 Virtual Reality in Orthopedic Rehabilitation

Aliaa Rehan Youssef and Mohammed Gumaa
Cairo University
Giza, Egypt

CONTENTS

INTRODUCTION

The incidence and prevalence of musculoskeletal conditions is high worldwide (Picavet and Hazes, 2003; Haralson and Zuckerman, 2009). With increased life expectancy, the prevalence of such conditions, and subsequently, the socioeconomic burden, are expected to rise; especially for age-related degenerative diseases such as osteoarthritis (OA) (Picavet and Hazes, 2003; WHO Scientific Group on the Burden of Musculoskeletal Conditions at the Start of the New Millennium, 2003; Haralson and Zuckerman, 2009; Woolf, Erwin and March, 2012; Chang *et al.*, 2019).

Patients with chronic musculoskeletal conditions may be referred for physical rehabilitation either as an integral part of conservative management or for pre- or post-operative care. Rehabilitation is challenged by patients' willingness to comply with treatment, especially when it is prolonged, expensive, and not easily accessible. Further, rehabilitation is mostly done in clinical settings that have controlled environment, which are far away from reality. Thus, despite the probability that patients may be receiving proper rehabilitation, yet treatment outcomes may not be fully transferrable to daily life encounters. Moreover, a proper patient-centered rehabilitation needs not only to consider the physical aspects of the condition, but also the direct integration between physical function and the motor control involved; which is a cornerstone in developing adequate coping strategies. Evidence suggests that chronic musculoskeletal disorders may be associated with central

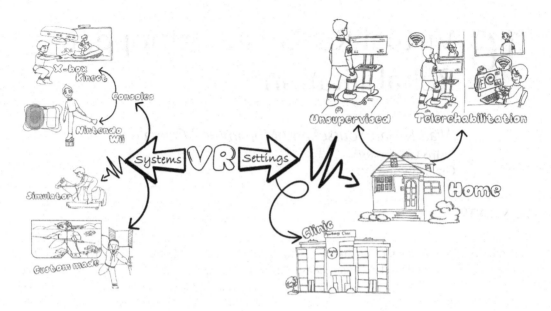

FIGURE 16.1   An illustration of VR use in orthopedic rehabilitation.

nervous system changes resulting in neurocognitive deficits (Grooms, Appelbaum and Onate, 2015; Tarragó *et al.*, 2016; Gandola *et al.*, 2017; Grooms *et al.*, 2017), that need to be addressed in order to prevent recurrence and improve treatment outcome.

Virtual reality (VR) may provide an alternative that simulates real world scenarios, thus transferring daily life events to clinical settings. Further, it requires the integration among various components of the neuromusculoskeletal system. Moreover, it is an amusing experience that may attract patients and increase their adherence to rehabilitation. In addition, it provides instantaneous feedback which may improve patient's response. Finally, it is a commercially available technology at an affordable price, and can be used remotely, which could reduce the work burden on therapists and allow patients to exercise within the convenience of their homes (Riener and Harders, 2012a, 2012b).

VR has been introduced over the past decades into the rehabilitation of pediatrics, geriatrics and patients with neurological diseases, burn, as well as musculoskeletal disorders (Morris, Louw and Grimmer-Somers, 2009; Ferreira Dos Santos *et al.*, 2016; Laver *et al.*, 2017; Ravi, Kumar and Singhi, 2017; Gumaa and Rehan Youssef, 2019). In orthopedic rehabilitation, the use of VR is still novel and evolving. It has been used in regular clinical settings (Lin *et al.*, 2007; McPhail *et al.*, 2016) and as an unsupervised home-based training or telerehabilitation (Burdea *et al.*, 2000). This technology could be used as a standalone or as an adjunctive treatment tool (McPhail *et al.*, 2016). It could be delivered using off-the-shelf consoles (such as Wii fit and Xbox), custom-made soft- and hardware systems, or simulator machines. The potential uses and route of administrating VR in orthopedic rehabilitation is summarized in Figure 16.1. For the sake of this chapter, evidence from clinical studies that investigated VR effectiveness in orthopedic rehabilitation will be discussed. This chapter is not intended to review the use of VR to enhance physical activity, prevent injury, or to assess patients' physical performance.

GENERALIZED MUSCULOSKELETAL CONDITIONS

VR has been used in rehabilitation of two generalized musculoskeletal conditions: fibromyalgia and rheumatoid arthritis. Preliminary evidence of VR effectiveness in the rehabilitation of those two conditions, compared to regular daily activities or home-based exercises, is promising.

Fibromyalgia is a challenging generalized musculoskeletal disease of an unknown etiology. Treatment aims to improve patients' quality of life (QoL). This could be achieved using a multi-disciplinary approach, especially pharmacological treatment and exercises. Specifically, aerobic conditioning and strength training have been recommended as corner stones in improving patients' symptoms. Further, the importance of encouraging patient's active participation and cognitive behavioral therapy are well emphasized (Hauser *et al.*, 2017; Macfarlane *et al.*, 2017). Thus, it would be expected that generalized amusing exergaming programs could positively impact patients' life. This has been proven in a few studies that compared VR exergaming to regular physical activities. In these studies, VR rehabilitation was given for a period ranging between 8 and 24 weeks. The aim of treatment was to improve physical performance and patient's daily physical activities by enhancing postural control and coordination, aerobic conditioning, strength and mobility as well as balance.

Results showed improved disease manifestations, QoL, and physical function (including strength, flexibility, coordination, and balance) (Collado-Mateo et al., 2017a, 2017b; Martin-Martinez *et al.*, 2019). Interestingly, perceived pain and QoL improved better, especially in women with poorer baseline QoL values (Villafaina, Collado-Mateo, Dominguez-Munoz, *et al.*, 2019). Moreover, VR was associated with improved electroencephalography (EEG) brain beta waves, especially in patients who had the disease symptoms for <17 years, which implies improved brain circulation (Villafaina, Collado-Mateo, Fuentes, *et al.*, 2019). All these studies used a custom-made hardware and software (VirtualEx-FM) exergaming system that tracked participants' motion using Microsoft Kinect and fed it into a personal computer. This system has a simple and easy interface that is controlled by a technician. It provides group therapy, which improves social interaction among patients. This particular VR system supports three different virtual environments to allow performing different motor training exercises. For example, warming up exercises require the patient to mimic the movement of an on-screen avatar, whereas coordination and movement control training are accomplished by having the participant reacting to an apple appearing on the screen. For mobility, balance, and coordination training, participants stepped on virtual footprints appearing on the screen. All games intensity and difficulty can be adjusted according to patients' performance. At the meantime, real time feedback is displayed on the screen, so that patients can reflect on how they execute movements.

Another challenging systemic disease is rheumatoid arthritis. Similar to fibromyalgia, generally treatment aims at controlling disease activity and manifestations as well as improving patient's function and QoL. While pharmacological treatment plays a substantial role in controlling inflammation and increasing disease remission (Lau *et al.*, 2019), evidence supports the crucial role of strengthening and conditioning exercises in improving symptoms and QoL (Kucukdeveci, 2020). A single cross-over study of a moderate quality showed that 12 weeks of home-based Nintendo Wii exergaming had effects similar to that of home-based strength, coordination, mobility and relaxation exercises (Zernicke *et al.*, 2016). In this study, the VR group selected exercises from commercially available Yoga, strengthening, balance and aerobic exercises. Preliminary evidence favors the potential use of VR exergaming as an alternative to home-based programs in this patients' population. However, more studies are needed to evaluate the various aspects of physical rehabilitation, including patient's adherence and satisfaction.

REGION-SPECIFIC MUSCULOSKELETAL CONDITIONS

VR has been used in the rehabilitation of chronic traumatic and degenerative diseases and disorders such as lateral ankle sprain, OA, neck and low back pain. It has also been used in post-operative rehabilitation especially following joint replacement. In the following section, VR rehabilitation for regional conditions will be discussed.

SPINE

The effectiveness of VR rehabilitation for chronic neck and back pain was explored in a few studies. VR was delivered using off-the-shelf consoles, custom-made systems and simulators. Overall preliminary evidence supports VR effectiveness that is similar to other traditional treatments, although patients' enjoyment and satisfaction tended to be greater with VR rehabilitation.

Neck pain is a prevalent disabling musculoskeletal condition. Exercises such as stabilization, endurance and strengthening, regardless to its type, are the most recommended treatment approaches (Cohen and Hooten, 2017; Fredin and Loras, 2017; Chou et al., 2018). To improve neck stability, mobility and control, the effectiveness of VR combined with kinematic training in improving neck range, movement velocity and accuracy was investigated in a couple of high quality randomized clinical controlled trials (RCTs). The VR program required patients to control a virtual Aeroplan appearing on a screen by moving the head in four directions (flexion, extension, right rotation, left rotation). In addition, patients in this group as well as a separate kinematic training control group patients were supplied with a laser pointer mounted on the head that can be projected on a poster for feedback. Kinematic training included supervised active neck movements, quick head movement in-between targets, static head positioning while moving the body, and smooth head movement following the target. Results favored VR adjunctive treatment in improving cervical flexion range, movement velocity, and accuracy compared to kinematic training alone and no-treatment controls, whereas kinematic training alone improved neck rotation. Self-reported exercise participation was similar in the two groups. After three-months, VR group showed a greater improvement in global perceived effect, while the kinematic training group had a greater movement velocity (Bahat et al., 2015, 2018; Gumaa and Rehan Youssef, 2019). Interestingly, when VR and kinematic training were given remotely (home-based), the VR group had a greater pain reduction as well as improved QoL, movement velocity, and accuracy that were sustained up to three months (Bahat et al., 2018). In these two previous studies, VR was delivered using an off-the-shelf hardware and customized software. The hardware was a head-mounted display with a three-dimensional (3D) built-in motion tracker (Wrap™1200VR by Vuzix, Rochester, New York). The interactive 3D virtual environment was developed using the Unity-pro software, version 3.5 (Unity Technologies, San Francisco) and the Vuzix Software development kit. It is worth noting that the outcomes in these two studies were assessed by the same VR system that was used for intervention; thus, the reported improvements in the VR group may be biased by the learning effect associated with patients' familiarity with the testing protocol (Gumaa and Rehan Youssef, 2019)

Similar to neck pain, exercise therapy is crucial for the management of chronic low back pain (Saragiotto et al., 2016; Gomes-Neto et al., 2017; Oliveira et al., 2018). VR rehabilitation of low back pain was delivered using an off-the-shelf Nintendo Wii console (Park, Lee and Ko, 2013; Kim et al., 2014; Zadro et al., 2019), custom-made VR games (Thomas et al., 2016), and horse-back riding simulator machine (Yoo et al., 2014). Treatment duration ranged from three days (Thomas et al., 2016) to eight weeks (Park, Lee and Ko, 2013; Yoo et al., 2014; Zadro et al., 2019). VR treatment was combined with other interventions such as electrophysical modalities (Park, Lee and Ko, 2013) or was given as a standalone intervention (Kim et al., 2014; Yoo et al., 2014; Thomas et al., 2016). VR training was delivered as supervised clinic-based (Park, Lee and Ko, 2013; Kim et al., 2014; Yoo et al., 2014; Thomas et al., 2016) or unsupervised home-based (Zadro et al., 2019). Although not explicitly mentioned, the aims of VR intervention were to improve muscle strength, neuromuscular control and balance, spine range of motion (ROM) as well as patients' QoL.

For clinic-based rehabilitation, after four weeks of Yoga-based VR, compared to Yoga-based stabilization exercises, VR significantly improved function and reduced pain, tenderness, disability and fear avoidance. Seemingly, the aim of treatment was to improve trunk stability. Notably, the methodological quality of this study was low due to bias in intervention delivery. Specifically, the VR group exercised for 30 minutes while the control group trained for double the time (one hour), which could have led to the development of fatigue that adversely affected outcome measures (Kim et al., 2014; Gumaa and Rehan Youssef, 2019). Another low-quality study, with apparently similar therapeutic goals, compared different sets of Nintendo Wii VR games to stabilization exercises and no-treatment controls. After eight weeks, findings supported similar effectiveness between the two interventions in reducing pain, increasing isometric back strength as well as improving the physical composite of

QoL. Yet, it should be stressed that only patients in the VR group showed a significant improvement in the mental subset of the QoL assessment, denoting an improved sense of well-being and the feeling of being energetic (Park, Lee and Ko, 2013). Although this was a small-scale study, with eight patients per each of the study arms, yet it provided preliminary evidence that VR may have favorable cognitive and mental effects, despite its clinical-effectiveness that is comparable to other interventions.

In an attempt to customize VR games to improve patients' neuromuscular control, pain and physical function, a third study investigated the feasibility and safety of a custom-made dodgeball VR game (using motion tracking). The game required the patient to play a dodgeball by controlling his/her avatar movements against four virtual opponent avatars. To improve VR immersion, participants wore Samsung 3D shutter glasses (Samsung Electronics Co Ltd) and were given auditory feedback such as crowd cheering, buzzers, and referee whistles. Patients were followed-up for three days only and were compared to no-treatment controls. The newly developed system showed similar results to that of the control group with regards to trunk flexion range gain and pain relief. Further, no adverse events were reported. Although this study was of a moderate quality, yet it was limited by its very brief duration (three consecutive days); which is quite short to expect meaningful clinical changes (Thomas et al., 2016; Gumaa and Rehan Youssef, 2019).

Moving from VR games to simulator machines, a moderate quality study that aimed at improving trunk muscles strength and reducing pain severity, showed a significant pain reduction in young men who received eight weeks of progressive training every other day on a horse-back riding simulator machine. Further, patients' isokinetic flexor and extensor strength as well as total muscle mass increased with less incorporation of fatty tissue. Horseback-riding simulator machine mimics the movement experienced during real horse riding, which enhances posture and balance, and subsequently, neuromuscular control. Horse-riding simulation exercises can progress in difficulty from regular walking to cantering (Yoo et al., 2014). It should be noted that in this study the simulator training was compared to no-treatment control patients. Exercise therapy, which is the gold standard for these cases, was not used as a comparator. Thus, there is a need for more non-inferiority studies including exercise as a fair comparator.

For home-based rehabilitation, an eight-week VR exergaming Wii Fit U flexibility training, a combined program of strengthening and aerobic exercises, was compared to patient's education and instructions with regards to improving patients' daily physical activities. The VR group showed immediate improvements in pain and function, and delayed (six months) improvements in self-efficacy and exercises engagement, with no adverse effects reported. However, these improvements were clinically insignificant (Zadro et al., 2019). Considering that the comparable control group did not receive any formal exercises nor they were followed up regularly by phone (as was the case with the VR group), it is difficult to conclude that home-based exergaming is not inferior to traditional exercise. Controlled non-inferiority trials as well as cost-effectiveness studies are needed to reach a solid conclusion.

LOWER EXTREMITY

VR was used for hip, knee and ankle rehabilitation whether for patients undergoing conservative or operative treatment for lower limb disorders or trauma. Available evidence supports VR telerehabilitation cost-effectiveness following total knee replacement (TKR) and the clinical efficacy of VR clinic-based ankle sprain rehabilitation. For other conditions and anatomic locations, evidence is still limited by the availability of clinical trials with rigorous methodological quality. Many of the published studies were pilot investigations that assessed VR feasibility and safety profile. On the other hand, well designed RCTs are limited. Currently, there are two registered well-designed RCTs protocols that are investigating the effectiveness of VR rehabilitation following total hip replacement (THR) and TKR (Negus et al., 2015; Eichler et al., 2017). We hope that these two studies will strengthen currently available evidence. In this section, available studies on VR rehabilitation for hip, knee and ankle are discussed.

VR for Hip Rehabilitation

The use of VR for hip rehabilitation is still novel and literatures are scarce. Two small-scale exploratory studies employed VR telerehabilitation for patients who undergone THR, no other hip diseases or surgeries were investigated. These studies investigated VR telerehabilitation feasibility and effectiveness. None of the studies was a randomized trial that directly compared telerehabilitation to face-to-face or unsupervised home exercises nor considered the economic aspects of rehabilitation. Thus, both studies could be considered as preliminary investigations rather than a robust RCT from which clear strong evidence could be reached.

Overall, preliminary evidence from available low-quality studies suggests that VR may have the advantage of being amusing to participants and may be beneficial for telerehabilitation. On the other hand, VR clinical effectiveness seems similar to traditional rehabilitation, yet as explained earlier, the evidence is still weak, limited and inconclusive. Considering that only VR telerehabilitation was investigated, it is worth mentioning that the effectiveness of VR telerehabilitation following THR, regardless to the method, is still controversial (Pastora-Bernal *et al.*, 2017; Wang *et al.*, 2019). Thus, the use of hip VR rehabilitation needs more studies before recommending its use for patients with hip pathology or following surgery. Following is a brief description of the available studies, so that the reader can get a glimpse on future directions in this research-rich area.

The first available study for hip VR telerehabilitation after THR was a pilot retrospective cohort study that investigated the effectiveness of the Virtual Exercise Rehabilitation Assistant, VERA (https://reflexionhealth.com/vera, Reflexion Health, San Diego, CA) in 16 patients who undergone THR. The VERA system has both assessment and rehabilitation modules. It is an FDA-cleared, and is a patient-centered, data-driven and value-based system. It consists of 3D motion tracking cameras and software that enables clinicians from assessing joint ROM as well as functional performance (using sit-to-stand and timed-up and-go tests). Further, clinicians can review and monitor patient-reported outcome measures. Rehabilitation features include clinician-prescribed physiotherapy protocols and an instructor avatar. Besides, VERA supports real-time tele-visits for close personalized rehabilitation.

In this study, patients received daily VR with a few interspersed face-to-face sessions until they achieved satisfactory progress (after an average of eight weeks). Compared to a historical cohort or literature referenced values (for patients who received traditional home- and/or outpatient-based physical therapy), VR telerehabilitation increased patients' satisfaction. On the other hand, the clinical effectiveness, adherence, and the incidence of adverse events were similar between the VR and control groups (Kuether *et al.*, 2019). Exercises were determined by the treating therapists according to a predetermined plan, with no sufficient details given. It should be noted that this study design was neither an RCT nor a matched-control retrospective cohort. Further, historical control received a variety of interventions (whether at home or clinic). Moreover, the THR cohort was a part of the study that also included patients who undergone TKR. The study conclusion was based on the whole study sample, and not just patients with THR as no subgroup analyses were done. Thus, the methodological quality of this study raises many concerns.

The second available study was a case series that included seven patients who undergone THR within 2–108 days. The primary goal of the study was to develop and validate a Kinect-based telerehabilitation system, rather than to investigate VR effectiveness. This system allows the therapist to record exercises using a moving avatar. Then, as the patient repeats the exercises, his/her own 3D avatar appears. Thus, patients can observe their own avatar's movement while receiving feedback on their performance in comparison with the physiotherapist's avatar. Patients were required to perform between one and four sessions within 1–21 days. Findings supported adequate system's capability for monitoring exercise execution accuracy. Further, participants reported an enjoyable and easy interaction with Kinect (Anton *et al.*, 2016).

VR FOR KNEE REHABILITATION

Knee rehabilitation using VR technology is the most abundant in literature, particularly its effectiveness after knee replacement. A couple of studies investigated VR effectiveness for knee OA (Lin et al., 2007; Kim, Kim and HwangBo, 2017) and only one study investigated its effectiveness post anterior cruciate ligament (ACL) repair (Baltaci et al., 2013).

VR rehabilitation following knee replacement was delivered either as face-to-face clinic-based or as home-based telerehabilitation. VR intervention was provided as an adjunctive to other exercises using off-the-shelf consoles or virtual telerehabilitation systems. It seems that VR telerehabilitation has the advantages of being accessible at home, attracting patients to participate in exercises, and saving the costs associated with regular clinic visits. This is in an agreement with the findings of a recent meta-analysis of technology-assisted rehabilitation following TKR (including telerehabilitation, game- or web-based therapy) that showed low to moderate quality evidence of a small clinical significance that these approaches particularly telerehabilitation, compared to traditional methods, improve pain and function following TKR. On the other hand, both approaches showed similar rates of re-hospitalization and adverse events (Wang et al., 2019). Following is a brief description of available clinical studies on VR clinic-based training and telerehabilitation for various knee conditions.

REHABILITATION AFTER TKR

After TKR, there is moderate quality evidence of improved pain, function and knee flexion range with adjunctive VR compared to adjunctive auto passive knee flexion exercise. These improvements were sustained at three- and six-months follow up. This evidence was driven from a study that compared a multimodal program combined with immersive rowing VR exercises (for 30 minutes) to auto-passive knee flexion (three sets for 30 repetitions/day) program. The multimodal program consisted of exercises (ankle movement, quadriceps strengthening, and passive knee flexion) combined with psychological intervention and pain management. Treatments were given day after day for four weeks. The details of used VR system were not given, except for its manufacture (Mide Technology Inc., Cangzhou, China), so that the exact nature of the received rehabilitation is not quite clear (Jin et al., 2018).

Also, as an adjunctive clinic-based intervention, progressive weightbearing (WB) biofeedback Nintendo Wii Fit Plus balance training combined with traditional exercises were given for six weeks and compared to traditional exercises alone. VR training consisted of bilateral stance, unilateral stance, sit-to-stand, and lunging exercises that were played using commercially available games (Ultimate Balance Challenge, Penguin Slide, Skiing, Table Tilt, Standing Knee Bend, Tree Pose, Chair Pose, Rowing Squat, Warrior Pose, and Lunge). Kinetic and kinematic analyses provided empirical good quality evidence supporting the similarity between VR and traditional exercises with regards to WB symmetry and knee extension moment during sit-to-stand transition. However, the VR group showed improved Sit-To-Stand test (which reflects lower limb strength, balance control, fall risk, and exercise capacity) that continued up to 26 weeks, whereas the traditional exercise group did not. Additionally, patients in the VR group showed a tendency towards improved walking speed (Christiansen et al., 2015). It should be noted that this study enrolled a small sample (12 patients in each arm) and the Wii balance board did not allow the use of a wide variety of dynamic training activities including walking, which was the activity assessed as an outcome measure in this study. This means that this study lacked training specificity, and thus a better response may have been obtained if a targeted training was provided.

Similarly, another study concluded that Wii Fit activities (focusing on lateral weight shift, multidirectional balance as well as static and dynamic postural control) using commercially available games (Ski Slalom, Tightrope Walk, Penguin Slide, Table Tilt, Hula Hoop, Balance Bubble, Deep Breathing, Half Moon, Torso Twist) in combination with stretching, strengthening and balance exercises were comparable to exercises alone in improving pain, function and knee range after TKR. However, the evidence from this study was limited by the small sample size (VR group: 27 patients; control group: 23 patients). Further, the selected VR activities were individualized and, thus, varied among patients (Fung *et al.*, 2012).

A third small scale short-follow up feasibility and safety study was conducted on 30 patients; 26 patients who undergone TKR (11 control patients receiving standard physiotherapy and 15 patients receiving VR exergames in addition to standard physical therapy) and four patients after ACL repair (two patients in each arm). Patients in the VR group received daily supervised knee flexion and extension exercises for ten minutes or until fatigue developed. VR was given via a Wii fit plus console that was connected to two remote plus controllers; one above and the other below the knee joint. This VR system was supported by custom-made software that graphically displayed the maximum, minimum and average knee angles reached, exercise time and the frequency of knee flexion motions performed. Exercises were given from the first postoperative day until discharge after approximately three days. Results confirmed the feasibility of using VR technology as well as patients' treatment acceptance. Further, short-term, knee function was similar in the VR and traditional rehabilitation groups. However, it should be emphasized that this was not a clinical study that investigated clinical effectiveness. Further, the results and conclusion were based on all enrolled patients; even those who did not undergo TKR (Ficklscherer *et al.*, 2016).

A pilot study investigated the effectiveness of a VR gait simulation exoskeleton assistive device (AVATAR-M, Shanghai Zhanghe Corporation, China) that moved the lower extremity in a fashion similar to walking cycle while allowing partial weight bearing adjusted to patients' preferred walking speed. In this study, patients either received AVATAR-M rehabilitation or walker-aided indoor ambulatory training. In addition, both groups received pain relief electrotherapy (neuromuscular electrical stimulation, NMES) and continuous passive motion (CPM). The VR group showed significantly better one-year sustained improvements of hospital for special surgery score (that evaluates pain, function, Knee ROM, muscle strength, flexion deformity, and instability), Berg score (that evaluates balance), ten-meter sitting-standing time, and six-minute walking distance. This implies that patients who received the robotic-assisted rehabilitation showed improved symptoms, gait steadiness, walking distance and speed. On the other hand, no differences were found with regards to knee kinesthesia, proprioception, or functional ambulation scores (Li *et al.*, 2014). It is worth mentioning that authors used a poorly valid scoring system for assessing knee joint kinesthesia and proprioception.

TKR TELEREHABILITATION

Evidence favoring virtual telerehabilitation following TKR is improving in strength and quality; moving from observational studies to well-designed RCTs. All studies used the VERA rehabilitation system. Available evidence supports its clinical- and cost-effectiveness. Initially, a non-controlled retrospective cohort study of 18 patients with TKR and 139 patients with uni-compartment knee replacement (UKR) proved VERA telerehabilitation feasibility and effectiveness. Patients' adherence to one-month rehabilitation was 76–78%. During the entire rehabilitation period, patients needed an average of three outpatient follow-up visits and two virtual visits. Patients spent an average of 10.8 hours performing exercises, with a daily average of 26.5 minutes, which accounts for approximately 21 patient-centered exercises. VERA usability, measured by a standardized ten-item system usability scale (that allows the evaluation of a wide variety of products and services), scored >90%; denoting that VERA rehabilitation system is effective, efficient, and satisfactory in attaining patients' treatment goals. Following therapy, patients reported an improved function as measured by walking and performance tests as well as self-reported questionnaires (Chughtai *et al.*, 2019).

Similar findings supporting the effectiveness of VERA rehabilitation system were obtained from a pilot retrospective cohort study that was conducted on 24 patients who undergone TKR.

Those patients were compared to historical controls or literature-based reference values as explained earlier. In this study, telerehabilitation was offered on a daily basis, in addition to 6–8 face-to-face postoperative physiotherapy sessions. Patients in the two groups were not different in acute care outcomes (post-operative complications, and hospitalization length). Adherence to VR varied from 12 to 92% depending on patients' ability to use the VERA system at home. Further, functional recovery exceeded that reported in literature. Moreover, patients highly recommended the VR telerehabilitation program to others. However, this study design was a pilot retrospective cohort that used a historical control treated at a different period of time. Further, the control group received a wide variety of treatments. Moreover, patient's distribution between the two study arms was unequal, thus the methodological quality raises a few concerns. Also, VR was combined with face-to-face visits, which may have had an effect on the perceived improvement. Finally, adherence calculation did not take into account the intervening face-to-face visits (Kuether et al., 2019). It is important to stress that these two studies although provided promising preliminary evidence, yet their retrospective nature and the associated inherited bias may have influenced patient's outcome measures.

A better quality RCT confirmed previous findings from observational studies regarding the cost-effectiveness of VERA rehabilitation following TKR. In one study, VERA telerehabilitation was compared to home-based traditional exercises after hospital discharge in 306 patients who undergone TKR. The VR group was allowed unrestricted use of VERA with its assessment and tele-visits features for 12 weeks. Results favored the VR telerehabilitation group as evident by the significantly reduced total post-discharge costs (by more than half); with a mean difference of $2,745. Patients receiving the VERA training had a significantly higher number of weekly physical therapy sessions, greater adherence to exercise program completion, less outpatient visits, and less re-hospitalization rates. On the other hand, the clinical effectiveness at 6 and 12 weeks supported previous findings that VR is non-inferior to traditional rehabilitation in terms of functional performance and knee ROM. Pain was not different between the two groups at 12 weeks, although the traditional program group had a significant pain reduction over the first six weeks greater than that of the VR group. Adverse events did not differ between the two groups, except for increased falling rate in the VR group; which was attributed to their increased participation in sports and recreational activities. In addition, patients who were assigned to the VR training reported less difficulty in sports and recreational activities participation. Finally, patients were satisfied with the VR program. It should be noted that this study considered only documented health delivery costs and did not account for the soft- and hardware development, technical support, nor other indirect costs. Further, rehabilitation program was patient-centered, resulting in some variability in delivered exercises among different patients (Bettger et al., 2019).

The promising evidence regarding VERA telerehabilitation is in an agreement with the general acceptance of home-based rehabilitation (whether monitored or unsupervised) as a non-inferior intervention, and in some occasions better, compared to face-to-face conventional rehabilitation. This was based on evidence from two systematic reviews and meta-analyses. One review and meta-analysis based on RCTs and non-RCTs supported the effectiveness of telerehabilitation (whether internet-based videoconferencing or interactive virtual- or telephone-based) as evident by increased patients' satisfaction and quadriceps strength compared to face-to-face conventional rehabilitation; although telerehabilitation did not differ with regards to clinical outcomes (pain, knee swelling, active range, hamstring strength as well as functional performance). This selective improvement was attributed to easy accessibility to treatment when it is performed at home (Shukla, Nair and Thakker, 2017). Although this is a logical explanation, however, it failed to explain the similarity between telerehabilitation and conventional therapy in the remaining outcomes of interest. This was further confirmed, as evidence from a second meta-analysis suggested that telerehabilitation, compared to face-to-face rehabilitation, results in better functional recovery as well as knee extension range and quadriceps strength. Yet it was similar in pain relief as well as knee flexion range gain and safety profile (Jiang et al., 2018).

VR REHABILITATION FOR OA

Surprisingly, despite the high prevalence of OA worldwide and the crucial role of exercise therapy in the conservative management of these cases, only two studies investigated VR rehabilitation effectiveness in patients with knee OA. These studies used custom-made systems or simulator machines and combined VR with strength training. Their findings supported the notion that VR effectiveness is comparable to traditional exercise, thus, they could be used interchangeably.

The first study was of moderate quality and investigated the effectiveness of custom-made interactive VR games (that required patients to press with their lower extremity on pedals in a closed-kinetic chain similar fashion as a target appeared on the screen) and compared them to traditional close kinetic chain exercises in patients with bilateral OA. After eight weeks of training, the two groups showed similar improvements in knee joint position sense, muscle strength, functional score, and walking speed. Yet, the traditional exercises group showed better improvement in muscle strength (Lin *et al.*, 2007). The second study was of low quality and used horse-back riding simulator training combined with strength training in elderly for eight-weeks. The details of simulator exercise were insufficiently described. This combined program was compared to strengthening exercises only. The two groups showed similar improvements in walking velocity and knee strength (Kim, Kim and HwangBo, 2017).

VR REHABILITATION AFTER ACL REPAIR

Again, as in OA rehabilitation, despite that high incidence of ACL in young adults and the well-proven crucial role of exercises and neuromuscular control in its rehabilitation, only one small-sized RCT (15 patients per arm) investigated VR effectiveness in young active adults after ACL endoscopic repair using hamstrings grafts. A twelve-week program of Nintendo Wii based rehabilitation was compared to a hybrid program of closed and open kinetic chain exercises. VR program included commercially available bowling and skiing games (in Wii sports), boxing, football and balance Board (within Sports Pro Series). Investigators selected these games for their potential to positively influence physical and functional movements, cognitive functioning, and driving. Each game was practiced for 15 minutes, for a total session duration of an hour, given three times per week. The control group received progressive exercises that included prone knee flexion hanging position, isometric strengthening, straight leg raising, squatting, and jogging. VR rehabilitation outcomes were similar to that of conventional rehabilitation with regards to dynamic balance, movement coordination, proprioception, and response time (Baltaci *et al.*, 2013).

VR FOR ANKLE REHABILITATION

For lateral ankle sprain (non-recurrent, one month after injury) and chronic ankle instability (CAI), moderate to high quality evidence showed that VR clinical effectiveness is similar to that of traditional exercises in improving balance; although it may be superior in increasing patients' enjoyment. On the other hand, VR superiority to wait-and-see or natural recovery is still inconclusive (Gumaa and Rehan Youssef, 2019). All available literature used VR to enhance balance; a neuromuscular control component that is believed to play a crucial role in the development and treatment of CAI (Delahunt *et al.*, 2018; Vuurberg *et al.*, 2018; Hertel and Corbett, 2019).

For lateral ankle sprain, the same research group conducted two studies comparing six weeks of VR, multimodal exercises, and no-treatment in terms of clinical and biomechanical outcomes. The VR group used commercially available Nintendo Wii games that focused on lateral shift (Ski Slalom, Penguin Slide) and multidirectional balance training (Table Tilt, and Balance Bubble). The first study showed similar responses in patients allocated to the three groups. All enrolled patients showed comparable improvements in patient-reported function, pain during walking, return to activity, and satisfaction, although, a small proportion of patients (*n* = 3) were not satisfied. Only the VR group had better pain reduction at rest (Punt *et al.*, 2016). Similarly, the second study showed no differences in gait parameters as the walking speed, cadence and step length improved in all patients. None of the groups showed improved dorsiflexion range, whereas the plantar flexion range improved in the two treated groups. Also, single-stance duration symmetry improved in the two treated groups, while only the Wii Fit group showed improved single support time (Punt *et al.*, 2017).

For CAI, two studies used commercially available VR systems. The first study was conducted in competitive soccer players. Patients received ten weeks of either VR exergaming (using X-box commercially available adventure balance games: Rally Ridge, River Rush, and 2000 Leaks), traditional balance training (using a mini trampoline and inflatable discs), or no treatment. Only, the two treated groups showed similar improvement in dynamic balance and comparable exercises compliance. However, the VR group enjoyed the treatment better (Vernadakis *et al.*, 2014).

Another perspective in evaluating VR rehabilitation effectiveness for CAI was to compare different VR games on the same outcome of interest: balance. One study showed that four weeks of training with Nintendo Wii Fit Plus in patients with CAI improved static and dynamic balance regardless to the type of games played. In this study, patients were randomly assigned to receive either commercially available VR-based balance training games (The Soccer Heading, Ski Slalom, Tight Rope Walk, Table Tilt, and Snowboard Slalom games) or VR-based strength training games (Lunges, Single-Leg Extensions, Sideways Leg Lifts, Single Leg Twists, and Rowing Squats). Both groups showed improved static and dynamic balance in selected directions. However, balance VR-training was better than VR strength training in improving overall static as well as dynamic overall and anteroposterior balance (Kim and Heo, 2015), which is an expected finding considering the similarity between the trained activity and the measured outcome. In this study, the rationale for selecting VR-based strengthening exercises as a comparator was not provided (Gumaa and Rehan Youssef, 2019). Probably authors should have considered assessing muscle strength as an outcome measure that is relevant to the training received in one of the study arms.

It should be noted that VR combined with robotic assisted telerehabilitation seems promising in treating paretic foot in patients with neurological diseases even after a single treatment session (Zeng *et al.*, 2018). To authors' knowledge, studies investigating the effectiveness of using this combined rehabilitation technology in orthopedics is lacking, which could be a new promising research avenue with important clinical applicability.

VR for Traumatic Fracture Rehabilitation

As an attempt to the examine the feasibility of using VR in the rehabilitation of lower extremity fractures, Wii Fit Plus balance board in combination with commercially-available games (that focus on weight shifting and multidirectional balance) were compared to regular care in treating patients with various lower extremity fractures, regardless to fracture site or fixation method. This study provided preliminary evidence that VR adjunctive usage in these cases is feasible and safe (McPhail *et al.*, 2016). However, this study was a non-randomized pilot exploratory investigation that aimed at gathering preliminary data to calculate sample size for future well-designed RCTs, thus, despite the promising conclusion, yet further validation is needed.

UPPER EXTREMITY

The use of VR in upper extremity orthopedic rehabilitation is still in its premature stage compared to neuro-rehabilitation of cases such as stroke and cerebral palsy; which is abundant in literature (Chen, Lee and Howard, 2014; Yates, Kelemen and Sik Lanyi, 2016; Lee *et al.*, 2018; Ahn and Hwang, 2019; Karamians *et al.*, 2019). Thus, evidence regarding VR effectiveness in this area is insufficient to draw a conclusion. There were a few attempts to develop VR-based rehabilitation systems for upper extremity musculoskeletal disorders, yet their clinical effectiveness needs further validation using high-quality clinical trials (Burdea *et al.*, 2000; Heuser *et al.*, 2007). Only VR effectiveness for home-based rehabilitation of subacromial impingement syndrome with scapular dyskinesia was investigated. For those patients, one study of moderate quality provided evidence supporting the effectiveness of six-week home-based exercises and VR Nintendo Wii bilateral shoulder elevation, boxing, bowling and tennis exergaming, compared to home-based exercises alone, in relieving pain (Pekyavas and Ergun, 2017). However, based on findings from patients with neurological disorders; we could expect that VR rehabilitation, especially after traumatic hand injury and probably frozen shoulder, may be promising. Proper RCTs are required to confirm or refute this hypothesis.

ILLUSTRATION OF THE APPLICATION FROM THE USERS PERSPECTIVE

Most of the studies that investigated patients' perspective showed evidence of increased satisfaction, adherence and enjoyment compared to conventional methods. VR systems provide instantaneous feedback regarding patients' performance; besides they record patients' training duration and progress. Further, VR may allow remote monitoring and optimization of treatment parameters by clinicians; without the need for regular in-clinic visits. Moreover, for home-based rehabilitation, VR systems may be readily accessible by patients, which may increase their exercising time (Palazzo *et al.*, 2016).

DISCUSSION AND FUTURE TRENDS/DEVELOPMENTS

Studies evaluating VR effectiveness in orthopedic rehabilitation are limited and lack rigorous methodology. The most abundantly studied orthopedic cases were fibromyalgia, low back pain, post-operative knee replacement, and lateral ankle sprain. Overall, VR clinical effectiveness seems similar to other conventional treatments. However, several studies favored VR in increasing patients' adherence to therapy and improving their satisfaction. Further, its remote use for home-based rehabilitation, whether unsupervised or through telerehabilitation, seems promising and may reduce work burden on therapists and the costs associated with face-to-face rehabilitation. Future studies are recommended to validate VR use for the rehabilitation of other common orthopedic conditions, in different age groups, and to conduct economic evaluation. More properly designed prospective non-inferiority and superiority trials that compare VR to traditional rehabilitation are recommended to confirm these findings. Further, economic studies are recommended to consider indirect treatment costs as well as the costs associated with VR innovation, system development, installation and maintenance.

Moreover, although a few studies evaluated VR rehabilitation clinical effectiveness and patients' satisfaction, yet only one study evaluated the effect of VR on the central nervous system. As central mechanism and neuropathic-etiology of pain have been claimed to contribute to the chronicity and recurrence of a few orthopedic diseases and dysfunctions, evaluation of VR effects on central neural mechanism using functional Magnetic Resonance Imaging (fMRI) and EEG may add another perspective on its clinical value in orthopedic rehabilitation.

Finally, engineers are encouraged to develop open source software that can be used with popular devices such as smartphones, commercially available VR goggles as well as Wii and Xbox Kinect consoles. Additionally, they are recommended to consider including training modules and patient's assessment tools to increase the feasibility and value of VR use in clinical settings. Also, they may consider combining VR with other technologies such as wearable sensors and 3D printed accessories to enrich data acquisition for assessment and to customize treatment of virtual environment. This may also help in obtaining real-world data that can be employed by machine learning and prediction algorithms to develop more effective rehabilitation protocols. It would be great if the patient's real environment could be captured, by an image acquisition method, and projected into the virtual environment, thus patients could be trained in an environment similar to that of real life, which could improve the patients' centeredness features of the rehabilitation and may allow an integrated rehabilitation that combines occupational therapy and vocational therapy with physical therapy. As technology are emerging every day, engineers, rehabilitation therapists and healthcare authorities are encouraged to collaborate in order to improve the existing VR technology, improve its immersion level, and increase its availability at affordable price to bridge the existing gap in rehabilitation.

ACKNOWLEDGMENT

We would like to thank Ms Hebatulkarim Gumah for designing the figure.

REFERENCES

Ahn, S. and Hwang, S. (2019) 'Virtual rehabilitation of upper extremity function and independence for stoke: a meta-analysis', *Journal of Exercise Rehabilitation*, 15(3), pp. 358–369. doi: 10.12965/jer.1938174.087.

Anton, D. *et al.* (2016) 'Validation of a Kinect-based telerehabilitation system with total hip replacement patients', *Journal of Telemedicine and Telecare*, 22(3), pp. 192–197. doi: 10.1177/1357633X15590019.

Bahat, H. S. *et al.* (2015) 'Cervical kinematic training with and without interactive VR training for chronic neck pain - a randomized clinical trial', *Manual Therapy*, 20(1), pp. 68–78.

Bahat, H. S. *et al.* (2018) 'Remote kinematic training for patients with chronic neck pain: a randomised controlled trial', *European Spine Journal*, 27(6), pp. 1309–1323.

Baltaci, G. *et al.* (2013) 'Comparison between Nintendo Wii Fit and conventional rehabilitation on functional performance outcomes after hamstring anterior cruciate ligament reconstruction: prospective, randomized, controlled, double-blind clinical trial', *Knee Surgery, Sports Traumatology, Arthroscopy*, 21(4), pp. 880–887. doi: 10.1007/s00167-012-2034-2.

Bettger, J. P. *et al.* (2019) 'Effects of virtual exercise rehabilitation in-home therapy compared with traditional care after total knee arthroplasty: veritas, a randomized controlled trial', *The Journal of Bone and Joint Surgery. American volume*. doi: 10.2106/JBJS.19.00695.

Burdea, G. *et al.* (2000) 'Virtual reality-based orthopedic telerehabilitation', *IEEE Transactions on Rehabilitation Engineering: A Publication of the IEEE Engineering in Medicine and Biology Society*, 8(3), pp. 430–432.

Chang, A. Y. *et al.* (2019) 'Measuring population ageing: an analysis of the global burden of disease study 2017', *The Lancet. Public Health*, 4(3), pp. e159–e167. doi: 10.1016/S2468-2667(19)30019-2.

Chen, Y., Lee, S.-Y. and Howard, A. M. (2014) 'Effect of virtual reality on upper extremity function in children with cerebral palsy: a meta-analysis', *Pediatric Physical Therapy: The Official Publication of the Section on Pediatrics of the American Physical Therapy Association*, 26(3), pp. 289–300. doi: 10.1097/PEP.0000000000000046.

Chou, R. *et al.* (2018) 'The global spine care initiative: applying evidence-based guidelines on the non-invasive management of back and neck pain to low- and middle-income communities', *European Spine Journal: Official Publication of the European Spine Society, the European Spinal Deformity Society, and the European Section of the Cervical Spine Research Society*, 27(Suppl 6), pp. 851–860. doi: 10.1007/s00586-017-5433-8.

Christiansen, C. L. *et al.* (2015) 'Effects of weight-bearing biofeedback training on functional movement patterns following total knee arthroplasty: a randomized controlled trial', *The Journal of Orthopaedic and Sports Physical Therapy*, 45(9), pp. 647–655. doi: 10.2519/jospt.2015.5593.

Chughtai, M. *et al.* (2019) 'The role of virtual rehabilitation in total and unicompartmental knee arthroplasty', *The Journal of Knee Surgery*, 32(1), pp. 105–110. doi: 10.1055/s-0038-1637018.

Cohen, S. P. and Hooten, W. M. (2017) 'Advances in the diagnosis and management of neck pain', *BMJ (Clinical Research ed.)*, 358, p. j3221. doi: 10.1136/bmj.j3221.

Collado-Mateo, D., Dominguez-Munoz, F. J., *et al.* (2017a) 'Effects of exergames on quality of life, pain, and disease effect in women with fibromyalgia: a randomized controlled trial', *Archives of Physical Medicine and Rehabilitation*, 98(9), pp. 1725–1731. doi: 10.1016/j.apmr.2017.02.011.

Collado-Mateo, D., Dominguez-Munoz, F. J., *et al.* (2017b) 'Exergames for women with fibromyalgia: a randomised controlled trial to evaluate the effects on mobility skills, balance and fear of falling', *Peer Journal*, 5, p. e3211. doi: 10.7717/peerj.3211.

Delahunt, E. *et al.* (2018) 'Clinical assessment of acute lateral ankle sprain injuries (ROAST): 2019 consensus statement and recommendations of the international ankle consortium', *British Journal of Sports Medicine*, 52(20), pp. 1304–1310. doi: 10.1136/bjsports-2017-098885.

Eichler, S. *et al.* (2017) 'Effectiveness of an interactive telerehabilitation system with home-based exercise training in patients after total hip or knee replacement: study protocol for a multicenter, superiority, no-blinded randomized controlled trial', *Trials*, 18(1), p. 438. doi: 10.1186/s13063-017-2173-3.

Ferreira Dos Santos, L. *et al.* (2016) 'Movement visualisation in virtual reality rehabilitation of the lower limb: a systematic review', *Biomedical Engineering Online*, 15(Suppl 3), p. 144. doi: 10.1186/s12938-016-0289-4.

Ficklscherer, A. *et al.* (2016) 'Testing the feasibility and safety of the Nintendo Wii gaming console in ortho-pedic rehabilitation: a pilot randomized controlled study', *Archives of Medical Science*, 12(6), pp. 1273–1278. doi: 10.5114/aoms.2016.59722.

Fredin, K. and Loras, H. (2017) 'Manual therapy, exercise therapy or combined treatment in the management of adult neck pain - a systematic review and meta-analysis', *Musculoskeletal Science & Practice*, 31, pp. 62–71. doi: 10.1016/j.msksp.2017.07.005.

Fung, V. *et al.* (2012) 'Use of Nintendo Wii Fit in the rehabilitation of outpatients following total knee replace-ment: a preliminary randomised controlled trial', *Physiotherapy*, 98(3), pp. 183–188. doi: 10.1016/j.physio.2012.04.001.

Gandola, M. *et al.* (2017) 'Functional brain effects of hand disuse in patients with trapeziometacarpal joint osteoarthritis: executed and imagined movements', *Experimental Brain Research*, 235(10), pp. 3227–3241. doi: 10.1007/s00221-017-5049-6.

Gomes-Neto, M. *et al.* (2017) 'Stabilization exercise compared to general exercises or manual therapy for the management of low back pain: a systematic review and meta-analysis', *Physical Therapy in Sport: Official Journal of the Association of Chartered Physiotherapists in Sports Medicine*, 23, pp. 136–142. doi: 10.1016/j.ptsp.2016.08.004.

Grooms, D., Appelbaum, G. and Onate, J. (2015) 'Neuroplasticity following anterior cruciate ligament injury: a framework for visual-motor training approaches in rehabilitation', *Journal of Orthopaedic & Sports Physical Therapy*, 45(5), pp. 381–393. doi: 10.2519/jospt.2015.5549.

Grooms, D. R. *et al.* (2017) 'Neuroplasticity associated with anterior cruciate ligament reconstruction', *Journal of Orthopaedic & Sports Physical Therapy*, 47(3), pp. 180–189. doi: 10.2519/jospt.2017.7003.

Gumaa, M. and Rehan Youssef, A. (2019) 'Is virtual reality effective in orthopedic rehabilitation? A system-atic review and meta-analysis', *Physical Therapy*, 99(10), pp. 1304–1325. doi: 10.1093/ptj/pzz093.

Haralson, R. H. 3rd and Zuckerman, J. D. (2009) 'Prevalence, health care expenditures, and orthopedic surgery workforce for musculoskeletal conditions', *JAMA*, 302(14), pp. 1586–1587. doi: 10.1001/jama.2009.1489.

Hauser, W. *et al.* (2017) 'Management of fibromyalgia: practical guides from recent evidence-based guide-lines', *Polish Archives of Internal Medicine*, 127(1), pp. 47–56. doi: 10.20452/pamw.3877.

Hertel, J. and Corbett, R. O. (2019) 'An updated model of chronic ankle instability', *Journal of Athletic Training*, 54(6), pp. 572–588. doi: 10.4085/1062-6050-344-18.

Heuser, A. *et al.* (2007) 'Telerehabilitation using the Rutgers Master II glove following carpal tunnel release surgery: proof-of-concept', *IEEE Transactions on Neural Systems and Rehabilitation Engineering: A Publication of the IEEE Engineering in Medicine and Biology Society*, 15(1), pp. 43–49. doi: 10.1109/TNSRE.2007.891393.

Jiang, S. *et al.* (2018) 'The comparison of telerehabilitation and face-to-face rehabilitation after total knee arthroplasty: a systematic review and meta-analysis', *Journal of Telemedicine and Telecare*, 24(4), pp. 257–262. doi: 10.1177/1357633X16686748.

Jin, C. *et al.* (2018) 'Virtual reality intervention in postoperative rehabilitation after total knee arthro-plasty: a prospective and randomized controlled clinical trial', *International Journal of Clinical and Experimental Medicine*, 11(6), pp. 6119–6124.

Karamians, R. *et al.* (2019) 'Effectiveness of virtual reality- and gaming-based interventions for upper extrem-ity rehabilitation poststroke: a meta-analysis', *Archives of Physical Medicine and Rehabilitation*. doi: 10.1016/j.apmr.2019.10.195.

Kim, K. J. and Heo, M. (2015) 'Effects of virtual reality programs on balance in functional ankle instability', *Journal of Physical Therapy Science*, 27(10), pp. 3097–3101.

Kim, S. K., Kim, S. G. and HwangBo, G. (2017) 'The effect of horse-riding simulator exercise on the gait, muscle strength and muscle activation in elderly people with knee osteoarthritis', *Journal of Physical Therapy Science*, 29(4), pp. 693–696. doi: 10.1589/jpts.29.693.

Kim, S.-S. *et al.* (2014) 'The effects of VR-based Wii fit yoga on physical function in middle-aged female LBP patients', *Journal of Physical Therapy Science*, 26(4), pp. 549–552. doi: 10.1589/jpts.26.549.

Kucukdeveci, A. A. (2020) 'Nonpharmacological treatment in established rheumatoid arthritis', *Best Practice & Research. Clinical Rheumatology*, p. 101482. doi: 10.1016/j.berh.2019.101482.

Kuether, J. *et al.* (2019) 'Telerehabilitation for total hip and knee arthroplasty patients: a pilot series with high patient satisfaction', *HSS Journal: The Musculoskeletal Journal of Hospital for Special Surgery*, 15(3), pp. 221–225. doi: 10.1007/s11420-019-09715-w.

Lau, C. S. *et al.* (2019) '2018 update of the APLAR recommendations for treatment of rheumatoid arthritis', *International Journal of Rheumatic Diseases*, 22(3), pp. 357–375. doi: 10.1111/1756-185X.13513.

Laver, K. E. *et al.* (2017) 'Virtual reality for stroke rehabilitation', *Cochrane Database of Systematic Reviews*, 11, p. CD008349. doi: 10.1002/14651858.CD008349.pub4.

Lee, S. H. *et al.* (2018) 'Virtual reality rehabilitation with functional electrical stimulation improves upper extremity function in patients with chronic stroke: a pilot randomized controlled study', *Archives of Physical Medicine and Rehabilitation*, 99(8), pp. 1447–1453.e1. doi: 10.1016/j.apmr.2018.01.030.

Li, J. *et al.* (2014) 'A pilot study of post-total knee replacement gait rehabilitation using lower limbs robot-assisted training system', *European Journal of Orthopaedic Surgery & Traumatology: Orthopedie Traumatologie*, 24(2), pp. 203–208. doi: 10.1007/s00590-012-1159-9.

Lin, D.-H. *et al.* (2007) 'Comparison of proprioceptive functions between computerized proprioception facilitation exercise and closed kinetic chain exercise in patients with knee osteoarthritis', *Clinical Rheumatology*, 26(4), pp. 520–528. doi: 10.1007/s10067-006-0324-0.

Macfarlane, G. J. *et al.* (2017) 'EULAR revised recommendations for the management of fibromyalgia', *Annals of the Rheumatic Diseases*, 76(2), pp. 318–328. doi: 10.1136/annrheumdis-2016-209724.

Martin-Martinez, J. P. *et al.* (2019) 'Effects of 24-week exergame intervention on physical function under single- and dual-task conditions in fibromyalgia: a randomized controlled trial', *Scandinavian Journal of Medicine & Science in Sports*, 29(10), pp. 1610–1617. doi: 10.1111/sms.13502.

McPhail, S. M. *et al.* (2016) 'Nintendo Wii Fit as an adjunct to physiotherapy following lower limb fractures: preliminary feasibility, safety and sample size considerations', *Physiotherapy*, 102(2), pp. 217–220. doi: 10.1016/j.physio.2015.04.006.

Morris, L. D., Louw, Q. A. and Grimmer-Somers, K. (2009) 'The effectiveness of virtual reality on reducing pain and anxiety in burn injury patients', *The Clinical Journal of Pain*, 25(9), pp. 815–826. doi: 10.1097/AJP.0b013e3181aaa909.

Negus, J. J. *et al.* (2015) 'Patient outcomes using Wii-enhanced rehabilitation after total knee replacement - the TKR-POWER study', *Contemporary Clinical Trials*, 40, pp. 47–53. doi: 10.1016/j.cct.2014.11.007.

Oliveira, C. B. *et al.* (2018) 'Clinical practice guidelines for the management of non-specific low back pain in primary care: an updated overview', *European Spine Journal: Official Publication of the European Spine Society, the European Spinal Deformity Society, and the European Section of the Cervical Spine Research Society*, 27(11), pp. 2791–2803. doi: 10.1007/s00586-018-5673-2.

Palazzo, C. *et al.* (2016) 'Barriers to home-based exercise program adherence with chronic low back pain: patient expectations regarding new technologies', *Annals of Physical and Rehabilitation Medicine*, 59(2), pp. 107–113. doi: 10.1016/j.rehab.2016.01.009.

Park, J.-H., Lee, S.-H. and Ko, D.-S. (2013) 'The effects of the Nintendo Wii exercise program on chronic work-related low back pain in industrial workers', *Journal of Physical Therapy Science*, 25(8), pp. 985–988. doi: 10.1589/jpts.25.985.

Pastora-Bernal, J. M. et al. (2017) 'Evidence of benefit of telerehabitation after orthopedic surgery: a systematic review', *Journal of Medical Internet Research*, 19(4), p. e142. doi: 10.2196/jmir.6836.

Pekyavas, N. O. and Ergun, N. (2017) 'Comparison of virtual reality exergaming and home exercise programs in patients with subacromial impingement syndrome and scapular dyskinesis: short term effect', *Acta Orthopaedica Et Traumatologica Turcica*, 51(3), pp. 238–242. Available at: file:///P:/research/PhD/conduction/quality assess stage/full articles/upper limb (2)/Pekyavas et al 2017.pdf. https://doi.org/10.1016/j.aott.2017.03.008.

Picavet, H. S. J. and Hazes, J. M. W. (2003) 'Prevalence of self reported musculoskeletal diseases is high', *Annals of the Rheumatic Diseases*, 62(7), pp. 644–650. doi: 10.1136/ard.62.7.644.

Punt, I. M. *et al.* (2016) 'Wii Fit exercise therapy for the rehabilitation of ankle sprains: its effect compared with physical therapy or no functional exercises at all', *Scandinavian Journal of Medicine & Science in Sports*, 26(7), pp. 816–823. doi: 10.1111/sms.12509.

Punt, I. M. *et al.* (2017) 'Effect of Wii Fit™ exercise therapy on gait parameters in ankle sprain patients: a randomized controlled trial', *Gait and Posture*, 58, pp. 52–58. doi: 10.1016/j.gaitpost.2017.06.284.

Ravi, D. K., Kumar, N. and Singhi, P. (2017) 'Effectiveness of virtual reality rehabilitation for children and adolescents with cerebral palsy: an updated evidence-based systematic review', *Physiotherapy*, 103(3), pp. 245–258. doi: 10.1016/j.physio.2016.08.004.

Riener, R. and Harders, M. (2012a) 'Introduction to virtual reality in medicine', in *Virtual Reality in Medicine*. Springer, pp. 1–12.

Riener, R. and Harders, M. (2012b) 'Virtual reality for rehabilitation', in *Virtual Reality in Medicine*. Springer, pp. 161–180.

Saragiotto, B. T. *et al.* (2016) 'Motor control exercise for chronic non-specific low-back pain', *The Cochrane Database of Systematic Reviews*, (1), p. CD012004. doi: 10.1002/14651858.CD012004.

Shukla, H., Nair, S. R. and Thakker, D. (2017) 'Role of telerehabilitation in patients following total knee arthroplasty: evidence from a systematic literature review and meta-analysis', *Journal of Telemedicine and Telecare*, 23(2), pp. 339–346. doi: 10.1177/1357633X16628996.

Tarragó, M. da G. L. *et al.* (2016) 'Descending control of nociceptive processing in knee osteoarthritis is associated with intracortical disinhibition', *Medicine*, 95(17), p. e3353. doi: 10.1097/MD.0000000000003353.

Thomas, J. S. *et al.* (2016) 'Feasibility and safety of a virtual reality dodgeball intervention for chronic low back pain: a randomized clinical trial', *The Journal of Pain: Official Journal of the American Pain Society*, 17(12), pp. 1302–1317. doi: 10.1016/j.jpain.2016.08.011.

Vernadakis, N. *et al.* (2014) 'The effect of Xbox Kinect intervention on balance ability for previously injured young competitive male athletes: a preliminary study', *Physical Therapy in Sport*, 15(3), pp. 148–155. doi: 10.1016/j.ptsp.2013.08.004.

Villafaina, S., Collado-Mateo, D., Dominguez-Munoz, F. J., *et al.* (2019) 'Benefits of 24-week exergame intervention on health-related quality of life and pain in women with fibromyalgia: a single-blind, randomized controlled trial', *Games for Health Journal*, 8(6), pp. 380–386. doi: 10.1089/g4h.2019.0023.

Villafaina, S., Collado-Mateo, D., Fuentes, J. P., *et al.* (2019) 'Effects of exergames on brain dynamics in women with fibromyalgia: a randomized controlled trial', *Journal of Clinical Medicine*, 8(7). doi: 10.3390/jcm8071015.

Vuurberg, G. *et al.* (2018) 'Diagnosis, treatment and prevention of ankle sprains: update of an evidence-based clinical guideline', *British Journal of Sports Medicine*, 52(15), p. 956. doi: 10.1136/bjsports-2017-098106.

Wang, X. *et al.* (2019) 'Technology-assisted rehabilitation following total knee or hip replacement for people with osteoarthritis: a systematic review and meta-analysis', *BMC Musculoskeletal Disorders*, 20(1), p. 506. doi: 10.1186/s12891-019-2900-x.

WHO Scientific Group on the Burden of Musculoskeletal Conditions at the Start of the New Millennium. (2003) The Burden of Musculoskeletal Conditions at the Start of the New Millennium, World Health Organization Technical Report Series. Switzerland.

Woolf, A. D., Erwin, J. and March, L. (2012) 'The need to address the burden of musculoskeletal conditions', *Best Practice & Research. Clinical Rheumatology*, 26(2), pp. 183–224. doi: 10.1016/j.berh.2012.03.005.

Yates, M., Kelemen, A. and Sik Lanyi, C. (2016) 'Virtual reality gaming in the rehabilitation of the upper extremities post-stroke', *Brain Injury*, 30(7), pp. 855–863. doi: 10.3109/02699052.2016.1144146.

Yoo, J. H. *et al.* (2014) 'The effect of horse simulator riding on visual analogue scale, body composition and trunk strength in the patients with chronic low back pain', *International Journal of Clinical Practice*, 68(8), pp. 941–949. doi: 10.1111/ijcp.12414.

Zadro, J. R. *et al.* (2019) 'Video-game-based exercises for older people with chronic low back pain: a randomized controlledtable trial (GAMEBACK)', *Physical Therapy*, 99(1), pp. 14–27. doi: 10.1093/ptj/pzy112.

Zeng, X. *et al.* (2018) 'Reviewing clinical effectiveness of active training strategies of platform-based ankle rehabilitation robots', *Journal of Healthcare Engineering*, 2018, p. 2858294. doi: 10.1155/2018/2858294.

Zernicke, J. *et al.* (2016) 'A prospective pilot study to evaluate an animated home-based physical exercise program as a treatment option for patients with rheumatoid arthritis', *BMC Musculoskeletal Disorders*, 17(1), p. 351. doi: 10.1186/s12891-016-1208-3.

17 Emerging Perspectives of Virtual Reality Techniques

Renee M. Hakim
University of Scranton
Scranton, Pennsylvania

Michael Ross
Daemen College
Amherst, New York

CONTENTS

INTRODUCTION

Advances in technology continue to progressively change the rehabilitation environment. Over the past several years, virtual reality (VR) systems, often combined with robotic applications, have been used to train individuals with loss of motor function (Levin, 2020). Clinical applications of this technology can serve to assist, augment, evaluate, and document the rehabilitation for a wide variety of movement disorders (Juras et al., 2019; De Keersmaecker et al., 2019; Felipe et al., 2020).

The focus of rehabilitation research using VR has been on interventions used to improve motor function (e.g., balance, gait, and activities of daily living (ADL)) in neurologic populations such as stroke, Parkinson's disease (PD), traumatic brain injury, spinal cord injury (SCI), multiple sclerosis, and cerebral palsy. Based on neuroplasticity (Kleim & Jones, 2008) and principles of motor control and motor learning (Levin et al., 2015), task-oriented repetitive movements using VR systems may provide the experience, timing, motivation and attention necessary to improve motor function (You et al., 2005; Cramer et al., 2011). These types of applications have not been comparably explored in patients with movement disorders resulting from orthopedic causes. In addition, VR applications may also provide benefits such as pain relief (Mallari et al., 2019) and fear abatement (Gumaa & Youssef, 2019) to further enhance motor outcomes. The purpose of this chapter is to review relevant evidence on neuroplasticity, describe VR rehabilitation applications (including VR-augmented robotic devices) to improve motor function, and review evidence in persons with neurologic and orthopedic disorders. In addition, new directions in research to enhance outcomes for patients recovering from orthopedic conditions will also be discussed.

VIRTUAL REALITY SYSTEMS AND REHABILITATION

VR systems are computer-based applications that provide interactive, multisensory stimulation environments similar to real-world objects and events. These systems engage the user in either immersive or non-immersive virtual environments while performing activities in real time (Levac et al., 2019; Rizzo & Kim, 2005). Fully immersive VR systems may use a head-mounted display, large screen projection, video capture (e.g., Interactive Rehabilitation & Exercise System (IREX)), or projection on a concave surround surface to create the sense of immersion (e.g., Balance Near Automatic Virtual Environment (BNAVE)). Nonimmersive VR systems typically include a computer screen with interface devices such as a computer mouse, joystick, force sensors, or haptic controllers (e.g., Cyberglove). Across all VR systems, the degree of immersion is related to the meaning of the task to the user (Sanchez-Vives & Slater, 2005).

In a rehabilitation setting, virtual environments allow motor performance to be objectively monitored as a user engages in challenging yet safe and ecologically valid tasks. Training parameters can be controlled with respect to stimulus delivery, feedback, and measurement of performance in real time (Levac et al., 2019). Training paradigms using immersive systems (e.g., IREX system, Reachin API, Crystal Eyes CE-2 immersive VR) have focused on reaching, grasping, and lifting (Jang et al., 2005), striking a virtual ball with a haptic device and knocking down a virtual array of bricks (Broeren et al., 2004). Nonimmersive systems (e.g., VR Rehabilitation System) have simulated virtual tasks such as putting envelopes into mailboxes (Holden & Dyar, 2002), hitting a nail with a hammer, and pouring liquid from a glass into a carafe (Piron et al., 2003, 2005). The outcomes measured have included impairment and functional level tests and measures such as manual dexterity (e.g., Box and Blocks Test, Perdue Pegboard), grip force (e.g., hand dynamometer), strength/control function of the affected upper extremity (UE) (e.g., Manual Function Test, Fugl-Meyer Arm Scale), and ADL (i.e., Structured Assessment of Independent Living Skills, Functional Independence Measure, Wolf Motor Function Test). With respect to dosage, VR interventions have varied widely with sessions lasting from 20 minutes to a maximum of 3.5 hours with frequencies ranging from three to five times per week and durations ranging from two to six weeks up to a total of 11–13 weeks (Rahman and Shaheen, 2011).

Insights for rehabilitation using VR-based practice were provided by Levac and colleagues (2019) with respect to learning and transfer of complex motor skills. They suggested that therapists begin by closely observing the quality of movement considering the differences in how patients with impairments move in nonimmersive (e.g., flat-screen VEs) or immersive (e.g., head mounted displays) virtual environments as compared to real environments. This may be particularly important when the goal is to eventually integrate virtual environments into unsupervised home-based practice. Skillful observations made by clinicians can guide their decisions to use verbal feedback, demonstration, or physical guidance in order to encourage movements that are relevant to real-world activities. Additionally, therapists may consider increasing the challenge in virtual environments beyond corresponding real-world tasks by taking advantage of VR attributes such as adding cognitive dual-task challenges with either visual or auditory modalities (Quadrado et al., 2017). In consideration of patients' functional status, lower-fidelity VEs may be more realistic options for patients with significant physical or cognitive limitations. Decisions about type of virtual display and interaction method should be made in consideration of patient goals, abilities, and nature of the practice setting, including the availability of patient supervision and monitoring (Levac et al., 2019).

To directly impact a patient's movement, VR applications may be combined with other types of technology such as robotics. Rehabilitation robotics have been used to facilitate movement training and promote recovery following central nervous system (CNS) injury (Mehrholz et al., 2018; Kahn et al., 2006).

These devices are typically guided by computer programs to provide varying amounts of assistance, sensory feedback, and information processing (Waldner et al., 2009). Robotic-assisted training devices enhanced with VR systems may provide virtual targets or complex gaming simulations. These integrated systems can monitor the specificity and frequency of sensory feedback (Rizzo & Kim, 2005). Because of programmable force-producing ability, robotic devices can mimic some features of a therapist's manual assistance, allowing patients to semi-autonomously practice their movement training. Most robotic devices are capable of providing varying amounts of assistance (from passive/guided movement to partial assistance) or resistance during paretic limb movement to progress training (Krebs & Volpe, 2013). Some devices may assist active movements at a single joint, while others may assist multiple segments of a UE or incorporate bilateral UEs during reaching movements (Krebs, 2012). Devices that contact a user only at the most distal aspect of a limb (with indirect influence on more proximal segments) are referred to as end-effector-based devices (e.g., haptic joystick, wrist/hand devices), while devices with a mechanical structure that directly contacts and controls the limb are referred to as exoskeleton-based devices (e.g., instrumented orthoses, multi-joint, segmented devices). The progression of therapy with robotic devices may be accomplished by changing assistance, resistance, amplitude, or precision of targeted movements (Schweighofer et al., 2012). In general, most robotic systems incorporate more than one mode option into a single device. A specific subset of robotics includes haptic devices which are designed to create touch sensations for the user. Haptic devices are particularly well-suited for interfacing with VR environments because they enhance the users' sensory experience. More than a dozen haptic interfaces are currently commercially available with a variety of rehabilitation applications (Fisch et al., 2003; Burdea, 2008).

FOUNDATIONS FOR TASK-ORIENTED REHABILITATION: NEUROPLASTICITY

Rehabilitation approaches to promote recovery may cause adaptive changes in the CNS in response to functional activity, the environment, and learning (Sasmita et al., 2018). These changes, also known as neuroplasticity, demonstrate the ability of the nervous system to respond to intrinsic and extrinsic stimuli by reorganizing its structure, function, and/or connections (Cramer et al., 2011). Neuroplasticity is manifested at many levels, from molecular, cellular, or body systems to behavioral (Cramer et al., 2011; Sasmita et al., 2018). Research in both animals and humans has shown that neuronal connections undergo continuous remodeling based on experience, with more intensive practice resulting in expansion of cortical mapping (Merzenich et al., 1996; Nudo et al., 1996; Plautz et al., 2000; Cramer et al., 2011).

The impact of CNS injury on neuroplasticity has been studied to a great extent, particularly in persons with stroke (Dimyan & Cohen, 2011; Koeneman et al., 2007). Studies of motor recovery after stroke indicated that many forms of neuroplasticity may occur simultaneously in parallel. Rehabilitation post-stroke produces a range of brain events such as a return to a normal degree of laterality (Buma et al., 2010) and the formation of new connections from neurons on the undamaged side of the brain to neurons in damaged areas of the midbrain and spinal cord (Chen et al., 2002). Adaptive changes may also occur following peripheral nervous system lesions. Following nerve damage in the UE, there is extensive functional reorganization in the somatosensory cortex (Chen et al., 2002; Davis et al., 2011; Hansson & Brismar, 2003; Lundborg, 2003; Florence & Kaas, 1995). In line with the research on CNS plasticity, recent studies support the view that training protocols designed to enhance sensory relearning substantially improve functional outcomes following surgical peripheral nerve repair (Rosen et al., 2014; Miller et al., 2012; Svens & Rosen, 2009; Rosén & Lundborg, 2003). Therefore, rehabilitation programs should target promotion of neuroplasticity following both central and peripheral nervous system lesions.

Grounded in neuroscience research, Kleim and Jones (2008) identified ten key principles of the training environment to promote experience-dependent neuroplasticity and optimize motor skill reacquisition. These principles include consideration of practice parameters such as repetition, intensity, and task-specificity to promote heightening of specific brain functions. The timing of training as well as the transfer of improvement to similar tasks may also impact plasticity and subsequent skill acquisition. In addition, personal characteristics such as age and whether the task is meaningful to the individual also matter.

Contemporary task-oriented rehabilitation approaches that emphasize high-repetition, intense training (Dobkin, 2008; Kwakkel et al., 1997; Proteau et al., 1994; Carr & Shepherd, 1987) of meaningful tasks in enriched environments provide motivation and attention which appear to be critical modulators of plasticity (Nithianantharajah & Hannan, 2006; Woldag & Hummelsheim, 2002; Winstein et al., 1999).

As training parameters greatly impact neuroplasticity and subsequent motor recovery (Cramer et al., 2011), activities that optimize motor control and motor learning principles may improve rehabilitation outcomes (Schmidt & Lee, 2011). Recent evidence supports the advantage of virtual rehabilitation as it may provide increased time in therapy (adjunctive or additional therapy sessions), while patients perform functional tasks using their upper and/or lower extremities (de Rooij et al., 2016). In addition, virtual rehabilitation can increase motivation to practice more often and/or more intensively using engaging virtual surroundings with the potential of positive rewards (e.g., increased scores). Furthermore, training parameters in virtual environments can be tailored by clinicians to meet individual patients' needs.

According to Levin and colleagues (2015), basic aspects of motor control including movement planning, kinematic redundancy, task-specificity, problem-solving, and experience can be implemented as guidelines for the organization of practice in VR training environments (Table 17.1). VR is also well-suited for the application of motor learning principles using controlled environments

---

TABLE 17.1

Principles of Motor Control and How They Can Be Incorporated into the Design of a VR Training Environment

Motor Control Principles	Virtual Environment Design Based on Motor Control Principles
The level of difficulty for motor task depends on both the speed and accuracy of the intended movement (i.e., Index of Difficulty, Fitts' Law), whereas movements made rapidly to small targets are more difficult than those made slowly to large targets	Virtual objects should be adjustable so that task difficulty can be graded according to Fitts' Law
The organization of movement (i.e., degrees of freedom, timing, and coordination of joint motions) is related to the location and distance of an object from the body with greater difficulty to the contralateral side (when the arm crosses midline) than to the ipsilateral side in patients with stroke	Virtual tasks should involve interacting with objects placed at different distances from the body as well as in different locations in the workspace (contralateral, midline, ipsilateral) to encourage the coordinated use of different combinations of arm and trunk segments
The organization of movement is related to the perceived distance of the object from the body, as well as visual cues of the arm and the interaction of the arm with the object	The VE should include three-dimensional visual cues such as perspective lines, shading, drop lines, and motion parallax (i.e., closer objects moving faster than further objects) to improve depth perception
The hand trajectory during the reaching phase and the positioning of the hand for grasping are related to the location, size, and orientation of the object to be grasped	Objects included in a VE should be of various shapes, sizes, and locations. Hand tracking and haptics may enhance this application
The organization of a reach-to-grasp movement depends on the intrinsic features of the object and the intention for its use	VE tasks involving grasping should have purposeful goals with specific objects
Meaningful feedback about movement features (such as quality, joint motion, and accuracy) is essential to improve motor ability	High-fidelity visual, auditory, and tactile feedback should be incorporated into VEs. Most VR applications provide continuous visual monitoring as well as game and time scores. Additional feedback such as KR about task success (precision) and KP about movement quality may be available

Note: VR, virtual reality; VE, virtual environment; 3D, three-dimensional; KR, knowledge of results; KP, knowledge of performance.

Adapted from: Levin MF, Weiss PL, Keshner EA. Emergence of virtual reality as a tool for upper limb rehabilitation: incorporation of motor control and motor learning principles. Phys Ther. 2015;95:415–425.

TABLE 17.2

Principles of Motor Learning and Their Application into the Design of a Virtual Reality Training Environment

Principles of Motor Learning	Application of Motor Learning Principles in the Virtual Reality Training Environment
Learning occurs through repetitive, diversified practice of meaningful tasks	Virtual tasks should involve multiple repetitions of different movements with goals that are meaningful to the user
Learning occurs when task difficulty is progressively increased based on the individual's ability	Difficulty levels (either adaptive or clinician-selected) should be associated with ongoing abilities of the user with decision rules to identify optimal times for increasing or changing task challenges
Training should include active problem-solving in order to cognitively engage the learner	Virtual tasks should be varied in their level of problem-solving to challenge cognitive processing and self-awareness
Learning is enhanced by an individual's motivation to improve	Virtual tasks including ambient features such as realistic scenes, direct feedback (e.g., scores), and high interaction should be used to motivate users
Sensory feedback that is related to the task performance is necessary for learning	Multimodal sensory feedback (e.g., haptic feedback from the fingers when an object is touched, visual cues, auditory cues) in the virtual environment should be explicitly linked to patient performance and progression through difficulty levels
Learning is enhanced when an individual receives positive feedback about task performance (KP) and outcome/results (KR)	Both KP and KR should be presented to the learner during successful task performance
Learning of maladaptive movement patterns can be avoided by providing specific feedback to limit movement compensations made during the task (KP)	KP of maladaptive performance should be presented to the user in a way that does not disrupt task performance

Note: KR, knowledge of results; KP, knowledge of performance.

Adapted from: Levin MF, Weiss PL, Keshner EA. Emergence of virtual reality as a tool for upper limb rehabilitation: incorporation of motor control and motor learning principles. Phys Ther. 2015;95:415–425.

which include repetitive and varied practice, progression of task difficulty, problem-solving or error correction, motivation, and salient sensory feedback (Table 17.2). Progression of task-oriented movements in a virtual environment may optimize a patient's abilities while "repetitively trying to achieve a goal" which is critical for motor learning (Schmidt & Lee, 2011). In light of the neuroplasticity literature and recent clinical findings, Levin and colleagues (2015) stated that one of the remaining challenges for the virtual rehabilitation community is to design training activities that are more strongly based on principles of motor control and motor learning than those currently available. Applications should include challenging functional activities linked to the skill level of the learner which provide relevant sensory feedback aimed at maximizing motor problem-solving, rather than merely repeating a prescribed movement or set of movements (Levin et al., 2015).

OVERVIEW OF VIRTUAL REALITY APPLICATIONS FOR NEUROLOGICAL REHABILITATION

With respect to clinical relevance, research using VR systems has included both immersive and non-immersive applications in patients with neurological disorders to target motor function outcomes, including balance, gait, and ADL. The following sections will review recent evidence regarding virtual rehabilitation applications, particularly with respect to improvement of motor function, in various neurologic patient populations.

STROKE

Stroke in a nonprogressive brain injury resulting from disruption of blood circulation to the brain by a blockage or rupture (ASA, 2020). Clinical presentation following stroke varies based on the location and extent of the lesion, ranging from mild to severe impairments and functional limitations (Langhorne et al., 2009). Compared with other neurological disorders, there is a large, expanding body of evidence examining the effects of using VR rehabilitation in patients with stroke (Laver et al., 2017; Juras et al., 2019; Saposnik and Levin, 2011). There is a considerable number of studies that focused on interventions aimed at improving motor function of the UE. Recently, Laver and colleagues (2017) updated a Cochrane review including randomized and quasi-randomized trials of VR in adults after stroke (total of 72 trials with 2,470 participants) with the primary outcomes of UE function and activity. A subgroup meta-analysis of 22 studies in patients with post-acute stroke (1,038 total) compared virtual rehabilitation with the same dose of conventional therapy and did not find statistically significant improvement of UE function (standardized mean difference (SMD), 0.07; 95% confidence interval (CI) [−0.05 to 0.20]; low-quality evidence). However, when virtual rehabilitation was used to supplement usual care by providing participants in the intervention group with a higher dose of therapy, statistically significant improvements in UE function occurred (SMD, 0.49; 95% CI [0.21–0.77], 10 studies, 210 participants, low-quality evidence). Benefits were greater for VR interventions specifically designed for rehabilitation than those using commercial gaming programs.

In a related systematic review, Karamians et al. (2019) investigated the efficacy of VR and gaming-based interventions for improving UE function post-stroke with examination of demographic and treatment-related factors that may moderate treatment response. On average, they found that VR or gaming interventions produced an improvement of 28.5% of the maximal possible improvement. Dose and severity of motor impairment did not significantly influence rehabilitation outcomes. Treatment gains were significantly larger overall (10.8%) when the training involved a gaming component compared with visual feedback alone. VR or gaming interventions showed a significant treatment advantage (10.4%) over active control treatments. Overall, they concluded that VR or gaming-based UE rehabilitation post-stroke appeared to be more effective than conventional methods to improve UE function. Additionally, a systematic review and meta-analysis of randomized controlled trials (RCTs) by Domínguez-Téllez et al. (2020) found that game-based VR interventions improved not only UE motor function ($n = 15$; Fugl–Meyer Assessment for UE; SMD = 1.53, 95% CI [0.51–2.54]) but also quality of life ($n = 15$; Functional Independence Measure, SMD=0.77, 95% CI [0.05–1.49]) in adults with stroke.

With respect to balance and mobility, a systematic review was conducted by Mohammadi and colleagues (2019) to determine the effect of VR on balance as compared to conventional therapy alone in persons post-stroke. A meta-analysis ($n = 13$; 348 participants) using Berg Balance Scale (BBS) scores found significant improvement in the VR (in combination with conventional therapy) group compared to the usual care control group (medium effect size of .64, CI [0.36–0.92]). Another systematic review by Iruthayarajah et al. (2017) focused on the effectiveness of VR interventions for improving balance in a chronic stroke population (≥6 months post). Overall, VR interventions [20 studies: Nintendo® Wii Fit balance board ($n = 7$), treadmill training and VR ($n = 7$), and postural training using VR ($n = 6$)] compared to conventional rehabilitation had significant improvements evaluating the BBS ($n = 12$; MD = 2.94 ± 0.57; $p < 0.001$) and Timed Up and Go (TUG) ($n = 13$; MD = 2.49 ± 0.57; $p < 0.001$). Similarly, Chen and colleagues (2016) reported moderate evidence (nine studies; PEDro Scores ranged four to nine) to support VR training as an effective adjunct to standard rehabilitation programs to improve balance in patients with chronic stroke. However, the effect of VR training in balance recovery was less clear in patients with acute or subacute stroke.

Li et al. (2016) analyzed 16 RCTs involving 428 participants with stroke who received VR interventions and reported marked improvements in BBS (MD = 1.46, 95% CI [0.09, 2.83]) and TUG scores (MD = 1.62, 95% CI [–3.07 to 0.16]) compared with controls.

With respect to dosage, de Rooij et al. (2016) analyzed studies of patients with mostly post-acute stroke and found that virtual rehabilitation was more effective when added to conventional therapy and dose-matched for time in therapy, compared to conventional treatment alone. Correspondingly, Corbetta and colleagues (2015) reviewed 15 trials involving 341 adults with stroke to determine if virtual rehabilitation improved walking speed, balance, and mobility more than the same duration of standard rehabilitation and if adding extra VR to usual care improved the benefits. Findings revealed that when VR replaced some or all of the standard rehabilitation, there were statistically significant benefits in walking speed (MD = 0.15 m/s; 95% CI [0.10–0.19]), balance (MD = 2.1 points on BBS; 95% CI [1.8–2.5]) and mobility (MD = 2.3 seconds on TUG, 95% CI [1.2–3.4]). When VR was added to standard rehabilitation, mobility showed a significant benefit (0.7 seconds on TUG; 95% CI [0.4–1.1]).

Concerning multiple outcomes, Lee et al. (2019) conducted a meta-analysis to examine whether VR training was effective for lower extremity (LE) function as well as UE and overall function in patients with chronic stroke. The effects of VR programs on specific outcomes were most effective for improving muscle tension, followed by muscle strength, ADL, joint range of motion, gait, balance, and kinematics. They concluded that VR training was effective in improving the function in patients with chronic stroke, corresponding to a moderate effect size (total effect size for VR programs = 0.440; UE function = 0.431, LE function = 0.424, overall function = 0.5450). Moreover, VR training showed a similar effect for improving LE compared with UE function. Relatedly, Gibbons and colleagues (2016) reviewed RCTs assessing effects of VR on LE outcomes in patients with stroke (n = 22, 552 participants). Significant differences in favor of VR groups were found for functional balance (SMD = 0.42, 95% CI [0.11–0.73]), gait velocity (WMD = 0.12, 95% CI [0.03–0.22]), cadence (WMD = 11.91, 95% CI [2.05–21.78]), and stride length (WMD = 9.79, 95% CI [0.74–18.84]) within the chronic population. They concluded that VR is equally as effective as conventional therapy; therefore, use in practice should be determined by affordability and patient/practitioner preferences.

Parkinson's Disease

Parkinson's disease (PD) is a neurodegenerative movement disorder characterized by slowness, tremor, rigidity, impaired posture, and imbalance with resulting functional limitations based on the stage of progression (Parkinson's Foundation, 2020). PD is optimally managed by a combination of medication and regular physical therapy (PT). A recent Cochrane Review by Dockx and colleagues (2016) was conducted to summarize the current best evidence for the effectiveness of VR interventions in the rehabilitation of patients with PD. The review included eight trials involving 263 persons with PD with evidence quality graded as low or very low. Most of the studies intended to improve motor function using commercially available devices, which were compared to conventional PT with interventions lasting between 4 and 12 weeks. Findings revealed that VR may lead to a moderate improvement in step and stride length (SMD = 0.69, 95% CI [0.30–1.08]; three studies; 106 participants; low-quality evidence) when compared to PT. Both VR and PT interventions may have similar effects on gait (SMD = 0.20, 95% CI [–0.14 to 0.55]; four studies; 129 participants; low-quality evidence), balance (SMD = 0.34, 95% CI [–0.04 to 0.71]; five studies; 155 participants; low-quality evidence), and quality of life (MD = 3.73 units, 95% CI [–2.16 to 9.61]; four studies; 106 participants). VR interventions did not lead to any reported adverse events, and exercise adherence did not differ between VR and other interventions.

Another more recent systematic review with meta-analysis (27 studies, n = 688 participants) was conducted by Triegaardt et al. (2020) to analyze the effectiveness of VR in rehabilitation of patients with PD. Findings revealed that VR training significantly ($p < 0.05$) improved motor function, balance and co-ordination, quality of life, and ADL. Meta-analysis (ten studies, n = 343 participants) determined that VR training led to greater improvement of stride length (SMD = 0.70; 95% CI [0.32–1.08], p = 0.0003) compared with active interventions and was equally effective for gait speed, balance and coordination, quality of life, and ADL. Compared with passive rehabilitation intervention, VR had greater effects on balance (SMD = 1.02, 95% CI [0.38-1.65], p = 0.002). Results from single RCTs showed that VR training was better than passive rehabilitation intervention for improving gait speed (SMD = 1.43, 95% CI [0.51–2.34], p = 0.002), stride length (SMD = 1.27, 95% CI [0.38–2.16], p = 0.005), and ADLs (SMD = 0.96, 95% CI [0.02–1.89]). The authors concluded that virtual rehabilitation improves a number of outcomes and may be considered for routine use in patients with PD.

With respect to balance and gait, Wang et al. (2019) evaluated the effectiveness of VR interventions for in persons with PD. A meta-analysis of 12 studies (median PEDro score = 6.4; 419 participants) demonstrated significant improvements in BBS (MD = 2.69; 95% CI [1.37–4.02]; p < 0.0001), TUG Test (MD = −2.86; 95% CI [−5.60 to −0.12]; p = 0.04), and stride length (MD = 9.65; 95% CI [4.31–14.98]; p = 0.0004) in patients with PD who received VR compared with controls. However, there was no significant difference in gait velocity and walk distance. A recent systematic review by Freitag et al. (2019) focused on the effects of VR on dual-task gait training in persons with PD. They concluded that VR dual-task gait training should be part of rehabilitation protocols for PD, although specific guidelines have not yet been established.

Similarly, Lei et al. (2019) analyzed 16 articles involving 555 participants with PD to examine outcomes of balance and gait. They found that VR rehabilitation training resulted in better than conventional or traditional rehabilitation training in three aspects: step and stride length (SMD = 0.72, 95% CI = 0.40,1.04, Z = 4.38, P < 0.01), balance function (SMD = 0.22, 95% CI = 0.01,0.42, Z = 2.09, P = 0.037), and mobility (MD = −1.95, 95% CI = −2.81, −1.08, Z = 4.41, P < 0.01) with no effect on the Dynamic Gait Index and gait speed. Adverse events were minimal and included mild dizziness (four patients) with one patient developing severe dizziness and vomiting. The authors concluded that VR rehabilitation training not only achieved the same effect as conventional rehabilitation training, it actually resulted in better performance on gait and balance in patients with PD. Consequently, when the effect of traditional rehabilitation training on gait and balance in patients with PD is not optimal, VR rehabilitation training can at least be used as an alternative therapy. However, more rigorous design of large-sample, multicenter RCTs are needed to provide a stronger evidence-based basis for verifying its potential advantages.

MULTIPLE SCLEROSIS

Multiple sclerosis (MS) is an inflammatory disease of the CNS in which the insulating cover of nerve cells in the brain and spinal cord are damaged (National MS Society, 2020). Individuals with MS display a widely variable clinical presentation based on the type, onset, and severity of this progressive neurological disorder. Many persons with MS experience motor dysfunction resulting in problems with balance, gait, and ADLs. A systematic review by Massetti and colleagues (2016) investigated the broad outcomes of previous studies on VR in persons with MS. Analysis of ten articles (low to moderate quality evidence) revealed varied VR systems, protocols, and outcome measures with dosage ranging from one to two sessions (three studies) to once or twice per week for 10–12 weeks (seven studies). Overall findings suggest that VR represents a motivational and effective alternative to traditional motor rehabilitation for patients with MS to improve multiple cognitive and/or motor deficits. Motor outcomes included improvements in UE movement/control, balance (BBS, TUG, Functional Reach, anticipatory control, and response), gait pattern, and speed and endurance. The authors reported the need to monitor factors associated with VR that may impact patients with MS such as excessive fatigue, game difficulty level, and high physical requirements. They concluded that additional research is needed to support rehabilitation protocols with VR and increase the effects of treatment.

Comparing a commercially available VR system, a recent RCT by Yazgan et al. (2019) included 47 persons with MS who were randomized into a VR group (Nintendo Wii Fit), balance training group, and a control group. Participants underwent an exercise program under the supervision of a PT on two days a week for eight weeks with outcome measures including balance (BBS, TUG), endurance (6 Six-Minute Walk Test), and fatigue (Fatigue Severity Scale). Findings revealed that all parameters evaluated in the VR and balance training groups showed statistically significant improvement after treatment with superior changes noted in the VR group compared with controls. Additionally, changes in BBS were greater in the VR group compared with balance training. The authors concluded that in comparison with no intervention, exergaming with Nintendo Wii Fit improves balance, increases functionality, reduces fatigue severity, and increases quality of life in people with MS.

SPINAL CORD INJURY

Injury to the spinal cord resulting from trauma, disease, or degeneration causes neurological damage that may include partial or complete loss of sensory function and/or motor control of arms, legs, and/or body (WHO, 2020). Clinical presentation of patients with SCI varies widely depending on the location and severity of damage (Shepherd Center, 2020). SCI is often associated with long-term functional limitations related to impairments of the sensorimotor system. A recent systematic review by de Araújo et al. (2019) explored the possible benefits and efficacy of VR-based rehabilitation in individuals with SCI. A total of 25 studies (including 482 participants, 47.6±9.5 years, 73% male) were analyzed with a large amount of heterogeneity noted in study design, VR protocols, and outcome measures. Only seven studies (28%) rated as excellent/good quality of evidence. However, substantial evidence for significant positive effects associated with VR therapy was found in most of the studies (88%), with no adverse events reported. Although the current evidence was limited, the findings suggested that VR-based rehabilitation in participants with SCI may lead to positive effects on aerobic function, balance, pain level, and motor function recovery with added benefits of improved psychological/motivational aspects. Further high-quality studies are needed to provide guidelines for clinical practice and to draw robust conclusions about the potential benefits of VR therapy for patients with SCI (de Araújo et al., 2019).

With a more narrow focus, Alashram and colleagues (2020) conducted a systematic review to investigate the effects of VR on balance ability in individuals with incomplete SCI and to identify an efficient training protocol. Five low-quality pilot studies met the inclusion criteria (PEDro scores ranged from two to three) with samples of less than 20 patients in all selected studies. The preliminary findings showed that VR training (12–20 sessions of 30–60 min) had a promising impact on balance ability in patients with incomplete SCI. However, the authors concluded that further high-quality evidence is strongly needed.

TRAUMATIC BRAIN INJURY

Traumatic Brain Injury describes a heterogenous, nonprogressive neurological condition resulting from an impact or jolt to the head or body or a penetrating object into the skull (BIAA, 2020). Extent of injury may vary widely from mild concussive symptoms to severe, life-threatening damage. Clinical presentation is based on the location and extent of the brain damage with subsequent impact on multiple body systems which may have long-lasting effects (BIAA, 2020). A recent review by Aida and colleagues (2018) evaluated a total of 11 studies (primarily of low-level evidence, with the exception of two RCTs). Overall, 10 of 11 studies demonstrated improvement with VR therapy, most frequently used to address gait or cognitive deficits. They concluded that while the current literature generally offers support for the use of VR in TBI recovery, there is a scarcity of strong evidence to support its widespread use. The increasing availability of immersive VR technology offers the potential for engaging therapy in TBI rehabilitation, but its utility remains uncertain given the limited studies available at this time.

CEREBRAL PALSY

Cerebral palsy is a group of disorders that affect movement, muscle tone, and posture caused by damage that occurs to the immature brain as it develops, most often before birth (Mayo Clinic, 2020). The clinical presentation of CP varies widely and may impact motor function in some or all areas of the body resulting in mild to severe functional limitations. The appeal of VR may be particularly useful in the rehabilitation of a disorder involving children and young adults. To investigate the effect of VR therapy on balance and walking in children with CP, Warnier et al. (2019) included 26 articles with meta-analyses and showed a significant result in favor of VR therapy for balance (SMD = 0.89, 95% CI [0.14–1.63]) and for walking (SMD = 3.10, 95% CI [0.78–5.35]). They concluded that VR therapy seems promising as an intervention for rehabilitation in children with CP. However, these results must be interpreted with caution due to differences in the interventions used, the lack of RCTs, and the relatively small groups.

Similarly, Wu and colleagues (2019) analyzed the effect of VR games on balance recovery in children with CP. Additionally, they evaluated the impact of VR game training parameters (i.e., duration, frequency, cycle, and total intervention time) on balance recovery. The results showed that VR games may improve the balance of children with CP (Hedge's $g = 0.29$; 95% CI [0.10–0.48]); however, no significant influence of the intervention parameters was shown in the subgroup analysis. They concluded that VR games played a positive role in the improvement of balance of children with CP, but these results should be viewed with caution owing to current methodological defects such as difference in outcome measures, heterogeneity of control groups, and intervention combined with other treatments.

Another systematic review by Ren and Wu (2019) focused on the rehabilitative effect of VR games for gross motor skills of children with CP. They included seven RCTs of children with CP (234 total) in a meta-analysis and found that VR games could improve gross motor skills (SMD = 0.37, 95% CI [0.06, 0.68], $p = 0.02$). The authors concluded that the VR game interventions can enhance gross motor skills of children with CP to some extent. However, they expressed the same methodological limitations as noted above.

With an emphasis on gait, Ghai and Ghai (2019) analyzed the influence of VR training, including relevant training dosage, on gait recovery in children with CP. They included 16 studies involving 274 children with CP. Findings revealed that 88% of the studies reported significant enhancements in gait performance after training with VR. Meta-analyses revealed positive effects of VR training on gait velocity (Hedge's $g = 0.68$), stride length (0.30), cadence (0.66), and gross motor function measure (0.44). Subgroup analysis reported a training duration of 20–30 minutes per session, ≤4 times per week across ≥8 weeks allowed maximum gains in gait velocity.

Overall, there is a variety of effective VR applications available for rehabilitation of persons with neurological disorders as noted above. Activities performed in virtual environments have been described as more interesting and enjoyable by children and adults which encourages higher numbers of repetitions (Lewis & Rosie, 2012). Recent supporting evidence has grown enough to establish VR as part of clinical practice guidelines for recovery of locomotor function in acute onset of CNS injury, especially following stroke (Hornby et al., 2020). With respect to clinical utility, a systematic review by Massetti et al. (2018) found a range of benefits associated with VR interventions, including improvement in motor functions, greater community participation, and improved psychological and cognitive function. Many studies provided support for the use of VR as part of a neurorehabilitation program in maximizing recovery. However, findings should be viewed with caution as there is inconclusive demonstration that the acquired skills from VE practice can be transferred to the real-world (Levac et al., 2019). Although virtual environments offer multiple benefits for patients and therapists compared to conventional interventions, practical and logistical factors such as the significant cost as well as the space and training required for equipment operation must be considered as therapists deliberate the use of VR in clinical practice (Glegg, 2017; Glegg et al., 2013; Levac et al., 2019).

In consideration of future applications, VR continues to emerge as a viable option for the reha-bilitation of patients with neurological disorders. Although the optimal type of system, training parameters, and dosage are yet to be determined, clinicians with access to VR systems may simu-late real-world activities in a safe, controlled environment. Virtual rehabilitation may be imple-mented for patients along the continuum of care from acute to chronic stages of recovery. As noted above, the role of VR as an adjunct to conventional therapy has the benefit of increasing the total amount of therapy delivered to the patient, which is linked to better outcomes as noted above. In addition, the growing availability of commercial gaming systems may not only provide opportuni-ties for patients to receive skilled interventions under the guidance and supervision of a clinician but may also present a physically active recreational opportunity for individuals at home with family, friends, or caregivers. As the presence of VR applications continues to grow in many environments, the opportunities for using this technology to promote health and wellness are boundless.

VIRTUAL REALITY AND PAIN MANAGEMENT

It is estimated that chronic pain affects 20% of the adult population, and pain-related consultations comprise a significant amount of all health care visits (Dahlhamer et al., 2018; Hoy et al., 2012, 2014). Pain relief as a quality outcome measure has influenced patients' expectations on the level of pain they should have and how it should be managed by the patient care team. Therefore, a cam-paign that was initiated in the 1990s advocated that pain be assessed as the "fifth vital sign," which linked patient satisfaction with pain control to hospital reimbursements (Lovecchio et al., 2019). This created a patient-care environment in which opioid analgesic medications were often overpre-scribed by physicians who treated patients for pain. Because of the opioid epidemic, a great deal of research has been done on a patient's pain-relief experience and the factors that may influence opioid use (Franklin et al., 2008; Lovecchio et al., 2019; Webster et al., 2007).

The alternative to opioids in many cases is non-opioid medications, such as non-steroidal anti-inflammatory drugs. However, there are concerns associated with nonsteroidal anti-inflammatory drugs, such as preventable adverse drug reactions that cause bleeding, heart attack, stroke, and renal damage (Davis and Robson, 2016). Thus, alternative pain management strategies are important to consider in an effort to mitigate some of these risks associated with opioid and non-opioid analgesic medications.

VR can assist in decreasing and managing pain in individuals either undergoing medical pro-cedures or dealing with chronic pain. In a recent systematic review and meta-analysis by Mallari et al. (2019), they determined that VR is an effective treatment for reducing acute pain both during and after a VR intervention and that VR is especially effective in mitigating medical procedure-related pain (e.g., episiotomy repair, dressing change for management of various types of wounds such as burns, ulcers, and necrotizing fasciitis). There is some evidence that suggests that VR can reduce chronic pain during the time of the intervention; however, further work is necessary to either support or refute whether VR is effective for long-term reductions in chronic pain. The authors concluded that immersive VR-based therapies should be considered as an adjunct to routine care to assist in reducing acute pain and potentially chronic pain that is experienced by individuals.

One clinical scenario that is particularly applicable to rehabilitation professionals is burn/wound care, which can be extremely painful for patients. Opioid medications are the analgesic of choice for many patients with severe burn injuries and other traumatic injuries. Unfortunately, patients often experience decreased analgesic effects with repeated administration of opioid medications; thus, over the course of care, it is not uncommon to see increased doses of opioid analgesic medica-tions being used to achieve the same analgesic effect prior to the patient developing drug tolerance. However, due to the risk of physical dependency and the side effects of opioid analgesic medica-tions, caution should be utilized and clinicians may be limited in terms of dosage levels and fre-quency of use.

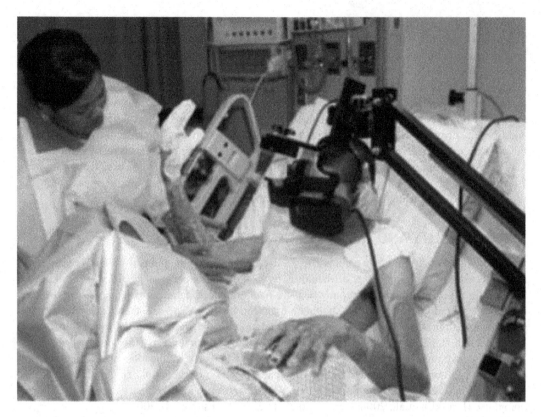

FIGURE 17.1   In this image, the patient is receiving immersive virtual reality to reduce the pain experienced during burn wound care to his right arm. The patient is using used robot-like arm mounted goggles, which does not require the patients to wear a helmet. He is controlling the mouse for the virtual reality software with his left hand. This design increased patient comfort and the number of patients who can use virtual reality (e.g., patients with bandaged wounds of the face and head as well as those who found the helmet too uncomfortable) (Maani et al., 2011). Photos and copyright Hunter Hoffman, UW, www.vrpain.com.

In consideration of clinical pain management, Mallari et al. (2019) examined studies that assessed pain and found that VR therapies in conjunction with medication usage significantly reduced pain during wound debridement or range of motion activities for the extremities that were affected by burns compared to medication usage alone (Carrougher et al., 2009; Maani et al., 2011; Morris et al., 2010). In an interesting study, Maani et al. (2011) assessed whether immersive VR could reduce the excessive pain experienced by of soldiers during wound debridement for combat-related burn injuries. These authors used a within-subject experimental design, with patients receiving half of their burn wound cleaning procedures (approximately six minutes) with individual medication regimens as determined by the treating physicians (e.g., fast acting opioids and/or ketamine) and the other half of their treatment supplemented with adjunctive immersive VR. The VR was facilitated through SnowWorld software, which allowed patients to follow a pre-determined "gliding" path through an icy three-dimensional virtual canyon (Figures 17.1 and 17.2). Each patient "looked" around the virtual environment (e.g., the sky when they looked up, a canyon wall when they looked left or right, and a flowing river when they looked down) and "threw" snowballs at snowmen, penguins, wooly mammoths, and jumping fish by pushing a mouse trigger button. Sound effects (e.g., a splash when a snowball hit the river) were also utilized and included background music by recording artist Paul Simon. The results revealed that patients reported significantly less pain when using VR during treatment; more specifically "worst pain," "pain unpleasantness," and "time spent thinking about pain" were all significantly influenced, especially in those patients who experienced the most

FIGURE 17.2 A screenshot of the game SnowWorld what patients see in the VR goggles during immersive VR pain distraction. SnowWorld is a three-dimensional computer graphic system that uses the imagery of an icy canyon with a river flowing through it as a backdrop for snowmen, penguins, woolly mammoth, fish, and snowfall. The object of the system is to distract the patient during wound treatment by allowing the participant to focus on throwing snowballs at objects while moving through the canyon (Maani et al., 2011). Image by Ari Hollander and Howard Rose, copyright Hunter Hoffman UW, www.vrpain.com.

pain. In addition to reducing pain, patients reported the VR experience during treatment was "pretty fun," which is in stark contrast to what they experienced without VR.

With respect to the mechanism of experiencing pain, Maani and colleagues (2011) noted that pain draws on our attention span and patients have a limited amount of attention available. They further speculated that VR draws upon our attention, leaving less attention span available to process pain signals that are experienced during treatment. They concluded that patients spent much less time thinking about their pain during wound care while in SnowWorld which led to significantly reduced pain during treatment.

VIRTUAL REALITY AND ORTHOPEDIC REHABILITATION

In addition to serving as a modality that can effectively manage pain in both acute and chronic conditions, VR continues to develop as a technology with a variety of potential benefits for many aspects of rehabilitation assessment, treatment, and research (Hakim et al., 2017). As noted above, the use of VR in neurorehabilitation for individuals with cerebral palsy, stroke, and PD has been studied extensively (Dockx et al., 2016; Laver et al., 2017; Ravi et al., 2017). Likewise, some recent clinical trials have also evaluated the effectiveness of VR in the orthopedic rehabilitation setting. This is especially important, since musculoskeletal disorders are one of the leading causes of chronic disability worldwide and there is a need to develop improved and more efficient rehabilitation methods.

(a) **(b)**

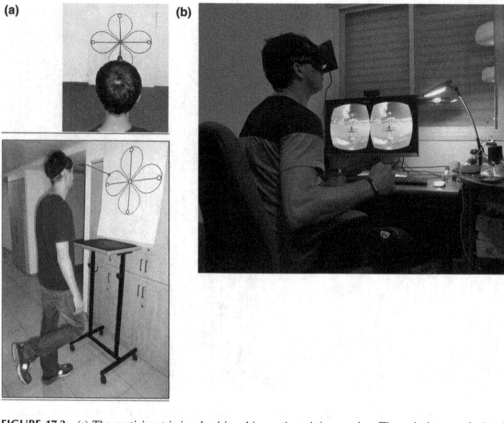

FIGURE 17.3 (a) The participant is involved in a kinematic training session. Through the use of a head-mounted laser beam aimed at a 70 cm × 70 cm poster, exercises were performed, such as following the lines with the laser or moving the laser quickly from one circle to another. Visual feedback regarding head motion was provided by the laser beam. (b) The participant is involved in a kinematic home training program using customized software with a virtual reality head-mounted display. The virtual airplane on the computer screen is controlled by head motion of the individual. Unlike kinematic training, the velocity of the motions of the head and neck can be controlled. Reprinted from Bahat HS, Croft K, Carter C, Hoddinott A, Sprecher E, Treleaven J. Remote kinematic training for patients with chronic neck pain: a randomised controlled trial. Eur Spine J. 2018;27:1309–1323, with permission from Springer Nature. The publication is available through Springer Nature at https://link.springer.com/article/10.1007/s00586-017-5323-0.

For the cervical spine, two high-quality studies that evaluated patients with chronic neck pain provided evidence of improved cervical flexion range of motion, movement velocity, and accuracy with VR in comparison with kinematic training and/or no-treatment; comparable improvements in pain and disability were noted in both groups (Bahat et al., 2015, 2018). In these studies, kinematic training involved using a head mounted laser beam projected onto a poster hung on the wall (Figure 17.3). Kinematic training focused on active neck movements to increase range of motion of the cervical spine, quick head movement to facilitate cervical spine motion control, static head positioning while moving the body, and smooth head movement. The VR training involved a head-mounted display and customized software with a three-dimensional virtual airplane controlled by head motion; three modules were developed that focused primarily on range of motion, velocity, and accuracy (Figure 17.3) that were facilitated and progressed by the patient's response to the visual stimuli that was provided.

Concerning shoulder impingement syndrome, Pekyavas and Ergun (2017) determined that a supervised clinic-based VR-based training program (two 45 minute treatments per week for six weeks) was more effective than a six week home based strengthening and stretching exercise program that was performed twice per week for 45 minutes per session in reducing complaints of shoulder pain and improving scapular dyskinesia. Patients in the VR groups used an off-the-shelf Nintendo Wii Fit gaming system that included the following activities: shoulder stretching and bilateral shoulder elevation, as well as boxing, bowling, and tennis games that were supplemented with a home-based progressive resistive exercise program with Theraband®. Patients in the home-based strengthening and stretching exercise group performed progressive resistive exercises with Theraband®, shoulder mobility activities, and capsular/pectoral muscle stretching exercises. The authors concluded that VR exergaming programs may have provide visual and sensory feedback on exercise which made an individual's performance more effective, which may have positively influenced outcomes.

A recent systematic review and meta-analysis by Gumaa and Youssef (2019) determined that the evidence of VR effectiveness is quite promising in patients with chronic neck pain and shoulder impingement syndrome. They also concluded that VR and exercises have similar effects in patients with ankle instability, anterior cruciate ligament reconstruction, knee osteoarthritis, and rheumatoid arthritis. For patients with fibromyalgia and low back pain, as well as those that have undergone total knee arthroplasty, the evidence of VR effectiveness compared with exercise is either inconclusive or absent.

Despite a lack of differences between VR and exercise, Gumaa and Youssef (2019) concluded that VR may assist with participant motivation and enhance enjoyment. This may have important clinical applications. For example, following anterior cruciate ligament reconstruction and total knee arthroplasty, regaining range of motion early after surgery is extremely important. Thus, if a patient is apprehensive or noncompliant due to a lack of motivation in performing their exercises to improve knee range of motion, rehabilitation may be facilitated with VR-based exercise to overcome some of the patient apprehension and concerns with motivation.

Likewise, fear avoidant behavior and kinesiophobia, especially in patients with spinal pain, are factors that are challenging to address during rehabilitation and can adversely influence outcome (Pincus et al., 2013). However, VR-based activities may be able to overcome some of these concerns by way of "distracting" the individual while they are performing exercises in a safe, entertaining, and enjoyable manner. VR activities can provide a fully immersive experience by having the individual "function" in a three-dimensional simulated environment where their beliefs of the real world and the potential influence of their symptoms are temporarily suspended. For example, most individuals with chronic low back pain that is associated with fear avoidant behavior would never imagine that their back could tolerate activities like running, wake boarding, surfing, horseback riding, or playing dodgeball (Figure 17.4). However, when properly dosed to achieve a positive effect for patients with chronic low back pain (e.g., 8–12 sessions over 10–12 weeks), VR activities can enhance outcome, especially when utilized with other aspects of graded exposure therapy. Once a patient has succeeded with performing these activities in a virtual three-dimensional environment, perhaps less strenuous ADL (e.g., walking, bending forward, prolonged standing, or sitting) will no longer evoke fear for the patient, which would likely significantly improve the progress of the patient's rehabilitation.

ROBOTIC ASSISTED-VIRTUAL REALITY AND ORTHOPEDIC REHABILITATION

While there is existing evidence of VR effectiveness with orthopedic disorders (Gumaa and Youssef, 2019), the literature pertaining to robotic-assisted VR applications is quite limited for orthopedic rehabilitation. On the contrary, most of the rehabilitation robotics research has focused on neurorehabilitation for patients with a history of stroke or SCI. More specifically, exoskeletons or end effector robots in combination with VR have shown advantages for neuromotor rehabilitation, especially for the upper extremities (Lo et al., 2010; Prange et al., 2006).

Using the principles of neurorehabilitation, Padilla-Castaneda et al. (2012, 2018) described a robotic-assisted training device integrated with a VR system that was designed for patients with

FIGURE 17.4 A screenshot of a virtual dodgeball intervention, developed using Vizard software (WorldViz™) (Thomas et al., 2016). Participants (left side of the images) were positioned 1.5 m in front of a 152.4 cm high definition three-dimensional television, which was mounted to the wall at approximately eye level. Movement of light-reflective markers for the head, upper arms, forearms, hands, trunk, pelvis, thighs, lower legs, and feet were tracked using a 10-camera Vicon Bonita system sampled at 100 Hz using TheMotion-Monitor software (Innovative Sports Training, Inc., Chicago, IL). Participants wore Samsung three-dimensional shutter glasses, which were required to provide the necessary three-dimensional effect with this particular television. The gaming environment (right side of the images) was in a basketball gymnasium in which the participant played dodgeball against four virtual opponents. In the virtual environment, the participant's avatar, viewed in the third person perspective, was located on one free-throw line and the four opponents were located on the free throw line of the opposite end of the court. Virtual balls were launched at the participant's avatar every 3.3 ± 0.3 seconds from one of four virtual opponents. In the top virtual reality image on the right, the opponent on the far left (encircled) threw the ball (encircled) at the participant's avatar. In the bottom virtual reality image, the participant had to flex their lumbar spine, hips, and knees in order to avoid the ball (ball is out of the image at this point). From https://www.youtube.com/watch?v=We0BrUeYrxo&feature=youtu.be.

orthopedic disorders affecting the distal UE. This rehabilitation device consisted of an enhanced end-effector haptic robotic arm that required grasping with a cylindrical type of power grip (Figure 17.5). It allowed patients to perform high intensity, task-oriented rehabilitation activities for the UE (primarily movements of forearm pronation/supination and elbow flexion/extension with the shoulder in an adducted and abducted position) by integrating VR games developed to assist with treatment following injury. The system simulated the actions of the patient's UE in an effort to improve range

FIGURE 17.5 A patient performing a rehabilitation session with the BRANDO system, which is comprised of a haptic interface with a six-degree of freedom robotic arm with three actuated degrees of freedom. The system is integrated with three different virtual reality games described in Figure 17.6. Reprinted from Padilla-Castaneda MA, Sotgiu E, Barsotti M, Frisoli A, Orsini P, Martiradonna A, Laddaga C, Bergamasco M. An orthopaedic robotic-assisted rehabilitation method of the forearm in virtual reality physiotherapy. J Healthc Eng. 2018;2018:7438609, with permission from Hindawi Limited. The publication is available through Hindawi Limited at https://doi.org/10.1155/2018/7438609.

of motion and motor control through. This was done by having the patient complete different virtual tasks through repetitive movements while providing proprioceptive and haptic force haptic feedback in three different training modes: passive, active, and resistive exercising. Three games were developed: one for pronation/supination of the forearm and two for the flexion/extension of the elbow with varying degrees of shoulder positions (Figure 17.6). The system was also able to evaluate the mobility of the joints of the UE and effectively dose the level of the rehabilitation session based on the patient's abilities. Feedback during gaming performance was provided in the form of sounds and performance scores. Kinematic measures of the patient's progress and knowledge of results were provided following completion of the activity. When UE muscular activity of two healthy individuals was evaluated via surface electromyography during a robotic-assisted training session, Padilla-Castaneda et al. (2012) concluded that the system can assign appropriate workloads during a typical treatment session. This is an important finding, as the forearm muscles are routinely involved in the performance of many ADL; thus, their rehabilitation with appropriately dosed therapeutic activities is key after injury to the UE.

In subsequent pilot testing on ten patients undergoing PT orthopedic rehabilitation for fractures of the UE that adversely influenced function, Padilla-Castaneda et al. (2018) determined that the system could provide a safe and controlled rehabilitation environment, but in a more systematic and feedback driven way than traditional PT. According to patient reports, changes in difficulty through the rehabilitation activities we well tolerated and the system was well accepted. The authors concluded that the proposed methodology was feasible for subsequent clinical trials. Additionally, based upon kinesiological observations and performance of the patients, the authors were able to

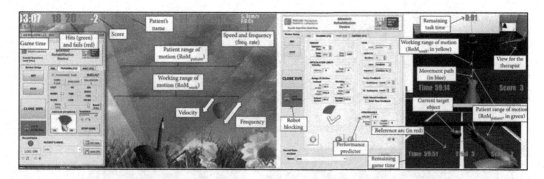

FIGURE 17.6   On the left is a screenshot of the Balloons game and on the right is the Bells game by Padilla-Castaneda et al. (2012, 2018). For the Balloons game (left), the subject is to avoid collisions with balloons that are gradually getting closer to the virtual hand by bursting them by orienting the pointed end of a pencil through movements of pronation and supination of the forearm while the shoulder is in an adducted position. The speed, size, and frequency rate of the balloons vary throughout the game. For the Bells game (top right), the subject is to touch and ring a bell with the index finger through movements of flexion and extension of the elbow while the shoulder is in an adducted position. The size, position, and sequence of the bells vary throughout the game. For the Balls game (bottom right), the subject is to hitting a tennis ball on a table with their hand through movements of flexion and extension of the elbow while the shoulder is in an abducted position. The size, position, and sequence of the balls vary throughout the game. Reprinted from Padilla-Castaneda MA, Sotgiu E, Barsotti M, Frisoli A, Orsini P, Martiradonna A, Laddaga C, Bergamasco M. An orthopaedic robotic-assisted rehabilitation method of the forearm in virtual reality physiotherapy. J Healthc Eng. 2018;2018:7438609, with permission from Hindawi Limited. The publication is available through Hindawi Limited at https://doi.org/10.1155/2018/7438609.

develop a predictive model to establish difficulty in rehabilitation parameters, which could systematically individualize rehabilitation training sessions. Unfortunately, since patients only participated in a single session, the impact on outcome in terms of motor performance and function was not described.

While the system described by Padilla-Castaneda et al. (2012, 2018) may be useful for restoration of range of motion and motor control for the forearm and elbow, it may not be helpful for restoring fine motor control of the hand and wrist. This is especially concerning in patients with orthopedic disorders of the distal UE (e.g., carpal tunnel syndrome, tendon rupture/repair, nerve crush/compression injury) who may display activity/participation restrictions characterized by difficulty with ADL such as handwriting or grasping objects. Thus, we developed a rehabilitation approach to explore robotic applications in patients with orthopedic disorders of the UE based on current best evidence in neurologic rehabilitation robotics and practice-dependent neuroplasticity (Hakim et al., 2017).

More specifically, task-oriented, functional training is typically the focus of rehabilitation programs for patients following stroke and other injuries involving the CNS. With respect to patients with orthopedic disorders of the wrist and hand; however, this type of task-specific training is typically not the focus of rehabilitation. Rather, the rehabilitation emphasis has been on management of impairments such as pain, weakness, and loss of motion through interventions such as modalities, edema management, splinting, progressive resistance exercises, stretching, and joint/soft tissue/neural mobilization (Green et al., 2016). As such, most patients with orthopedic disorders do not receive high intensity, task-oriented training as part of a typical plan of rehabilitation care. However, as there is a continued emphasis on improving functional outcomes in patients with orthopedic disorders, a reconsideration of these prior rehabilitation approaches is warranted.

In order to achieve both power and precision grasp training that may be required to restore fine motor control of the hand and wrist, an end-effector robot such as a haptic joystick may be utilized. Unlike most of the robotic systems that have been used for neurologic rehabilitation, end-effector robotic systems do not provide assistance or support to the patient's entire UE. Therefore, the patient must have adequate control of the UE in order to grasp and appropriately control the joystick handle without physical assistance from a clinician or external support. High-intensity repetition training of grasp and/or manipulation tasks may be then be accomplished in the either the clinical or home setting. Thus, we used on an end-effector based, haptic robotic device that contacted the distal aspect of the individual's UE.

The following section describes the application of a commercially available VR-enhanced haptic robotic device using a task-oriented rehabilitation approach (i.e., handwriting computer program) for a patient with an orthopedic disorder of the wrist and hand to improve functional outcomes. The feasibility of this approach, as well as the acceptance of the device and training program, is discussed.

CASE REPORT DESCRIBING THE APPLICATION OF A VIRTUAL REALITY-ENHANCED ROBOTIC DEVICE IN ORTHOPEDIC REHABILITATION

The patient was a 73-year-old woman who had a history of right wrist trauma (secondary to a fall in 16 years prior) and multiple surgeries (including total wrist arthroplasty in four years prior with a recent revision in one year prior) who presented with chief complaints of right wrist/hand pain and loss of motion that resulted in increasing difficulty with functional mobility for the past several months. Her primary goal was to improve mobility, in particular as it related to her penmanship, which had been adversely affected since her surgeries. She had seen multiple rehabilitation specialists with only very limited effects. The patient's past medical history included rheumatoid arthritis; her past surgical history included right wrist/hand tendon repair; and contracture reduction. She was a retired teacher who enjoyed golfing and photography.

Tests and measures conducted before and after training consisted of pain assessment (numeric pain rating scale with 0 being "no pain" and ten being the "worst pain imaginable"), sensory screening (i.e., visual acuity, UE sensation assessment), impairment level testing (i.e., grip/pinch strength, wrist/hand active/passive range of motion), and functional testing (i.e., Jebsen Hand Function Test, Manual Ability Measure-36). At the time of initial examination, the patient had right wrist/hand pain (three out of ten) and intact sensation, limited active range of motion (right wrist flexion and extension decreased by 43 and 8 degrees, respectively, versus left), reduced grip strength (right = 2.7 kg, left = 7.3 kg,), and decreased hand function (Jebsen Hand Function Test = 47 seconds on the right; Manual Ability Measure-36 = 60%) (Jebsen et al., 1969; Chen and Bode, 2010).

For our rehabilitation protocol, we were specifically interested in a commercially available, relatively low-cost haptic device that is accessible to practicing clinicians. Thus, we selected an end-effector robotic device, with a VR-integrated haptic stylus (i.e., Geomagic Touch), that allowed objectively monitored training in the context of a virtual environment that also provides specific feedback on motor performance (Figure 17.7). Over a four week period, the patient participated in eight training sessions (90 minutes each) that included high-repetition fine motor training using a VR-enhanced haptic robotic device (i.e., Geomagic Touch).

A custom-designed software program was developed specifically for our protocol in conjunction with faculty and graduate students from the computing sciences department to provide targeted retraining of handwriting for both printing (serial task) and cursive (continuous task) letters using a haptic stylus shaped like a large pen linked to a virtual environment (i.e., letter-shaped maze on the screen). She was instructed to move a virtual pen smoothly and accurately through the letter maze without "touching" the walls (Figure 17.8). If the walls were "touched," haptic/tactile feedback was given, a dissonant alarm sounded and an error score was recorded. Thus, the distance between the walls of the path provides bandwidth feedback on performance errors – any movements within the walls are considered correct, while touching the wall is considered incorrect. With progression, she was challenged by making the path more narrow with levels I–III (bandwidth feedback) and/ or requiring her to move more quickly (i.e., in accordance with Fitt's Law) (Shumway-Cook and Woollacott, 2001). Completion of each maze/trial resulted in a time and error score.

FIGURE 17.7 Geomagic Touch haptic device equipped with a VR-integrated haptic stylus.

With respect to outcomes, the patient progressed from writing 52 words (three letters each) with an overall average of 3.4 errors and 24.6 seconds elapsed time per trial during the first session (Level 1: average per trial = 25.0 seconds, 2.2 errors for 6 trials; Level 2: average per trial = 24.7 seconds, 2.6 errors for 15 trials; Level 3: average per trial = 24.0 seconds; 5.3 errors for four trials) to 113 words completed on the final session with an overall average of 2.6 errors and 12.3 seconds elapsed time per trial (Level 1: average per trial = 12.0 seconds, 2.2 errors for 15 trials; Level 2: average per trial = 11.6 seconds, 1.7 errors for 29 trials; Level 3: average per trial = 13.3 seconds; 4.0 errors for 75 trials). She achieved overall adherence of 100% (12 hours total) to the massed practice training protocol with no adverse events and her goal of improved penmanship was achieved. At the time of final examination, subtle improvements were noted with impairments and function (Jebsen Hand Function Test = 44 seconds on right; Manual Ability Measure-36 = 63%). Patient feedback regarding the device/computer program noted problems that included dissatisfaction with the completion indicator for each letter, difficulty moving the stylus to the starting position, and a lack of fluidity when moving the stylus; however, the ease of use, virtual handwriting task, and elapsed time/error scores provided motivation for improvement and enjoyment.

The use of a VR-enhanced haptic robotic device allowed high repetition, task-specific training with augmented extrinsic feedback (i.e., visual, auditory, tactile) for a patient with a chronic orthopedic disorder of the wrist/hand. This paradigm shift focused on functional training rather than impairment reduction which is typical in orthopedic rehabilitation. The training protocol, which was safe, feasible, and enjoyable for this patient, allowed her to achieve her primary goal of improved penmanship. Her feedback also informed the ongoing development of an updated software program which will infer more natural stroke paths for each letter in the virtual environment. Future research using rehabilitation robotics for task-specific training in an orthopedic population appears warranted to determine the optimal training parameters, outcome measures, and diagnostic groups who will benefit most to improve function.

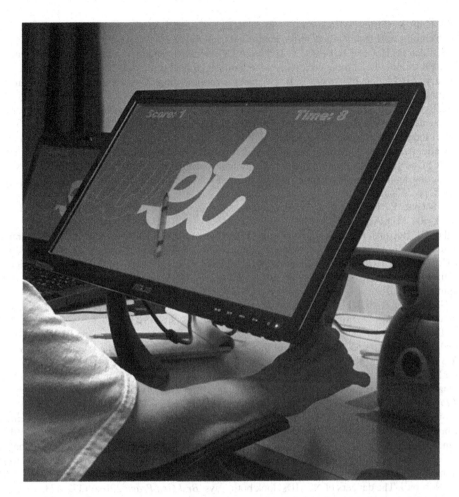

FIGURE 17.8 Configuration and view of the computer screen for the letter maze used for the haptic robotic handwriting training program.

CONCLUSIONS

Technological advances in the fields of VR and robotics provide exciting new options for the rehabilitation of patients with neurologic and orthopedic dysfunction. Based on the principles of neuroplasticity, these devices can serve to supplement traditional rehabilitation paradigms by providing high intensity, task-specific training of functional tasks. In addition, quantitative data can be generated to allow progression of interventions and tracking of patient outcomes. Future research in the form of randomized clinical trials should focus on determining the optimal approach to training using functional paradigms to contribute to improved overall patient outcomes.

REFERENCES

About Stroke. www.stroke.org. https://www.stroke.org/en/about-stroke. Accessed April 13, 2020.

Aida J, Chau B, Dunn J. Immersive virtual reality in traumatic brain injury rehabilitation: a literature review. *NeuroRehabilitation*. 2018;42(4):441–448. doi: 10.3233/NRE-172361.

Alashram AR, Padua E, Hammash AK, Lombardo M, Annino G. Effectiveness of virtual reality on balance ability in individuals with incomplete spinal cord injury: a systematic review. *J Clin Neurosci*. 2020;72:322–327. doi: 10.1016/j.jocn.2020.01.037. Epub 2020 Jan 16.

Bahat HS, Croft K, Carter C, Hoddinott A, Sprecher E, Treleaven J. Remote kinematic training for patients with chronic neck pain: a randomised controlled trial. *Eur Spine J.* 2018;27:1309–1323.

Bahat HS, Takasaki H, Chen X, Bet-Or Y, Treleaven J. Cervical kinematic training with and without interactive VR training for chronic neck pain: a randomized clinical trial. *Man Ther.* 2015;20:68–78.

Brain Injury Association of America. Brain injury overview. https://www.biausa.org/brain-injury/about-brain-injury/basics/overview. Accessed April 13, 2020.

Broeren J, Rydmark M, Sunnerhagen K. Virtual reality and haptics as a training device for movement rehabilitation after stroke: a single-case study. *Arch Phys Med Rehabil.* 2004;85(8):1247–1250.

Buma FE, Lindeman E, Ramsey NF, Kwakkel G. Functional neuroimaging studies of early upper limb recovery after stroke: a systematic review of the literature. *Neurorehabil Neural Repair.* 2010;24(7):589–608.

Burdea GC. Chapter 25: The role of haptics in physical rehabilitation. In: Otaduy MA, Peters AK. Haptic rendering foundations, algorithms, and applications, Boca Raton, FL: CRC Press/Taylor & Frances; 2008, pp. 517–529.

Carr JH, Shepherd RB, editors. *Movement science: foundations for physical therapy in rehabilitation.* Rockville, MD: Aspen; 1987.

Carrougher GJ, Hoffman HG, Nakamura D, et al. The effect of virtual reality on pain and range of motion in adults with burn injuries. *J Burn Care Res.* 2009;30(5):785–791. doi: 10.1097/BCR.0b013e3181b485d3.

Chen CC, Bode RK. Psychometric validation of the manual ability measure-36 (MAM-36) in patients with neurologic and musculoskeletal disorders. *Arch Phys Med Rehabil.* 2010; 91(3):414–420. doi: 10.1016/j.apmr.2009.11.012.

Chen P, Goldberg D, Kolb B, Lanser M, Benowitz L. Inosine induces axonal rewiring and improves behavioral outcome after stroke. *Proc Natl Acad Sci USA.* 2002;99(13):9031–9036.

Chen L, Lo WL, Mao YR, Ding MH, Lin Q, Li H, Zhao JL, Xu ZQ, Bian RH, Huang DF. Effect of virtual reality on postural and balance control in patients with stroke: a systematic literature review. *Biomed Res Int.* 2016;2016:7309272. doi: 10.1155/2016/7309272. Epub 2016 Dec 7.

Corbetta D, Imeri F, Gatti R. Rehabilitation that incorporates virtual reality is more effective than standard rehabilitation for improving walking speed, balance and mobility after stroke: a systematic review. *J Physiother.* 2015;61(3):117–124. doi: 10.1016/j.jphys.2015.05.017. Epub 2015 Jun 18.

Cramer SC, Sur M, Dobkin BH, O'Brien C, Sanger TD, Trojanowski JQ, Rumsey JM, Hicks R, Cameron J, Chen D, Chen WG. Harnessing neuroplasticity for clinical applications. *Brain.* 2011;134(6):1591–1609.

Dahlhamer J, Lucas J, Zelaya, C, et al. Prevalence of chronic pain and high-impact chronic pain among adults - United States, 2016. *MMWR Morb Mortal Wkly Rep.* 2018;67:1001–1006. http://dx.doi.org/10.15585/mmwr.mm6736a2external icon.

Davis A, Robson J. The dangers of NSAIDs: look both ways. *Br J Gen Pract.* 2016;66:172–173.

Davis KD, Taylor KS, Anastakis DJ. Nerve injury triggers changes in the brain. *Neuroscientist.* 2011;17(4):407–422.

de Araújo AVL, Neiva JFO, Monteiro CBM, Magalhães FH. Efficacy of virtual reality rehabilitation after spinal cord injury: a systematic review. *Biomed Res Int.* 2019;2019:7106951. doi: 10.1155/2019/7106951.

De Keersmaecker E, Lefeber N, Geys M, Jespers E, Kerckhofs E, Swinnen E. Virtual reality during gait training: does it improve gait function in persons with central nervous system movement disorders? A systematic review and meta-analysis. *NeuroRehabilitation.* 2019;44(1):43–66. doi: 10.3233/NRE-182551.

De Rooij IJM, van de Port GL, Meijer JWG. Effect of virtual reality training on balance and gait ability in patients with stroke: systematic review and meta-analysis. *Phys Ther.* 2016;96:1905–1918.

Dimyan, MA, Cohen LG. Neuroplasticity in the context of motor rehabilitation after stroke. *Nat Rev Neurol.* 2011;7(2):76–85.

Dobkin BH. Training and exercise to drive poststroke recovery. *Nat Clin Pract Neurol.* 2008;4(2):76–85.

Dockx K, Bekkers EM, Van den Bergh V, et al. Virtual reality for rehabilitation in Parkinson's disease. *Cochrane Database Syst Rev.* 2016;12:CD010760.

Domínguez-Téllez P, Moral-Muñoz JA, Salazar A, Casado-Fernández E, Lucena-Antón D. Game-based virtual reality interventions to improve upper limb motor function and quality of life after stroke: systematic review and meta-analysis. *Games Health J.* 2020;9(1):1–10. doi: 10.1089/g4h.2019.0043.

Felipe FA, de Carvalho FO, Silva ÉR, Santos NGL, Fontes PA, de Almeida AS, Garção DC, Nunes PS, de Souza Araújo AA. Evaluation instruments for physical therapy using virtual reality in stroke patients: a systematic review. *Physiother.* 2020;106:194–210. doi: 10.1016/j.physio.2019.05.005.

Fisch A, Mavroidis C, Melli-Huber J, Bar-Cohen Y. Chapter 4: Haptic devices for virtual reality, telepresence, and human-assistive robotics. In: Bar-Cohen Y, Breazael C. *Biologically inspired intelligent robots*, Bellingham, WA: SPIE Press; 2003, pp. 73–102.

Florence SL, Kaas JH. Large-scale reorganization at multiple levels of the somatosensory pathway follows therapeutic amputation of the hand in monkeys. *J Neurosci*. 1995;15:8083–8095.

Franklin GM, Stover BD, Turner JA, et al. Early opioid prescription and subsequent disability among workers with back injuries: the disability risk identification study cohort. *Spine*. 2008;33:199–204.

Freitag F, Brucki SMD, Barbosa AF, Chen J, Souza CO, Valente DF, Chien HF, Bedeschi C, Voos MC. Is virtual reality beneficial for dual-task gait training in patients with Parkinson's disease? A systematic review. *Dement Neuropsychol*. 2019;13(3):259–267. doi: 10.1590/1980-57642018dn13-030002.

Ghai S, Ghai I. Virtual reality enhances gait in cerebral palsy: a training dose-response meta-analysis. *Front Neurol*. doi: 10.3389/fneur.2019.00236. eCollection 2019.

Gibbons EM, Thomson AN, de Noronha M, Joseph S. Are virtual reality technologies effective in improving lower limb outcomes for patients following stroke - a systematic review with meta-analysis. *Top Stroke Rehabil*. 2016;23(6):440–457.

Glegg S. Virtual rehabilitation with children: challenges for clinical adoption [from the field]. *IEEE Pulse*. 2017;8:3–5. doi: 10.1109/MPUL.2017.2750858.

Glegg SM, Holsti L, Velikonja D, Ansley B, Brum C, Sartor D. Factors influencing therapists' adoption of virtual reality for brain injury rehabilitation. *Cyberpsychol Behav Soc Netw*. 2013;16:385–401. doi: 10.1089/cyber.2013.1506.

Green JB, Deveikas C, Ranger HE, et al. Chapter 10: Hand, wrist and digit injuries. In: Magee DJ, Zachazewski JE, Quillen WS, Manske RC. *Pathology and intervention in musculoskeletal rehabilitation*. 2nd ed. Maryland Heights, MO: Elsevier; 2016, pp. 344–436.

Gumaa M, Youssef AR. Is virtual reality effective in orthopedic rehabilitation? A systematic review and meta-analysis. *Phys Ther*. 2019;99:1304–1325.

Hakim RM, Tunis BG, Ross MD. Rehabilitation robotics for the upper extremity: review with new directions for orthopaedic disorders. *Disabil Rehabil Assist Technol*. 2017;12:765–771.

Hansson T, Brismar T. Loss of sensory discrimination after median nerve injury and activation in the primary somatosensory cortex on functional magnetic resonance imaging. *J Neurosurg*. 2003;99:100–105.

Holden MK, Dyar T. Virtual environment training: a new tool for rehabilitation. *Neurol Rep*. 2002;26(2):62–71.

Hornby TG, Reisman DS, Ward IG, Scheets PL, Miller A, Haddad D, Fox EJ, Fritz NE, Hawkins K, Henderson CE, Hendron KL. Clinical practice guideline to improve locomotor function following chronic stroke, incomplete spinal cord injury, and brain injury. *J Neurol Phys Ther*. 2020;44(1):49–100.

Hoy D, Bain C, Williams G, March L, Brooks P, Blyth F, et al. A systematic review of the global prevalence of low back pain. *Ann Rheum Dis*. 2012;64(6):2028–2037.

Hoy DG, Smith E, Cross M, Sanchez-Riera L, Buchbinder R, Blyth FM, et al. The global burden of musculoskeletal conditions for 2010: an overview of methods. *Ann Rheum Dis*. 2014;73(6):982–989.

Iruthayarajah J, McIntyre A, Cotoi A, Macaluso S, Teasell R. The use of virtual reality for balance among individuals with chronic stroke: a systematic review and meta-analysis. *Top Stroke Rehabil*. 2017;24(1):68–79. Epub 2016 Jun 16.

Jang SH, You SH, Hallett M, Cho YW, Park CM, Cho SH, Lee HY, Kim TH. Cortical reorganization and associated functional motor recovery after virtual reality in patients with chronic stroke: an experimenter blind preliminary study. *Arch Phys Med Rehabil*. 2005;86(11):2218–2223.

Jebsen, RH, Taylor, N, Trieschmann, RB, Trotter, MJ, Howard, LA. An objective and standardized test of hand function. *Arch Phys Med Rehabil*. 1969;50:311–319.

Juras G, Brachman A, Michalska J, Kamieniarz A, Pawłowski M, Hadamus A, Białoszewski D, Błaszczyk J, Słomka KJ. Standards of virtual reality application in balance training programs in clinical practice: a systematic review. *Games Health J*. 2019;8(2):101–111.

Kahn LE, Lum PS, Rymer WZ, Reinkensmeyer DJ. Robot-assisted movement training for the stroke-impaired arm: does it matter what the robot does? *J Rehabil Res Dev*. 2006;43(5):619.

Karamians R, Proffitt R, Kline D, Gauthier LV. Effectiveness of virtual reality- and gaming-based interventions for upper extremity rehabilitation poststroke: a meta-analysis. *Arch Phys Med Rehabil*. 2019. doi: 10.1016/j.apmr.2019.10.195. [Epub ahead of print].

Kleim JA, Jones TA. Principles of experience-dependent neural plasticity: implications for rehabilitation after brain damage. *J Speech Lang Hear Res*. 2008;51:S225–S239.

Koeneman EJ, Swenson P, Shepard B, Perez M, Koeneman JB. Neurological reorganization after brain injury: stimulation by robotic therapy. *J Histotechnol* 2007;30(4):229–233.

Krebs HI. Robotic technology and physical medicine and rehabilitation. *Eur J Phys Rehabil Med*. 2012;48(2):319–324.

Krebs HI, Volpe BT. Rehabilitation robotics. *Handb Clin Neurol*. 2013;110:283–294.

Kwakkel G, Wagenaar RC, Koelman TW, Lankhorst GJ, Koetsier JC. Effects of intensity of rehabilitation after stroke: a research synthesis. *Stroke.* 1997;28(8):1550–1556.

Langhorne P, Coupar F, Pollock A. Motor recovery after stroke: a systematic review. *Lancet Neurol.* 2009;8:741–754.

Laver K, Lange B, George S, Deutsch J, Saposnik G, Crotty M. Virtual reality for stroke rehabilitation. *Cochrane Database Syst Rev.* 2017;11:CD008349.

Lee HS, Park YJ, Park SW. The effects of virtual reality training on function in chronic stroke patients: a systematic review and meta-analysis. *Biomed Res Int.* 2019;18:7595639. doi: 10.1155/2019/7595639.

Lei C, Sunzi K, Dai F, Liu X, Wang Y, Zhang B, He L, Ju M. Effects of virtual reality rehabilitation training on gait and balance in patients with Parkinson's disease: a systematic review. *PLoS One.* 2019 Nov 7;14(11):e0224819. doi: 10.1371/journal.pone.0224819. eCollection 2019.

Levac DE, Huber ME, Sternad D. Learning and transfer of complex motor skills in virtual reality: a perspective review. *J NeuroEngineer Rehabil.* 2019;16:121. https://doi.org/10.1186/s12984-019-0587-8.

Levin MF. What is the potential of virtual reality for poststroke sensorimotor rehabilitation? *Expert Rev Neurotherapeut.* 2020;20(3):195–197. doi: 10.1080/14737175.2020.1727741.

Levin MF, Weiss PL, Keshner EA. Emergence of virtual reality as a tool for upper limb rehabilitation: incorporation of motor control and motor learning principles. *Phys Ther.* 2015;95:415–425.

Lewis GN, Rosie JA. Virtual reality games for movement rehabilitation in neurological conditions: how do we meet the needs and expectations of the users? *Disabil Rehabil.* 2012;34(22):1880–1886.

Li Z, Han XG, Sheng J, Ma SJ. Virtual reality for improving balance in patients after stroke: a systematic review and meta-analysis. *Clin Rehabil.* 2016;30(5):432–440. doi: 10.1177/0269215515593611. Epub 2015 Jul 3.

Lo AC, Guarino PD, Richards LG et al., Robot-assisted therapy for long-term upper-limb impairment after stroke. *N Engl J Med.* 2010;362(19):1772–1783.

Lovecchio F, Premkumar A, Stepan JG, Albert TJ. Fighting back: institutional strategies to combat the opioid epidemic: a systematic review. *HSS J.* 2019;15:66–71.

Lundborg G. Nerve injury and repair - a challenge to the plastic brain. *J Periph Nerv Sys.* 2003;8(4):209–226.

Maani CV, Hoffman HG, Morrow M, et al. Virtual reality pain control during burn wound debridement of combat-related burn injuries using robot-like arm mounted VR goggles. *J Trauma.* 2011;71(1 Suppl):S125–S130. doi: 10.1097/TA.0b013e31822192e2.

Mallari B, Spaeth EK, Goh H, Boyd BS. Virtual reality as an analgesic for acute and chronic pain in adults: a systematic review and meta-analysis. *J Pain Res.* 2019;12:2053–2085. Published 2019 Jul 3. doi: 10.2147/JPR.S200498.

Massetti T, da Silva TD, Crocetta TB, Guarnieri R, de Freitas BL, Bianchi Lopes P, Watson S, Tonks J, de Mello Monteiro CB. The clinical utility of virtual reality in neurorehabilitation: a systematic review. *J Cent Nerv Syst Dis.* 2018. doi: 10.1177/1179573518813541.

Massetti T, Trevizan IL, Arab C, Favero FM, Ribeiro-Papa DC, de Mello Monteiro CB. Virtual reality in multiple sclerosis - a systematic review. *Mult Scler Relat Disord.* 2016;8:107–112. doi: 10.1016/j.msard.2016.05.014. Epub 2016 May 21.

Mayo Clinic. Cerebral palsy symptoms and causes. https://www.mayoclinic.org/diseases-conditions/cerebral-palsy/symptoms-causes. Accessed April 13, 2020.

Mehrholz J, Pohl M, Platz T, Kugler J, Elsner B. Electromechanical and robot-assisted arm training for improving activities of daily living, arm function, and arm muscle strength after stroke. *Cochrane Database Syst Rev.* 2018;9. Art. No.:CD006876. doi: 10.1002/14651858.CD006876.pub5.

Merzenich MM, Wright BA, Jenkins W, Xerri C, Byl N, Miller S, Tallal P. Cortical plasticity underlying perceptual, motor, and cognitive skill development: implications for neurorehabilitation. *Cold Spring Harb Symp Quant Biol.* 1996;61:1–8.

Miller LK, Chester R, Jerosch-Herold C. Effects of sensory reeducation programs on functional hand sensibility after median and ulnar repair: a systematic review. *J Hand Ther.* 2012;25(3):297–307.

Mohammadi R, Semnani AV, Mirmohammadkhani M, Grampurohit N. Effects of virtual reality compared to conventional therapy on balance poststroke: a systematic review and meta-analysis. *J Stroke Cerebrovasc Dis.* 2019;28(7):1787–1798. doi: 10.1016/j.jstrokecerebrovasdis.2019.03.054. Epub 2019 Apr 25.

Morris LD, Louw QA, Crous LC. Feasibility and potential effect of a low-cost virtual reality system on reducing pain and anxiety in adult burn injury patients during physiotherapy in a developing country. *Burns.* 2010;36(5):659–664. doi: 10.1016/j.burns.2009.09.005.

National Multiple Sclerosis Society. What is MS? https://www.nationalmssociety.org/What-is-MS. Accessed April 13, 2020.

Nithianantharajah J, Hannan AJ. Enriched environments, experience dependent plasticity and disorders of the nervous system. *Nat Rev Neurosci.* 2006;7(9):697–709.

Nudo R, Wise B, SiFuentes F, Milliken G. Neural substrates for the effects of rehabilitative training on motor recovery after ischemic infarct. *Science.* 1996;272(5269):1791–1794.

Padilla-Castaneda MA, Sotgiu E, Barsotti M, Frisoli A, Orsini P, Martiradonna A, Laddaga C, Bergamasco M. An orthopaedic robotic-assisted rehabilitation method of the forearm in virtual reality physiotherapy. *J Healthc Eng.* 2018;2018:7438609. Epub 2018 Aug 1.

Padilla-Castaneda MA, Sotgiu E, Frisoli A, Bergamasco M. A robotic & virtual reality orthopedic rehabilitation system for the forearm. *Stud Health Technol Inform.* 2012;181:324–328.

Parkinson's Foundation. What is Parkinson's? https://www.parkinson.org/understanding-parkinsons/what-is-parkinsons. Accessed April 13, 2020.

Pekyavas NO, Ergun N. Comparison of virtual reality exergaming and home exercise programs in patients with subacromial impingement syndrome and scapular dyskinesis: short term effect. *Acta Orthop Traumatol Turc.* 2017;51:238–242.

Pincus T, Kent P, Bronfort G, Loisel P, Pransky G, Hartvigsen J. Twenty-five years with the biopsychosocial model of low back pain—is it time to celebrate? A report from the twelfth international forum for primary care research on low back pain. *Spine shila Pa 1976).* 2013;38:2118–2123. https://doi.org/10.1097/BRS.0b013e3182a8c5d6.

Piron L, Tonin P, Atzori AM, Zucconi C, Massaro C, Trivello E, Dam M. The augmented-feedback rehabilitation technique facilitates the arm motor recovery in patients after a recent stroke. *Stud Health Technol Inform.* 2003;94:265–267.

Piron L, Tonin P, Piccione F, Iaia V, Trivello E, Dam M. Virtual environment training therapy for arm motor rehabilitation. *Presence.* 2005;14:732–740.

Plautz EJ, Milken GW, Nudo RJ. Effects of repetitive motor training on movement representations in adult squirrel monkeys: role of use versus learning. *Neurobiol Learn Mem.* 2000;74(1):27–55.

Prange, GB, Jannink MJ, Groothuis-Oudshoorn CG, Hermens HJ, and Ijzerman MJ. Systematic review of the effect of robot-aided therapy on recovery of the hemiparetic arm after stroke. *J Rehab Res Dev.* 2006;43(2):171.

Proteau L, Blandin Y, Alain C, Dorion A. The effects of the amount and variability of practice on the learning of a multi-segmented motor task. *Acta Psychol.* 1994;85(1):61–74.

Quadrado VH, Silva TDD, Favero FM, Tonks J, Massetti T, Monteiro CBM. Motor learning from virtual reality to natural environments in individuals with Duchenne muscular dystrophy. *Disabil Rehabil Assist Technol.* 2017;10:1–9.

Rahman SA, Shaheen AA. Virtual reality use in motor rehabilitation of neurological disorders: a systematic review. *Middle East J Sci Res.* 2011;7:63–70.

Ravi DK, Kumar N, Singhi P. Effectiveness of virtual reality rehabilitation for children and adolescents with cerebral palsy: an updated evidence-based systematic review. *Physiotherapy.* 2017;103:245–258.

Ren Z, Wu J. The effect of virtual reality games on the gross motor skills of children with cerebral palsy: a meta-analysis of randomized controlled trials. *Int J Environ Res Public Health.* 2019;16(20):3885. doi: 10.3390/ijerph16203885.

Rizzo AA, Kim GJ. A SWOT analysis of the field of virtual reality rehabilitation and therapy. *Presence-Teleop Virt Envir.* 2005;14:119–146.

Rosén B, Lundborg G. Early use of artificial sensibility to improve sensory recovery after repair of the median and ulnar nerve. *Scand J Plast Reconstr Surg Hand Surg.* 2003;37(1):54–57.

Rosén B, Vikström P, Turner S, McGrouther DA, Selles RW, Schreuders TA. Enhanced early sensory outcome after nerve repair as a result of immediate post-operative re-learning: a randomized controlled trial. *J Hand Surg [Br].* 2014;7:1–9.

Sanchez-Vives M, Slater M. From presence to consciousness through virtual reality. *Nat Rev Neurosci.* 2005;6(4):332–339.

Saposnik GS, Levin M. Virtual reality in stroke rehabilitation a meta-analysis and implications for clinicians. *Stroke.* 2011;42:1380–1386.

Sasmita AO, Kuruvilla J, Ling APK. Harnessing neuroplasticity: modern approaches and clinical future. *Int J Neurosci.* 2018. doi: 10.1080/00207454.2018.1466781.

Schmidt RA, Lee TD. *Motor control and learning: a behavioral emphasis.* 5th ed. Champaign, IL: Human Kinetics; 2011.

Schweighofer N, Choi Y, Winstein C, Gordon J. Task-oriented rehabilitation robotics. *Am J Phys Med Rehabil.* 2012;91(11):S270–S279.

Shepherd Center. Spinal cord injury rehabilitation. https://www.shepherd.org/patient-programs/spinal-cord-injury/about. Accessed April 12, 2020.

Shumway-Cook A, Woollacott M. Chapter 16: Normal reach, grasp, and manipulation. In: *Motor control: theory and practical applications.* 2nd ed. Philadelphia, PA: Lippincott Williams-Wilkins; 2001, pp. 477–501.

Svens B, Rosen B. Early sensory re-learning after median nerve repair using mirror training and sense substitution. *J Hand Ther.* 2009;14(3):75–82.

Thomas JS, France CR, Applegate ME, Leitkam ST, Walkowski S. Feasibility and safety of a virtual reality dodgeball intervention for chronic low back pain: a randomized clinical trial. *J Pain.* 2016;17:1302–1317.

Triegaardt J, Han TS, Sada C, Sharma S, Sharma P. The role of virtual reality on outcomes in rehabilitation of Parkinson's disease: meta-analysis and systematic review in 1031 participants. *Neurol Sci.* 2020;41(3):529–536. doi: 10.1007/s10072-019-04144-3. Epub 2019 Dec 6.

Waldner A, Tomelleri C, Hesse S. Transfer of scientific concepts to clinical practice: recent robot-assisted training studies. *Funct Neurol.* 2009;24(4):173–177.

Wang B, Shen M, Wang YX, He ZW, Chi SQ, Yang ZH. Effect of virtual reality on balance and gait ability in patients with Parkinson's disease: a systematic review and meta-analysis. *Clin Rehabil.* 2019;33(7):1130–1138. doi: 10.1177/0269215519843174. Epub 2019 Apr 24.

Warnier N, Lambregts S, Port IV. Effect of virtual reality therapy on balance and walking in children with cerebral palsy: a systematic review. *Dev Neurorehabil.* 2019;1:1–17. doi: 10.1080/17518423.2019.1683907. Epub ahead of print.

Webster BS, Verma SK, Gatchel RJ. Relationship between early opioid prescribing for acute occupational low back pain and disability duration, medical costs, subsequent surgery and late opioid use. *Spine.* 2007;32:2127–2132.

Winstein CJ, Merians AL, Sullivan KJ. Motor learning after unilateral brain damage. *Neuropsychology.* 1999;37(8):975–987.

Woldag H, Hummelsheim H. Evidence-based physiotherapeutic concepts for improving arm and hand function in stroke patients: a review. *J Neurol.* 2002;249:518–528.

World Health Organization. Spinal cord injury. https://www.who.int/news-room/fact-sheets/detail/spinal-cord-injury. Accessed April 13, 2020.

Wu J, Loprinzi PD, Ren Z. The rehabilitative effects of virtual reality games on balance performance among children with cerebral palsy: a meta-analysis of randomized controlled trials. *Internat J Envirol Res Pub Health.* 2019;16(21):4161.

Yazgan YZ, Tarakci E, Tarakci D, Ozdincler AR, Kurtuncu M. Comparison of the effects of two different exergaming systems on balance, functionality, fatigue, and quality of life in people with multiple sclerosis: a randomized controlled trial. *Mult Scler Relat Disord.* 2019;21;39:101902. doi: 10.1016/j.msard.2019.101902.

You SH, Jang SH, Kim Y, Hallett M, Ahn SH, Kwon YH, Kim JH, Lee MY. Virtual reality-induced cortical reorganization and associated locomotor recovery in chronic stroke: an experimenter-blind randomized study. *Stroke.* 2005;36(6):1166–11671.

18 Non-immersive Virtual Reality as a Tool for Treating Children with Cerebral Palsy

Lígia Maria Presumido Braccialli and Ana Carla Braccialli
Universidade Estadual Paulista Júlio de Mesquita Filho – UNESP
Marília, São Paulo, Brazil

Michelle Zampar Silva
University of São Paulo – USP
Ribeirão Preto, São Paulo, Brazil

CONTENTS

INTRODUCTION

Cerebral palsy (CP) is the most common condition of childhood disability, with an estimated incidence of 2–2.5/1,000 live births in developed countries (Shimony et al., 2008); and in Brazil, 7 per 1,000 live births (Lima and Fonseca, 2004), and has been described as a group of movement and posture disorders that cause activity limitations, attributed to non-progressive changes occurring in the fetal or infant brain. Motor disorders are often accompanied by sensory and cognitive changes in communication, perception, behavior, and/or seizures (Bax, et al., 2005; Brasil, 2013). These children present different motor conditions, but invariably they have a deficit in postural control, restriction in manual skills and in reaching objects, and impaired fingers dexterity. These children's ability and manual motor performance depend on the relevance and importance of the proposed task. Motor limitation is one of the factors that corroborates non-participation and engagement in activities, even everyday life activities (Codgno, Braccialli and Presumido Braccialli, 2018).

Despite the importance of engaging children with CP in physical activities in their natural environment, in general, these children, even the ones who can walk, are sedentary and spend little time participating in physical activities, either at school or at home (Li et al., 2017). Children and adolescents with disabilities have a lower level of participation in activities in different environments, with a lower range of activity and less social interaction in comparison with their peers without disabilities (Schreuer, Sachs and Rosenblum, 2014; Silva, 2019). The level of participation of these children seems to be associated with gross motor function, more effective adaptive behavior, age and family guidance (Palisano et al., 2011). In relation to Brazilian children with physical disabilities, the level of participation is related to the environment, with greater participation at home (83.16%), gradually decreasing in the school setting (67.37%) and in the community (56,67%) (Silva, 2019).

Due to the conditions presented by people with CP, the literature indicates the need to implement therapies which (1) are family-centered, that is, all decisions and actions are taken together with the child and the family; (2) have prioritized functional objectives; and (3) are capable of promoting the participation and inclusion of children in all activities developed in different settings, home, school, and community. To this end, modifications should be made, when necessary, to maximize function and provide children with opportunities to learn and move in natural environments and engage in meaningful activities (Longo, De Campos and Palisano, 2019).

From this perspective, the use of virtual reality tools has been a strategy used to simultaneously stimulate the participation of this population in different settings and enable the performance of physical activity that may assist in the acquisition of new skills and development. The use of Exergaming can serve two purposes: to encourage participation in physical activities and to promote friendship and social cohesion with peers (Williams and Ayres, 2020).

Virtual reality training programs for children and adolescents with CP offer the possibility to perform, in a virtual environment, activities that cannot be carried out in the real world, either due to the limitations inherent to their own disability or due to safety (Braccialli, Silva and Rebelo, 2019). In this context, the use of VR can bring advantages: (1) the possibility of adapting the virtual environment relatively easily to meet the therapeutic goals and physical abilities of the user; (2) dangerous activities may be performed in a safe environment with lower risks of injury; (3) ease of real-time performance feedback; (4) motivational; (5) possibility of increasing or decreasing task complexity; (6) tasks can be performed in a collaborative setting; (7) user performance is recorded in a database, which facilitates stracking development; (8) ecological validity; (9) sensory input control; (10) content control; (11) possibility of extended practice at home, school, or other community setting; (12) flexibility of the learning setting (Braccialli, Rebelo and Pereira, 2012); and (13) allows the execution of a greater number of repetitions of body movements per session in comparison with conventional motor intervention techniques (Cavalcante Neto, Steenbergen and Tudella, 2019).

RECOMMENDATIONS FOR USING VIRTUAL REALITY WITH CHILDREN WITH CEREBRAL PALSY

Based on a search for renowned studies in the area, we synthesized recommendations for using programs with virtual reality tools with children and adolescents with CP. Recommendations were made regarding: (1) population; (2) indications and benefits of using VR based on ICF components; (3) characteristics of VR programs for children and adolescents with CP; and (4) clinical implications.

POPULATION

Virtual reality as a tool for rehabilitation has been used with children and adolescents with CP of both sexes, aged between 4 and 17 years. The reviews indicate studies conducted with children and adolescents with different topographic distributions (diplegia, hemiplegia, and quadriplegia), with different types of tone (spasticity, dyskinesia, ataxia, and hypotonia) and classified into the five levels of the Gross Motor Function Classification System (Chen, Fanchiang and Howard, 2018; Ravi, Kumar and Singhi, 2017; Monge Pereira et al., 2014). However, there is no indication as to whether the type of CP in relation to topographic distribution and tone and the severity of gross motor skill impairment would be related to the positive or negative results obtained with the use of this resource.

INDICATIONS AND BENEFITS OF USING VR BASED ON ICF COMPONENTS

There is scientific evidence that VR provides positive results for this population in the different ICF components: Body structure and function; Activity and participation; Environmental and personal factors. We highlight that most studies have analyzed outcomes related to body structure and function and that the strongest evidence of benefits is in this component. Few studies have been conducted regarding the impact of VR on the components Activity and participation; Environmental and personal factors.

For the component Body structure and function, research has indicated that the use of VR results in positive motor function outcomes (Ghai and Ghai, 2019; Massetti et al., 2018; Ravi, Kumar and Singhi, 2017; Teo et al., 2016; Massetti et al., 2014; Wang and Reid, 2011; Snider, Majnemer and Darsaklis, 2010); in range of motion (Snider, Majnemer and Darsaklis, 2010; Tatla et al., 2013), in upper limb motor function (Chen, Fanchiang and Howard, 2018; Massetti et al., 2018; Ravi, Kumar and Singhi, 2017; Monge Pereira et al., 2014; Snider, Majnemer and Darsaklis, 2010), in postural control (Chen, Fanchiang and Howard, 2018; Monge Pereira et al., 2014), in balance (Ravi, Kumar and Singhi, 2017), in physical capacity (Weiss, Tirosh and Fehlings, 2014; Tatla et al., 2013; Mitchell et al., 2012), and walking (Ghai and Ghai, 2019; Monge Pereira et al., 2014; Tatla et al., 2013).

Regarding range of motion, studies investigating ankle dorsiflexion kinematics have found greater range of motion in this joint during VR programs in comparison with conventional exercises (Snider, Majnemer and Darsaklis, 2010; Tatla et al., 2013).

Training with Wii Sports on Nintendo and EyeToy on Playstation 2 has been effective in improving the physical capacity and performance of children with CP (Mitchell et al., 2012) in addition to increasing energy expenditure (Tatla et al., 2013) and decreasing sedentary behavior, so common in this population (Weiss, Tirosh and Fehlings, 2014).

Current studies indicate strong evidence that VR training has positive effects on postural control and balance (Chen, Fanchiang and Howard, 2018). Those programs carried out with Nintendo Wii and Xbox Kinect were effective in improving balance in the population with CP (Ravi, Kumar and Singhi, 2017).

Current findings suggest that a VR training program has positive influence on gait performance in relation to spatiotemporal parameters, i.e. gait speed, cadence, and stride length (Ghai and Ghai, 2019). Improvement in gait parameters may be a result of multisensory biofeedback and motivation during virtual training when compared to conventional training (Tatla et al., 2013).

Evidence shows that well-structured and systematized immersive and non-immersive VR programs favor upper limb functional performance in the following aspects: quality of movement, selective motor control, postural control, and coordination (Monge Pereira et al., 2014) and the number of upper limb movements (Chen, Fanchiang and Howard, 2018).

As for motor skill enhancement, VR programs can contribute to acquisition due to increased engagement and motivation during the game, which favors the repeated practice of a specific task in a virtual environment (Chen, Fanchiang and Howard, 2018). A six-week training with Nitendo Wii has resulted in an increase in the average Gross Motor Function Measurement score (Ravi, Kumar and Singhi, 2017). Other studies have indicated improved visual-perceptual processing, functional mobility, fine motor skills (Snider, Majnemer and Darsaklis, 2010), and sitting ability in children with levels IV and V in the Gross Motor Classification System – GMFCS (Ravi, Kumar and Singhi, 2017). To achieve these goals, the virtual environment must have ecological validity and enable the adjustment of the task difficulty, in addition to offering multisensory biofeedback.

Research about the effect of VR programs on activity and participation of children and adolescents with CP is scarce. However, considering that by lowering barriers and increasing facilitators, children with CP can gradually improve body structure and function components and decrease activity limitations, and consequently increase school and community participation (Chen, Fanchiang and Howard, 2018). Some reports have indicated that virtual reality therapy maximized hand function in children during daily living activities, although study participants had good cognitive ability to understand virtual reality intervention (Ravi, Kumar and Singhi, 2017).

However, there is still no consensus on the transfer and generalization of trained skills in the virtual world to the real world. Some authors have suggested that the degree of immersion in the virtual environment influences transfer and generalization of retained and learned activity in the virtual environment to the real environment (Standen and Brown, 2006; Sveistrup, 2004).

VR programs have been found to have positive effects on personal factors such as motivation, volition, and perceptions of self-efficacy in comparison with no intervention or alternative treatment methods (Snider, Majnemer and Darsaklis, 2010). Evidence shows that children with CP are more motivated to practice motor tasks in a virtual environment compared to conventional therapy. However, the type of activity interferes with motivation levels (Tatla et al., 2013).

VR environments must present the following characteristics to enable increasing volition of children with CP: (1) be an unpredictable environment, requiring continued concentration and readiness from the child and allow for activity variation; (2) be challenging, but fit the user's skills, neither too difficult nor too easy; and (3) be a competitive environment that encourages the participant to set concrete goals (Harris and Reid, 2005).

PROGRAM CHARACTERISTICS

VR systems have been used primarily in rehabilitation clinics and laboratories. However, emerging approaches, which provide child and family-centered care and seek participation in different contexts, have proposed home-based VR programs through telerehabilitation strategies (Ravi, Kumar and Singhi, 2017; Golomb et al., 2010) and the school environment during Physical Education classes (Silva and Braccialli, 2017; Braccialli et al., 2016). Using VR as a systematic home exercise program can maximize intervention (Chen, Fanchiang and Howard, 2018).

A variety of technologies have been used to implement immersive, semi-immersive, and non-immersive, interactive, motivating, and appropriate virtual environments for this population. We highlight the use of non-immersive systems such as videogames with commercial games such as Nintendo Wii (Chen, Fanchiang and Howard, 2018; Ravi, Kumar and Singhi, 2017), PlayStation EyeToy (Weiss, Tirosh and Fehlings, 2014; Chen, Fanchiang and Howard, 2018, Ravi, Kumar and Singhi, 2017; Tatla et al., 2013), Xbox Kinect (Chen, Fanchiang and Howard, 2018; Weiss, Tirosh and Fehlings, 2014; Ravi, Kumar and Singhi, 2017), as well as more complex and immersive systems such as Motek's CAREN (Weiss, Tirosh and Fehlings, 2014; Ravi, Kumar and Singhi, 2017), IREX® (Monge Pereira et al., 2014; Ravi, Kumar and Singhi, 2017), and Gesture Xtreme (Weiss, Tirosh and Fehlings, 2014; Chen, Fanchiang and Howard, 2018; Tatla et al., 2013; Monge Pereira et al., 2014) in addition to robot-assisted virtual rehabilitation – NJIT-RAVR (Monge Pereira et al., 2014). Associated with these or other systems, accessories such as keyboards, adapted or non-adapted mouse, joysticks, trackers, screens, haptic devices such as data gloves, head-mounted display, head trackers, eye tracking, and motion capture suits can be attached.

Although the user's sense of virtual presence increases from the simplest to the most complex systems, it is still not possible to state whether there is a relationship between degree of immersion and better performance. All systems have advantages and disadvantages of use. The non-immersive virtual reality system enables partial immersion in the virtual world because the user feels predominantly in the real world (Tori, Kirner and Siscoutto, 2006). Non-immersive systems, such as video games, have advantages over immersive systems due to their low price, affordability, technical support, easy installation, and no additional modification is required (Monge Pereira et al., 2014). Immersive systems still pose limitations such as the presence of wires, which limit the user's movements, difficulties to implement in clinical practice (Jannink et al., 2008); depth perception and insufficient tactile feedback; inadequate association between vision and action, which may lead to different performances than those in physical reality (Wang et al., 2011); possibility of side effects such as dizziness, stomachache, headache, eye fatigue, dizziness, and nausea (Holden, 2005).

However, these systems have high ecological validity, which allows them to safely teach tasks for people with different disabilities (Craig, Sherman and Will, 2009). Although there is no consensus, some authors indicate that learning activities performed in a virtual environment are retained (Mirelman et al., 2011) and transferred to the real environment (Adelola et al., 2009; Meriams et al., 2006) and that the training transfer from the virtual environment to the real environment is greater if the individual is immersed in the training environment (Sveistrup, 2004).

We highlight that the currently available systems, regardless of the degree of immersion, allow to program activities predictably and with personalized stimuli for the user's skill level and learning strategy.

The number of weeks of training proposed by the authors ranged from 1 to 20 weeks. In studies reviewed by Mitchell et al., training sessions were conducted between 5 days and 20 weeks; Tatla et al. found trainings from four to six weeks, and Chen et al. from 4 to 20 weeks (Chen, Fanchiang, Howard, 2018, Tatla et al., 2013; Mitchell et al., 2012). The analysis carried out in a study on the influence of VR training on the gait of children with CP indicated that the duration of training influences performance. Thus, these authors suggest that VR training should be at least eight weeks long (Ghai and Ghai, 2019). Scientific evidence shows that longer training programs are needed to produce clinically significant changes in strength and functional performance in children with CP. Some studies suggested that training programs with children with CP should have a minimum duration of 12 weeks, since lower levels of physical fitness in this population should be considered (Ross et al., 2016).

Regarding the duration of the session, investigations indicated from 20 to 120 minutes, and that the time is related to the previously established therapeutic or leisure goals. For gait and gross motor function training, the indication is training sessions lasting 20–30 minutes, which are more effective at improving spatiotemporal gait parameters than longer sessions with 40–45 minutes training (Ghai and Ghai, 2019). For upper limb function training, the suggestion is that longer sessions had a better effect than low-dose practice, sessions were more effective and had better therapeutic effect when they lasted from 80 to 90 minutes. However, for postural control and walking more intensive training had no positive impact on performance (Chen, Fanchiang and Howard, 2018). Improvement of upper limb function requires intensive training of a functional task for neural modification and reorganization to occur (Adamovich et al., 2009).

As for the weekly frequency of sessions, there is a range of indications from once a week to seven days a week (Ghai and Ghai, 2019; Chen, Fanchiang and Howard, 2018; Ravi, Kumar and Singhi, 2017; Tatla et al., 2013; Mitchell et al., 2012). For gait training, a frequency of ≤4 sessions per week was more efficient than ≥5 weekly sessions for improving gait speed (Ghai and Ghai, 2019).

CLINICAL RECOMMENDATIONS

To develop a program using virtual reality for children and adolescents with CP, we initially propose to establish therapeutic goals to be achieved together with the family, and if possible, with the patient. To this end, during the assessment and definition of therapeutic objectives process, we use Goal Attainment Scaling (GAS), which is based on partnership among family, patient and therapist to establish therapeutic goals that must be specific, measurable, acceptable, relevant and timing (Santos, 2018; McDougall and King, 2007).

In the next phase, we choose, together with the patient, the games to be used and the rules to be followed during the interventions. The child's ability to choose their own level of intensity and their game can contribute to reduce boredom and encourage adherence, and the game can and should be adjusted or adapted to meet the different needs and skills of the player (Williams and Ayres, 2020).

In addition, the therapist using VR for therapeutic purposes, prior knowledge of the resource and an analysis of the game to be worked with is recommended. This analysis can be contextualized from the ICF perspective describing the use of VR in the different components.

On the other hand, most patients included in the studies carried out with virtual reality were classified as levels I and II in the GMFCS and were hemiplegic type, and the games and the performed interventions are poorly described. The proposed programs are usually standardized and fail to consider the specific skills and characteristics of each child (Bonnechère et al., 2014).

Studies by Almeida et al. (2013) and Silva (2014) analyzed games by Nintendo Wii and X-BOX consoles, which can help in the therapeutic planning for children with CP classified in the GMFCS with levels I, II, III, and IV, hemiparetic, diparetic, quadriparetic types. Based on these previous studies, we propose the use of games such as sports, dance platform, action, action adventure, and when necessary we suggest the association of therapeutic resources such as small or large trampoline, exercise mat, shin pad, swiss ball, to facilitate or hinder the execution of the movement. The type of the game, the accessories and resources to be used in therapy depend on the pre-established goals and the motor skills of the child with CP.

Children with more severe gross motor impairment, GMFCS III and IV, benefit from sports games that are easily adapted to be performed in a sitting position. This game type is used when the therapeutic goal is found in the components of the ICF Structure and Function and Activity and Participation. This game type increases the player's energy expenditure, improves physical capacity (Lanningham-Foster et al., 2009), and range of motion. These games allow the use of accessories, such as tennis rackets attached to the joystick to increase immersion, and associate other therapeutic resources, such as the trampoline to hinder the execution of the activity depending on the patient's motor skill level.

Some games can be played individually or multiplayer, adding the activity allows participation enables interactivity and cooperation and/or competitiveness; generating greater socialization and participation in the virtual environment interactive with another player in the virtual setting, this contributes the inclusion of the child CP with the environment that he lives dealing with contexts and technologies in dialogues that own also manages to carry out in practice, as a typical child.

The game inspired in the real world allows the participant of the activity to encourage the player to solve problems to reach the goal and decision making. If a risk occurs in the proposed situation in any part of the game, the game allows a new attempt to be made, stimulating persistence and bringing greater concentration and attention to the player, quick thinking, strategy, and spatial stimuli of the situation.

During the transport of activity from the virtual world to real movements of external perception, it stimulates the use of all parts of the body, stimulates the range of movement and the strengthening of muscles in the upper limbs, weight gain in the lower limbs, stimulates the balance, fine and coarse coordination, works range of motion, temporal space, kinesthetic, and rhythm.

The fact that some video games allow the use of real accessories, such as tennis racket attached to the joystick, skate attached to the Balance Fit Platform, provides greater credibility when playing, leaving the imaginary for something concrete, this is for a better experience in the game, even more often it is about children who never had the chance to have a real experience in that situation. Adding rehabilitation resources such as small or large trampoline, mat, shin, Swiss ball and bandage can be a pleasant and intensify the typical therapy, making the experience more difficult and rich, making a real training.

Dance games require greater motor skills for the child to participate and therefore are indicated for children classified in the GMFCS level I; they also bring therapeutic benefits in the components of the ICF Structure and Function and Activity and Participation. This type of game provides children with typical development with objective and self-directed challenges, interaction among people, promotes intrinsic motivation, involvement, and pleasure in participating in the physical activity, in addition to improving physical capacity (Gao, Podlog and Huang, 2013).

The types of games action, action adventure, and platform games are more suitable for children with levels I and II of the GMFCS and they stimulate skills in the components of the ICF Structure and Function and Activity and Participation. Action games have a set of characteristics: (1) fast pace with time restriction for motor planning; (2) high degree of perceptual and motor load, requires working memory, planning and setting of goals; (3) constant alternation between one state of attention and another; and (4) high degree of distraction. Due to these characteristics, the action video game robustly improves the domains of attention, spatial cognition, and perception in children (Bediou et al., 2018).

The use of active games, or called "exergaming," because of their early and long-term adherence to physical activity, may contribute due to motivational, psychological, and social factors involved during playing (Trout and Christie, 2007; Reynolds and Koh, 2010; Lieberman et al., 2011). Several advantages to the user in the virtual environment are verified: mobility promotion, persistence with the programmed task, interaction and social skills building, increased confidence and self-esteem (Hannah and Stuart, 2012). Exergames allow to monitor body mass index, track exercises during the program, and compare player performance over time. Studies have found that Wii Fit Age optimizes participation in physical activity and increases energy expenditure by addressing diverse skills and interests (Bruin et al., 2010; Lieberman et al., 2011).

Regarding the parameters of use, VR training is recommended for children with CP from four years and older. This type of resource can be used for group or individual therapies, but preferably as complementary therapy in different clinical, home, school, and community contexts, especially when the child's participation in recreational activities and play is established.

When proposing any physical activity program, with or without the use of virtual reality games, we highlight that the impact of the exercise depends on the intensity, duration, and frequency of execution.

For gait training, balance training, and postural control, a training session lasting 20–30 minutes, from ≤4 times a week for ≥8 weeks is suggested (Ghai and Ghai, 2019). However, when the goal is muscle strengthening and improvement of functional skills, especially of the upper limbs, the suggestion is longer sessions, 90 minutes, performed daily. There is scientific evidence that training for muscle strengthening in children with CP requires programs with sessions of 40–50 minutes, at least three times a week (with better effect when performed five times a week), lasting 8–12 weeks, as the effect on physical adaptation of exercise occurs in approximately six to eight weeks (Park and Kim, 2014; Kraemer et al., 2002).The increase in muscle strength occurs especially in the first weeks of training due to neural mechanisms (Park and Kim, 2014); however, it has been reported that the cross-sectional area increase of the muscle occurs after six weeks of progressive resistance training in individuals with CP (Gillett et al., 2016). Based on this information and the fact that the physical fitness level of these children is lower, it is recommended that a training program for muscle strength gain has a minimum duration of 12 weeks for significant changes in strength and functional performance to occur (Ross, MacDonald and Bigouette, 2016).

A training program for individuals with CP that aims to strengthen muscles must also consider a combination of factors other than the parameters of duration and intensity. Factors such as the used range of motion, movement speed, and contraction mode, concentric or eccentric, may be decisive for the adaptation of muscle fascicles after training (Gillett et al., 2016).

The program should be planned based on the following perspectives:

1. Creation of the avatar by the child to mirror themselves in games that would be possible and not possible to perform due to their physical impairment, providing opportunity to relate to real situations the child has probably never experienced before;
2. Use of virtual games which are part of the child's daily life, for example, excluding snow games when used by Brazilian children;
3. Use of video game accessories (steering wheel, racket, fishing rod, and others), control and platform to make the activity more appealing;
4. Use of therapeutic materials associated with the program with virtual reality, such as exercise mat, trampoline, swiss ball, and elastic bandage;
5. Verbal commands to guide the participant to persist in the task;
6. Use of the video game ranking to encourage the participant to improve their score by challenging themselves and beating their fellow participants;
7. Transposition of virtual environment activity to real environment;
8. When possible, choose to play multiplayer games as playing in groups is an encouraging factor and advantage. It also favors participation in real environments.

CONCLUSIONS

The use of non-immersive virtual reality has the potential to be a tool to assist in the rehabilitation of children with CP in different contexts and situations, in addition to being an affordable and commercially available resource. The effectiveness of this therapy depends on proper assessment, objectives, and planning. The literature has indicated positive outcomes regarding the use of this resource when the therapeutic objective is inserted in the ICF Structure and Function component. Further studies need to be conducted to verify whether the use of games has positive outcomes for objectives in the ICF Activity and Participation component.

REFERENCES

Adamovich, S. V., et al. (2009). Sensorimotor training in virtual reality: A review. *Neuro Rehabilitation v. 25, n.* 1, p. 1–21.

Adelola, I. A., Cox, S. L., Rahman, A. (2009). Virtual environments for powered wheelchair learner drivers: Case studies. *Technology and Disability* v. *21, n.* 3, p. 97–106.

Almeida, V. S., Oliveira, N. De A., Santos, L. A., Braccialli, L. M. P. (2013). Estudo da demanda para o uso do videogame na reabilitação e habilitação física de deficientes. *VIII Encontro Da Associação Brasileira De Pesquisadores Em Educação Especial*, Londrina, Nov.

Bax, M., et al. (2005). Proposed definition and classification of cerebral palsy. *Developmental Medicine and Child Neurology v.* 47, p. 571–576.

Bediou, B., Adams, D. M., Mayer, R. E., Tipton, E., Green, C. S., Bavelier, D. (2018). Meta-analysis of action video game impact on perceptual, attentional, and cognitive skills. *Psychological Bulletin v. 144, n.* 1, p. 77–110.

Bonnechère, B., et al. (2014). Can serious games be incorporated with conventional treatment of children with cerebral palsy? A review. *Research in Developmental Disabilities v. 35, n.* 8, p. 1899–1913.

Braccialli, L. P., Almeida, V. S., Silva, F. C. T., Silva, M. S. (2016). Video game na escola e na clinica: Auxiliar da inclusão. *Journal of Research in Special Educational Needs v. 16, n.* 1, p. 1078–1081.

Braccialli, L., Rebelo, F., Pereira, L. (2012). Can virtual reality methodologies improve the quality of life of people with disabilities? In: Francisco Rebelo, Marcelo M Soares (Org). *Advances in Usability Evaluation part II.* 1st ed.: CRC Press, v.1, p. 20–29.

Braccialli, L. M. P., Silva, M. Z., Rebelo, F. (2019). The impact of virtual reality on the learning and rehabilitation of people with special needs. In: Anthony S Thomas. (Org.). *New studies on video games and heath.* 1st ed.: Nova, *v. 1*, p. 1–20.

Brasil (2013). Ministério da Saúde. Secretaria de Atenção à Saúde. Departamento de Ações Programáticas Estratégicas. Diretrizes de atenção à pessoa com paralisia cerebral/Ministério da Saúde, Secretaria de Atenção à Saúde, Departamento de Ações Programáticas Estratégicas. – Brasília: Ministério da Saúde, 80 p.

Bruin, E. D., et al. (2010). Use of virtual reality technique for the training of motor control in the elderly. *Z GerontolGeriatrv v.* 43, p. 229–234.

Cavalcante Neto, J. L., Steenbergen, B., Tudella, E. (2019). Motor intervention with and without Nintendo® Wii for children with developmental coordination disorder: Protocol for a randomized clinical trial. *Trials v. 20*, p. 794.

Chen, Y., Fanchiang, H. D., Howard, A. (2018). Effectiveness of virtual reality in children with cerebral palsy: A systematic review and meta-analysis of randomized controlled trials. *Physical Therapy v. 98, n.* 1, p. 63–77.

Codgno, F., Braccialli, A. C., Presumido Braccialli, L. M. (2018). Change inmanual dexterity of student with cerebral palsy with the use of adequate school furniture. *Revista Brasileira de Educação Especial v. 24, n.* 4, p. 501–516.

Craig, A., Sherman, W., Will, J. D. (2009). *Developing virtual reality applications: Foundations of effective design.* Retrieved from http://books.google.com/books?hl=en&lr=&id=2P91gPYr5Kk C&oi=fnd&pg=PR11&dq=Developing+virtual+reality+applications&ots=k keXyMxXef&sig=CXSNzVkOx08wjPCHCAi02RqauL0

Da Silva, F. C. T. (2019). Nível De Atividade Física, Participação E Qualidade De Vida De Pessoas Com Deficiência Física Em Diferentes Contextos. *Tese de doutorado em Educação. Universidade Estadual Paulista.*

Gao, Z., Podlog, L., Huang, C. (2013). Associations among children's situational motivation, physical activity participation, and enjoyment in an active dance video game. *Journal of Sport and Health Science v. 2*, p. 122–128.

Ghai, S., Ghai, I. (2019) Virtual reality enhances gait in cerebral palsy: A training dose-response meta-analysis. *Frontiers in Neurology v. 10*, p. 236.

Gillett, J. G. B., et al. (2016). The impact of strength training on skeletal muscle morphology and architecture in children and adolescents with spastic cerebral palsy: A systematic review. *Research in Developmental Disabilities v. 56*, p. 183–196.

Golomb, M. R., McDonald, B. C., Warden, S. J., Yonkman, J., Saykin, A. J., Shirley, B. (2010). In-home virtual reality videogame telerehabilitation in adolescents with hemiplegic cerebral palsy. *Archives of Physical Medicine and Rehabilitation v. 91*, p. 1–8.

Hannah, R. M., Stuart, T. S. (2012). Interactive videogame technologies to support independence in the elderly: A narrative review. *Games for Health Journal: Research, Development, and Clinical Applications v. 1, n. 2*, p. 139–152.

Harris, K., Reid, D. (2005). The influence of virtual reality play on children's motivation. *The Canadian Journal of Occupational Therapy v. 72, n. 1*, p. 21–29.

Holden, M. K. (2005). Virtual environments for motor rehabilitation: Review. *CyberPsychology & Behavior v. 3, n. 8*, p. 187–211.

Jannink, M. J. A., et al. (2008). A low-cost video game applied for training of upper extremity function in children with cerebral palsy: A pilot study. *CyberPsychology & Behavior v. 11, n. 1*, p. 27–32.

Kraemer, W. J., et al. (2002). Detraining produces minimal changes in physical performance and hormonal variables in recreationally strength-trained men. *Journal of Strength and Conditioning Research v. 16, n. 3*, 373–382.

Lanningham-Foster, L., Foster R. C., McCrady, S. K., Jensen, T. B., Mitre, N., Levine, J. A. (2009). Activity-promoting video games and increased energy expenditure. *The Journal of Pediatrics v. 154, n. 6*, p. 819–823.

Li, R., et al. (2017). Children with physical disabilities at school and home: Physical activity and contextual characteristics. *International Journal of Environmental Research and Public Health v.14, n. 7*, 687.

Lieberman, D. A., et al. (2011). The power of play: Innovations in getting: A science panel. *Circulation v. 123*, p. 2507–2516.

Lima, C. L. F. A., Fonseca, L. F. (2004). Paralisia cerebral. *Rio de Janeiro: Guanabara-Koogan* 1. ed., v. 1, p. 37.

Longo, E., De Campos, A. C., Palisano, R. (2019). Let's make pediatric physical therapy a true evidence-based field! Can we count on you? *Brazilian Journal of Physical Therapy v. 23, n. 3*, p. 187–188.

Massetti, T., da Silva, T. D., Crocetta, T. B., Guarnieri, R., de Freitas, B. L., Bianchi Lopes, P., de Mello Monteiro, C. B. (2018). The clinical utility of virtual reality in neurorehabilitation: A systematic review. *Journal of Central Nervous System Disease v. 10*. doi: 10.1177/1179573518813541

Massetti, T., et al. (2014). Motor learning through virtual reality in cerebral palsy – A literature review. *Medical Express v. 1, n. 6*, 302–306.

Meriams, A. S., Poizner, H., Boian, R., Burdea, G., Adamovich, S. (2006). Sensorimotor training in a virtual reality environment: Does it improve functional recovery poststroke? *Neurorehabilitation and Neural Repair v. 20, n. 2*, p. 252–267.

Mirelman, A., Maidan, I., Herman, T., Deutsch, J. E., Giladi, N., Hausdorff, J. M. (2011). Virtual reality for gait training: Can it induce motor learning to enhance complex walking and reduce fall risk in patients with Parkinson's disease? *The Journals of Gerontology. Series A, Biological Sciences and Medical Sciences v. 66, n. 2*, p. 234–240.

Mitchell, L., Ziviani, J., Oftedal, S., Boyd, R. (2012). The effect of virtual reality interventions on physical activity in children and adolescents with early brain injuries including cerebral palsy. *Developmental Medicine & Child Neurology v. 54*, p. 667–671.

Monge Pereira, E., Molina, R. F, Alguacil, D. I. M., Cano, De La Cuerda, R., De Mauro, A., Miangolarra, P. J. C. (2014). Empleo de sistemas de realidad virtual como método de propiocepción en parálisis cerebral: Guía de práctica clínica. *Neurología v. 29*, p. 550–559.

Palisano, R. J., et al. (2011). Determinants of intensity of participation in leisure and recreational activities by children with cerebral palsy. *Developmental Medicine & Child Neurology v. 53*, p. 142–149.

Park, E.-Y., Kim, W.-H. (2014). Meta-analysis of the effect of strengthening interventions in individuals with cerebral palsy. *Research in Developmental Disabilities v. 35, n. 2*, p. 239–249.

Ravi, D. K., Kumar, N., Singhi, P. (2017). Effectiveness of virtual reality rehabilitation for children and adolescents with cerebral palsy: An updated evidence-based systematic review. *Physiotherapy v. 103*, p. 245–258.

Reynolds F., Koh L. (2010). Physical and psychosocial effects of Wii video game use among older women. *Society v. 8*, p. 85–98.

Ross, S. M., MacDonald, M., Bigouette, J. P. (2016). Effects of strength training on mobility in adults with cerebral palsy: A systematic review. *Disability and Health Journal v. 9*, n. 3, p. 375–384.

Schreuer, N., Sachs, D., Rosenblum, S. (2014). Participation in leisure activities: Differences between children with and without physical disabilities. *Research in Developmental Disabilities v. 35*, n. I.1, p. 223–233.

Shimony, J. S., et al. (2008). Imaging for diagnosis and treatment of cerebral palsy. *Clinical Obstetrics and Gynecology v. 51*, n. 4, p. 787–799.

Silva, M. Z. (2014). Participação e Qualidade De Vida De Crianças Com Paralisia Cerebral: programa de intervenção com realidade virtual. Trabalho de Conclusão de Curso, graduação em fisioterapia, UNESP FFC.

Silva, F. C. T., Braccialli, L. M. P. (2017). Exergames como recurso facilitador da participação de aluno com deficiência física nas aulas de educação física: percepção do aluno. *Revista Cocar* (ONLINE) *v. 11*, p. 184–208.

Snider, L., Majnemer, A., Darsaklis, V. (2010). Virtual reality as a therapeutic modality for children with cerebral palsy. *Developmental Neurorehabilitation v. 13*, n. 2, p. 120–128.

Standen, P., Brown, D. (2006). Virtual reality and its role in removing the barriers that turn cognitive impairments into intellectual disability. *Virtual Reality v. 10*, n. 3, p. 241–252. Available from: http://www.springerlink.com/index/R350U12H6077725N.pdf

Sveistrup, H. (2004). Motor rehabilitation using virtual reality. *Journal of Neuroengineering and Rehabilitation v. 1*, n. 1, p. 10. Available from: http://www.pubmedcentral.nih.gov/articlerender.fcgi?artid=546406&tool=pmcentrez&rendertype=abstract

Tatla, S. K., Sauve, K., Virji-Babul, N., Holsti, L., Butler, C. Van Der, Loos, H. F. M. (2013). Evidence for outcomes of motivational rehabilitation interventions for children and adolescents with cerebral palsy: An American academy for cerebral palsy and developmental medicine systematic review. *Developmental Medicine & Child Neurology v. 55*, p. 593–601.

Teo, W. P., et al. (2016). Does a combination of virtual reality, neuromodulation and neuroimaging provide a comprehensive platform for neurorehabilitation? – A narrative review of the literature. *Frontiers in Human Neuroscience v. 10*, p. 284.

Tori, R., Kirner, C., Siscoutto, R. (2006). Fundamentos de realidade virtual e aumentada. In: Tori R, *VIII Symposium on Virtual Reality, Editora SBC* - Sociedade Brasileira de Computação, Porto Alegre.

Trout, J., Christie, B. (2007). Interactive video games in physical education. *Journal of Physical Education, Recreation & Dance v. 78*, p. 29–34.

Wang, C. Y., Hwang, W. J., Fang, J. J., Sheu, C. F., Leong, I. F., Ma, H. I. (2011). Comparison of virtual reality versus physical reality on movement characteristics of persons with Parkinson's disease: Effects of moving targets. *Archives of Physical Medicine and Rehabilitation v. 92*, n. 8, p. 1238–1245.

Wang, M., Reid, D. (2011). Virtual reality in pediatric neurorehabilitation: Attention deficit hyperactivity disorder. *Autism and Cerebral Palsy Neuroepidemiology v. 36*, p. 2–18.

Weiss, P., Tirosh, E., Fehlings, D. (2014). Role of virtual reality for cerebral palsy management. *Journal of Child Neurology v. 29*, n. 8, p. 1119–1124.

Williams, W. M., Ayres, C. G. (2020). Can Active Video Games Improve Physical Activity in Adolescents? A Review of RCT. *International journal of environmental research and public health, v. 17*, n. 2, p. 669–679.

19 Virtual Reality Applications in the Context of Low-Vision Rehabilitation

Marie-Céline Lorenzini and Walter Wittich
University de Montreal
Montreal, Canada

CONTENTS

LOW VISION: DEFINITION AND THE ROLE OF REHABILITATION

Visual impairment is globally prevalent across the lifespan, affecting around 314 million people worldwide (Foster et al., 2008) and includes blindness and low vision (Alabdulkader and Leat, 2010). Low vision corresponds to mild or moderate visual impairment that is not correctable with glasses, contact lenses, or surgical interventions and interferes with normal everyday functioning (Corn and Lusk, 2010). The World Health Organization has established criteria for low vision which are used in the International Classification of Diseases (World Health Organization, 2018). Based on an estimate of visual loss in terms of impairment (describes how the eye/visual system functions), low vision is defined as a best-corrected visual acuity worse than 0.5 logMAR (Snellen fraction equivalent of 6/18 or 20/60) but equal to or better than 1.3 logMAR (6/120 or 20/400) in the better eye, or visual field loss corresponding to less than 20° in the horizontal or vertical plane in the better eye with best standard correction. In industrialized countries, older adults (75+ age group) constitute the fastest growing segment of the population with low vision (Margrain, 1999, Watson, 2001), whereby visual impairment has been ranked third (Vos et al., 2016), after anaemia and hearing loss, among conditions that cause people older than 70 years to require assistance in activities of daily living (Scott et al., 1999).

Low vision is associated with well-identified adverse outcomes, including increased challenges with daily activities, an increased likelihood of social isolation and depression, as well as reduced life expectancy (Agency for Healthcare Research and Quality, 2004). In degenerative diseases of the retina, scotomas (blind regions) in the central visual field considerably challenge reading performance (Rubin, 2013). One major goal of people with central visual field deficits is to improve their ability to read text (Elliott et al., 1997). One of the main methods of achieving such improvement is through the provision of and training in the use of low-vision aids, a non-invasive approach that can attenuate the effects of the visual impairment without correcting the physiological causes of the disease (Arya et al., 2010).

Low-vision rehabilitation represents the primary intervention for individuals with chronic, disabling visual impairment (Markowitz, 2006). The goal is to improve activities of daily living of individuals with reduced visual function by optimizing the use of their remaining sight through the provision of appropriate refractive correction, training in the use of vision assistive equipment and compensatory strategies (Binns et al., 2012). The vision rehabilitation process typically includes the prescription of assistive devices and the provision of training in the use of low-vision aids.

Among individuals with low vision presenting for rehabilitation, approximately 25% have considerable peripheral field loss, while the remaining 75% have predominantly central vision loss (Owsley et al., 2009). As such, it is not surprising that the recommendation of magnification devices is among the most common forms of intervention in a low-vision rehabilitation program (Robillard and Overbury, 2006, Hooper et al., 2008), specifically for individuals with central vision loss due to macular pathology. Commonly prescribed magnifying low-vision aids for individuals with central visual defects include over-correction with reading spectacles with convex lenses, handheld or stand magnifiers, as well as closed-circuit televisions (Virtanen and Laatikainen, 1993). A 2012 systematic review on the effectiveness of vision rehabilitation services concluded that there was meaningful evidence that low-vision rehabilitation services improved clinical and functional abilities (Binns et al., 2012). More specifically, there is established evidence that the use of low-vision aids by individuals with low vision can improve functional reading ability (Binns et al., 2012).

Electronic low-vision devices represent a considerable component of potential low-vision rehabilitation solutions in the future. Electronic magnification technologies have been a boon and benefitted a large portion of the low-vision population, allowing a vast number of people worldwide to maintain their ability to read (Wolffsohn and Peterson, 2003a), read faster, and for longer and at smaller print sizes than with traditional optical magnifiers (Papageorgiou et al., 2007, Trauzettel-Klosinski, 2010). However, many problems, such as glare from the screen, non-optimal contrast settings, and light-sensitivity can cause poor image quality. Another negative point is that most electronic magnification technologies are by design limited to performing a single type of task, usually at a specific distance, while people living with low vision require a number of devices to assist with a variety of tasks at multiple distances (Culham et al., 2004). The color quality, sharpness, and high contrast are further important factors and have to be faithfully reproduced for an optimized visual experience.

Ideally, devices should offer a wide and continuous range of magnification options, and the usability to adjust image light intensity and contrast, while maintaining low weight, portability, and esthetic acceptability. Given the current massive technology evolution, the rapid miniaturization of cameras, advanced image processing capabilities, and improved display electronics, novel head-mounted displays using virtual reality have become available and are now a viable alternative to traditional assistive technologies for individuals with vision impairment (Wolffsohn and Peterson, 2003b, Hwang and Peli, 2014). Simulator displays using virtual reality are also more recently applied as a tool for the assessment and training of individuals with vision impairment.

TERMINOLOGY RELATED TO SIMULATOR OR VIRTUAL REALITY DISPLAYS

The most advanced immersive experience corresponds to a psychological state where the individual ceases to be aware of his or her own physical state. An immersive virtual environment is a real or artificial interactive computer-generated scene in which the user can be completely absorbed (experiencing a sense of 'presence'). To create a feeling of total immersion, ideally all five senses need to be solicited so that the digital environment can be perceived as real. Thus, depending on the technology features selected, simulator displays may offer different degrees of the immersion experience. The simplest visual immersion systems involve computer-generated wide-screen movies, whereas head-mounted displays have the potential to provide more complete visual immersion depending on visual field offered. It is also possible to add a more complete sensory immersion through auditory and/or tactile stimulation, depending on the application.

Head-mounted displays can be classified based on how visual information is conveyed, as well as their usability, and optical design (i.e., images are magnified with classic lenses in front of eyes or using pupil-forming systems to shift the image directly on the retina) (Cakmakci and Rolland, 2006, Spitzer et al., 2015). Based on how visual information is conveyed, head-mounted displays can be classified into monocular, bi-ocular, or binocular systems. The simplest solution to implement is a monocular display, presenting images to one eye only. Traditionally, this type of system is used as a low-vision aid (Hwang and Peli, 2014). Second, bi-ocular displays present the same image to both eyes. This solution is traditionally employed in military applications to improve peripheral vision (movement detection) (Spitzer et al., 2015) and also in low vision (van Rheede et al., 2015). The third solution is the most ecological one but also the more technically complex and consists in binocular displays. It is based on retinal disparities (each eye receives its own image that comes from a different field of view) and generally requires two cameras. However, artificial binocular vision can be created using one camera only and a software shifting images as if they were provided from two different areas of space (such as binocular vision) and provide a smaller visual field than two cameras. Given their high price, binocular displays for low-vision application remain rare and are mostly used in action video games for entertainment.

Head-mounted displays can also be classified based on usability in virtual reality (also known as immersive reality) devices, whereby the users' eyes are covered by opaque screens and see-through displays. Immersive displays eliminate the direct path between the user's eyes and their real environment and have traditionally been used for entertainment applications (video games), in which users are immersed in a simulated environment. In contrast, augmented reality displays super-impose images, monocularly or binocularly, directly on top of the user's fields of view, without occluding it. Virtual information that does not otherwise exist in the environment is presented to the user.

In recent decades, simulator displays using virtual reality have become a substitute sustainable solution to traditional low-vision aids (Wolffsohn and Peterson, 2003b, Hwang and Peli, 2014) and have more recently been applied as a tool for assessment and training for individuals with vision impairment (Apfelbaum et al., 2007, Seki and Sato, 2011, Bowman and Liu, 2017).

APPLICATIONS OF VIRTUAL REALITY IN THE CONTEXT OF LOW VISION

HEAD-MOUNTED DISPLAYS USED AS A VISUAL PROSTHETICS

With massive technology development, the rapid miniaturization of cameras, advanced image processing, and improved display electronics in recent decades, novel head-mounted displays have become available and are now an alternative to traditional low-vision aids (Wolffsohn and Peterson, 2003b, Hwang and Peli, 2014) for individuals with vision impairment. Head-mounted devices for individuals with low vision include a display in front of the user's eyes, using a frontal camera to capture live video and incorporate image processing software to present digitally enhanced visual information (Ehrlich et al., 2017). This signal is projected in (near) real time to the user, typically through a pair of micro-displays positioned in front of the eyes.

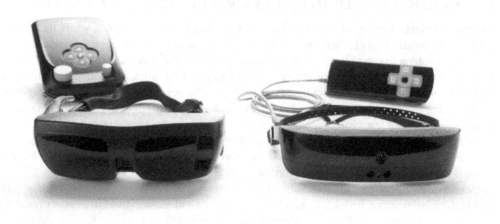

FIGURE 19.1   eSight 2 and eSight 3 head-mounted displays.

Advantages of head-mounted displays are that they enable hands-free magnification for resolution tasks at far, intermediate, and near distances; provide autofocus and variable magnification to facilitate viewing; and offer video inversion as well as contrast and brightness enhancement. Given their features, such assistive technologies have the potential to be used for various activities, instantly switching from near vision (reading a book) to intermediate vision (observing the person who enters the room) or far distance (looking out the window), in a wide range of environments. Head-mounted displays were initially developed for military application, but decreased costs have made them accessible to industrial and entertainment (video games) use. This family of technologies has the potential to improve patients' vision through the association of image processing and wearable visual display systems (Ehrlich et al., 2017). Interest in head-mounted displays has expanded over the years. The first device generations designed for individuals with low vision were the Low Vision Enhancement System (Massof and Rickman, 1992, Ballinger et al., 2000) and the Joint Optical Reflective Display (Enhanced Vision Systems, Huntington Beach, CA). They exhibited technical challenges (size, weight, restricted visual field, low resolution, lag time) (Massof and Rickman, 1992, Culham et al., 2004) that limited their full integration into the environment of low-vision users.

Recently, head-mounted displays have made a comeback. Devices currently commercially available include the eSight Eyewear (eSight Corp., Toronto, ON, Canada) (see Figure 19.1 that displays eSight generations 2 and 3), NuEyes Pro Smartglasses (NuEyes USA, Newport Beach, CA), CyberTimez Cyber Eyez (Cyber Timez, Winchester, VA), Evergaze seeBOOST (Evergaze, LLC, Richardson, TX), IrisVision (Visionize, L.L.C., Berkeley, CA), and the redesigned Enhanced Vision Systems Jordy (Enhanced Vision Systems, Huntington Beach, CA). These new systems use more modern-color micro-displays or cell phone displays, and higher-resolution color video cameras.

Head-mounted displays can vary by their positioning to the user's eyes, visual field size, the presence of stereopsis, type and range of illumination, levels of resolution, and user interface options. These visual system parameters are key components and need to be considered in the development and assessment of head-mounted displays (Ehrlich et al., 2017).

Visual Parameters

Difficulties with contrast sensitivity are commonly reported among individuals with low vision and highly contribute to reductions in their ability to function visually (Colenbrander and Fletcher, 2006). For example, reading performance and face recognition are generally negatively affected by reduced contrast sensitivity. Therefore, optimal contrast-enhancement strategies are essential in the development of head-mounted displays designed for this population. Most video magnifiers incorporate contrast stretching, consisting of mapping pixel intensities above or below a criterion value depending on the intensity, and linearly rescaling the pixel intensities. As another contrast-enhancement strategy, many of the current head-mounted low-vision enhancement systems use edge enhancement to compensate for reduced contrast sensitivity. This strategy selectively stretches the contrast along sharp luminance gradients at edges of objects and features (Peli et al., 2004).

Color perception through a head-mounted display is another important functional feature. Currently, none of the available displays is able to provide the exhaustive range of colors that people without visual impairment can perceive under photopic conditions. Adverse effects on color perception remain all the more important to consider as individuals with visual impairment can have additional reductions in color vision (Rassi et al., 2016).

Considering gaze position, as well as the size and location of the pupil, needs to be considered in the design and choice of a head-mounted display for specific applications. Moreover, individual differences exist in the interpupillary distance and distance between the temple and the eyes, these features play an important role in user comfort while wearing the device. Therefore, personalized attention is necessary, starting with the initials steps that consider the proper selection and setting of the head-mounted display.

In healthy eyes, the monocular visual field is approximately 140° in the horizontal plane and 110° in the vertical plane, while and binocular visual field spans around 190°. In individuals with visual field limitations, optics solutions exists to enlarge (scanners) or/and relocate the visual fields depending on the needs of users (Ehrlich et al., 2017).

Another visual parameter to consider concerns the management of depth perception (stereopsis). Vergence eye movements (i.e., convergence and divergence) play an important role in depth perception, whereby any imbalance can induce eyestrain or ocular fatigue (asthenopia), double vision (diplopia) or blurred vision. To accommodate such binocular vision-related symptoms, optical solutions such as additional convex lenses or multifocal-displays can be included in the device.

Finally, head motion needs to be addressed in the context of head-mounted displays because it may be challenging. Head motion occurs when tracking a moving object, while eye movements intend to keep the image stationary on the retina, both resulting in the visual vestibulo-ocular reflex. A heavy device reduces the velocity of the head movements, resulting in a lag between the image and the initiation of head movement, thus affecting the quality of the visual experience (Ehrlich et al., 2017). Moreover, enhanced image motion, sometimes exaggerated through magnification, increases the risk of cybersickness and other symptoms of visual discomfort (i.e., disorientation, vertigo, headaches, and eyestrain) (Peli, 1996) because the camera included in a head-mounted low-vision enhancement system moves with the users' head movements.

The experience of cybersickness is important to consider as it is inherent to the technology (Peli, 1996, Cobb et al., 1999, LaViola Jr, 2000). Cybersickness has been mainly explained by sensory conflict theory (Kolasinski, 1995, Cobb et al., 1999, LaViola Jr, 2000), given the mismatch between velocities recorded by the visual system when viewing an angularly magnified image, and those recorded by the vestibular system (Kolasinski, 1995). Physiologically the vestibulo-ocular reflex enables individuals to adapt to a certain level of magnification and beyond that, the system cannot compensate the mismatch, causing cybersickness symptoms appear (Demer et al., 1989). Proper selection and correct adjustment of these visual parameters are essential to reach optimal visual functioning in daily activities.

Effects on Functional Vision

The research literature highlights that head-mounted displays involving virtual reality have the potential to improve a wide variety of visual performances, and such devices have generated increasing interest in recent years. Important benefits, including improved print reading, contrast sensitivity, and magnification control, have been reported for both adults (Ortiz et al., 1999, Ballinger et al., 2000) and children (Geruschat et al., 1999) with low vision. Previous research reported that the early generation of head-mounted displays had the potential to improve visual search, night time travel and dark adaptation deficits in individuals with visual field loss, given the variable magnification, and the ability to magnify/minify portions of the visual field (Peli et al., 2007). Their greater magnification range significantly improved utility at far and intermediate distances (Culham et al., 2004, Pelaez-Coca et al., 2009) than traditional optical aids (Massof et al., 1995). In addition, it was documented that reading with a conventional optical aid, such as a hand-held magnifier, was slower than with head-mounted displays (Peterson et al., 2003), whereas a multicenter randomized controlled trial documented patient-reported functional improvement while using head-mounted displays (Stelmack et al., 2008).

Since then, head-mounted low-vision aids have made considerable technical progress and their use has shown the immediate improvement of basic visual functions, such as visual acuity and contrast sensitivity (Crossland et al., 2019, Yoon et al., 2019). Other head-mounted displays have been developed for improving binocular vision (Lodato and Ribino, 2018, Coco-Martin et al., 2019). Meaningful positive effects on activities of daily living had already been documented with previous generations of head-mounted displays (Harper et al., 1999, Culham et al., 2004) and have been replicated in the context of reading (Wittich et al., 2018, Kammer et al., 2019) and face recognition (Wittich et al., 2018). The most studied functional vision performance is reading speed (Wittich et al., 2018, Yoon et al., 2019) because of its relevance for individuals with low vision.

More comprehensively, a multicenter prospective trial investigated the effect of head-mounted displays on 51 novice eSight Eyewear users for three months (Wittich et al., 2018). Visual functions (i.e., distance and near visual acuity, contrast sensibility), as well as functional vision (i.e., visual activities of daily living and reading), showed the greatest benefit from device use (Wittich et al., 2018). Improvements were mainly obtained for reading ability, followed by visual information processing (such as face perception) and visual motor skills (such as pouring a liquid into a cup). A recent feasibility study (Lorenzini and Wittich, Under review-a) concluded that eSight usage and training yielded improvements within the first two weeks in functional vision on the reading, visual information, mobility, and visual motor sub-scales of the Veterans Affairs Low Vision Visual Functioning Questionnaire 48 and continued to improve during the first three months.

While immediate improvement of visual functions has been documented, research on head-mounted low-vision aids reports the importance of usage and/or training, especially to improve complex vision skills, such as mobility and visual confidence (users' ability to evaluate the reliability of their perception) (Zhao et al., 2017, Kinateder et al., 2018). Although evidence of functional benefits of head-mounted displays use exist, a device abandonment rate of around 15% has been reported (Wittich et al., 2018, Lorenzini and Wittich, Under review-a).

Facilitators and Barriers

Head-mounted displays are a new class of magnifying low-vision aid, and little is currently known about which factors may be related to their non-use. The decision-making process around the (non-) use of assistive technologies is multifactorial. Four variable categories (personal, device-related, environmental, and interventional) have been previously identified in the literature (Wessels et al., 2003, Federici et al., 2016, Lorenzini and Wittich, 2019), in line with the bio-psycho-social model of the International Classification of Functioning, Disability and Health (World Health Organization, 2001).

Applying this multifactorial paradigm, a cross-sectional study (Lorenzini et al., 2019) determined predictors of the continued use of a new generation of head-mounted low vision device, eSight Eyewear, in visually impaired individuals. Device owners who perceived more positive impact of the device on their quality of life, those reporting higher satisfaction with the device, and those expressing the absence of headaches associated with their device use, were consistently more likely to continue using this device for a variety of visual activities. In this study, the reported reasons for not using eSight Eyewear were mainly associated with ergonomics, whereby 21% were linked to discomfort and 29% cited its heavy weight. Usability and acceptably by device users is essential as this technology continues to develop. According to qualitative research with head-mounted displays, light weight, cosmetically acceptable devices with controls that are easy to use are the users' priorities (Crossland et al., 2019, Jeganathan et al., 2019).

In a prospective study (Lorenzini and Wittich, Under review-b), the majority of users who abandoned their head-mounted displays were renters who decided not to buy the device at the end of the rental period because they did not experience reading improvement. Interestingly, several demographic variables such as living situation, eye disease onset, report of other impairments, or health conditions did not emerge as usage predictors (Lorenzini et al., 2019, Lorenzini and Wittich, Under review-b). Anticipated factors such as having experience with video magnifiers and/or other head-mounted displays (Lorenzini et al., 2019, Lorenzini and Wittich, Under review-b) and the current use of several low-vision aids did not predict device use (Lorenzini et al., 2019). According to the qualitative comments provided by device users, in most cases, the wearing of the eSight device in public generated positive curiosity from other people. This curiosity was associated with increased device use and was not focused on a possible visual impairment but rather observers thought that the device user was engaging in a video or virtual reality game (Lorenzini et al., 2019).

In some virtual environments, cybersickness and related-symptoms could be an important usability issue potentially affecting the adoption of technology (Cobb et al., 1999). So far, limited data are available on cybersickness experienced by individuals with low vision using head-mounted displays, as well as on the possible connection with device abandonment. Given that the absence of headaches was reported as a robust predictor of eSight device use, cybersickness was studied more systematically in a prospective study (Lorenzini and Wittich, Under review-a), using the Simulator Sickness Questionnaire (Kennedy et al., 1993). The results indicated that cybersickness was not considered as a widespread usability issue. Few participants were severely affected by symptoms of cybersickness and visual discomfort. This may be explained by the semi-immersive condition offered by eSight Eyewear and by the need for sensory adaptation (Cobb et al., 1999).

In order to provide evidence-based practice recommendations, further studies exploring factors related to head-mounted display use remain essential. They should contribute to better screening of users that could benefit from individualized attention during device training and low-vision rehabilitation provision in order to decrease the likelihood of device abandonment and maintain/increase social participation, functional independence, and quality of life. While virtual reality systems have been mainly developed for prosthesis purposes as low-vision aids, applications oriented toward evaluation, training, and education have been explored in the field of visual impairment as well.

A TOOL FOR EVALUATING AND TRAINING INDIVIDUALS WITH LOW VISION

Even though less common than prosthetic vision applications, the use of virtual reality for evaluating and training of orientation and mobility skills (i.e., safe and effective travel) and other important vision-related functions have been explored in the field of visual impairment rehabilitation.

Orientation and Mobility and Other Functional Vision Competencies

The peer-reviewed literature provides evidence that the proprioceptive signals associated with walking can increase navigation efficiency within a virtual environment (Ruddle and Lessels, 2006), and can improve the sense of immersion (Apfelbaum et al., 2007). Based on these findings, a virtual-reality locomotion simulator using a stationary surface that simulates real walking was explored in a sample of ten individuals with low vision (Apfelbaum et al., 2007). The goal was to assess how these individuals utilized the available sensory information during active mobility and to identify how to optimally enhance this information with a low-vision aid. Using a treadmill-based virtual display environment, significant improvement of participants' accuracy in heading assessment while walking was obtained, compared to standing.

The role of training in orientation and mobility rehabilitation is to prepare people with visual impairment in a safe training environment. There has been abundant literature on training orientation and mobility skills in functionally blind individuals who have to rely on sensory substitution cues. It has been reported that blind people could successfully explore a new virtual acoustic and/or haptic space (Connors et al., 2014) and that training in a virtual acoustic environment could increase orientation and mobility skills compared to training in a real environment (Seki and Sato, 2011). In comparison, little work has been done to explore the training of visual skills in people with low vision in virtual reality. Recently, a study (Bowman and Liu, 2017) concluded that virtual reality-based orientation and mobility training could be as efficient as real street training for optimizing street safety in individuals with low vision. Participants were able to improve their street crossing timing in real streets after learning orientation and mobility skills in virtual streets, demonstrating a positive transfer of their virtual reality training to real streets.

Many essential sensory cues from the real environment that influence orientation and mobility, such as depth perception, vibration, and air flow, are technically too sophisticated to be simulated (Bowman and Liu, 2017). While virtual reality cannot be considered as physical reality, it offers a standardized, measurable, and secure environment for learning and practicing orientation and mobility skills. More extensively, real-world experiments may be challenging in the assessment of functional vision in individuals with low vision, especially for elderly participants due to the associated physical effort, possible comorbidities and other limitations such as age-related change in proprioception and balance (Vos et al., 2016). Considering these issues, virtual reality environments have the advantage of allowing researchers to conduct experiments in a simulated world, which may induce behaviour from the participants comparable to that expected in real situations (Gourlay et al., 2000). With this consideration in mind, a virtual environment platform was recently developed to assess functional vision in healthy individuals and patients with low vision, measuring reading and visual search behavior, while monitoring other visual parameters associated with the success of these tasks (i.e., distance and near contrast, distance and near visual angle, time taken for completion of task and error rate) (Gopalakrishnan et al., 2020).

With contrast and lighting being important variables in object or scene recognition, another study (Klinger et al., 2013) tested these parameters using an immersed virtual environment application. The authors used the SENSIVISE tool to assess in which manner scene and objects recognition are affected while performing simulated tasks that are comparable to indoor conditions of real life. When compared to normally sighted participants, those with low vision (i.e., blurred vision and central vision loss) were able to recognize landmarks in an indoor environment, however, with longer reaction time. In addition to the various applications of virtual reality that are centered on visual performance assessment for rehabilitation purposes, public health education and promotion have shown interest in the field of visual impairment.

Visual Impairment Simulation for Educational Purposes

Sighted individuals often lack knowledge about the impact visual impairment on everyday tasks. When family members and friends of persons with low vision are aware of and understand the nature and extent of the vision impairment and its consequences on functional abilities, they can contribute to improved independence (Cimarolli and Boerner, 2005). In order to enhance the understanding of informal caregivers about challenges met by individuals with disabilities in daily life, virtual reality has been explored. The use of virtual reality for educational or learning purposes has mainly focused on spatial perception and orientation among persons living with blindness, in order to assess and improve their mobility skills. A virtual reality-based application, the SENSIVISE tool, was developed to raise awareness about low vision by simulating several visual conditions (i.e., central scotoma, blurred or tunnel vision) in order to demonstrate their impact on activities of daily living (Boumenir et al., 2014). The application was designed to provide sighted users with the opportunity to navigate and interact within a virtual apartment with several rooms and to experience the challenges of having low vision while completing daily activities. When vision was blurred or when the peripheral visual field was reduced (i.e., tunnel vision), it took participants longer to complete these tasks. Additionally, compared to the other visual conditions, times to navigate and interact were longer with a central scotoma when the tasks required participants to see and resolve detail.

Simulated impairment enabled normally sighted participants to understand and experience the perceptual effort and the difficulties caused by reduced vision (Boumenir et al., 2014). However, depending on the visual tasks and the type of visual impairment, simulated impairment using this virtual reality application likely does not adequately reflect the lived experience of individuals with low vision. When investigating whether the visual experience of individuals with advanced peripheral vision loss due to retinitis pigmentosa (characterized by tunnel vision) can be appropriately simulated for normally sighted individuals, Apfelbaum and colleagues (Apfelbaum et al., 2007) concluded that head-mounted field restriction was insufficient and inadequate for simulating tunnel vision. Considering the range of current and potential applications of virtual reality for low-vision rehabilitation, several priorities and innovations related to technology development as well as clinical implications still need to be addressed.

FUTURE PRIORITIES AND POSSIBILITIES

PERSPECTIVES RELATED TO TECHNOLOGIES

The main challenges faced by industrial developers of head-mounted low-vision aids are the effect of magnification within a limited visual field (Ehrlich et al., 2017), the negative impact of magnification on the resolution in first-generation head-mounted displays, and the effect of digital magnification on image contrast (Deemer et al., 2018). A number of technical aspects have already been identified in order to optimize the virtual reality experience. Future research will need to explore binocular vision, gaze control, cybersickness, and other associated symptoms further, as well the benefits of mainstream and personalized devices.

Binocular Vision

The majority of head-mounted displays utilize a single camera and deprive the user of binocular vision. This limitation can create a mismatch between the perceived versus the actual location of objects. Head-mounted display systems using two cameras enable stereoscopic vision, helping people to localize objects, thereby potentially enhancing mobility (Coco-Martin et al., 2019) and restoring binocular vision by correcting certain visual field defects (Lodato and Ribino, 2018). Binocular vision can be challenging when magnifying within a region of interest because it also magnifies binocular disparity. In addition, it increases the likelihood of diplopia in individuals with fragile binocular vision (Tarita-Nistor et al., 2006). Considerable innovation will be necessary to insure binocular disparity under conditions of magnification. The question about the advantages provided by binocular versus bi-ocular vision still needs to be raised. For instance, the effect of head-mounted displays on binocular inhibition or summation (i.e., binocular function is worse or better than monocular function, respectively (Faubert and Overbury, 2000)) and their potential benefits should be explored.

Optimization of the Virtual Reality Experience Using Eye Tracking and Gaze Control

Virtual and augmented reality systems are generating a demand for high-performance eye-tracking systems in head-mounted displays to improve visual performance (Deemer et al., 2018). A recent study (Aguilar and Castet, 2017) tested a gaze-controlled system that magnified a portion of text while maintaining global viewing of the entire document. Reading speed improvements were obtained compared to traditional uniformly applied magnification. An eye-tracker application made it possible to artificially fill in a scotoma using remapping strategies while the image was stretched around the scotoma (Gupta et al., 2018). With an integrated eye-tracking system in the future, it might be feasible to use eye movements to relocate the magnification bubble (i.e., limited central area of visual field that is magnified while the remaining peripheral visual information is presented without added magnification) on the display so that it always remains superimposed on the region of the retina used for fixation. However, poor fixation stability in individuals with central vision impairment and eye-tracking calibration issues remain challenging for accurate fixation and visual search (Deemer et al., 2018). More research and development is required to determine the most effective way to include eye tracking and gaze control with magnification within head-mounted displays when performing various tasks.

Reducing Experience of Cybersickness and Other Associated Symptoms

Device users that report the presence of headaches are more likely to abandon their head-mounted displays (Lorenzini et al., 2019). This finding points at the importance of further exploring this and other symptoms of cybersickness that can occur during head-mounted display use. Additionally, the experience of motion sickness (i.e., nausea) associated with increased levels of magnification can be minimized by reducing head motion. A proposed strategy to reduce motion sickness consists of magnifying image motion within the magnification bubble while presenting visual information outside the bubble without added magnification. However, the limitations of dynamic visual acuity on resolution also limit the user's performance (Harper et al., 1999). Another potential solution is the minimization of imbalances between head and image motion beyond the range of vestibulo-ocular reflex adaptation. Including features for converting angular magnification to linear magnification in head-mounted displays and integrating accelerometers and gyroscopes into new generation of devices all have the potential to control these symptoms for low-vision device users. Further exploration of the mechanisms behind cybersickness and its associated predictors remains essential, given the increasing variability in head-mounted display types and the limited literature available on their link with this condition in individuals with low vision.

Orienting Toward Mainstream and Personalized Devices

Future head-mounted displays may dynamically adjust certain functions (e.g., magnification, contrast) and allow for personalized functionality to adapt to the changing needs of people with low vision. Such features are specifically important for individuals whose visual impairment is not stable and may progress or change over time (e.g., in diabetic retinopathy or macular degeneration). The next generation of head-mounted displays may also use image processing algorithms individually in order to compensate for each type of vision impairment without conceding performance trade-offs (e.g., resolution, visual field, binocularity) to accommodate system limitations (Ehrlich et al., 2017).

In parallel with the requirement to personalize functionality, there is a need to develop affordable mainstream head-mounted displays that can be used by people with low vision. The high cost of current commercially available head-mounted displays has been reported as the most significant barrier by users (Eslambolchilar et al., 2019). Basic accessibility settings provided by some virtual reality software improve performance for users with low vision. Several virtual reality devices offer brightness levels superior to traditional domestic lighting (i.e., 300–500 lumens) and have the potential to include integrated cameras that provide the opportunity to magnify the real environment, even without any special adaptations. A case study (Powell et al., 2020) concluded that mainstream virtual reality devices could be accessible (displaying high levels of contrast and illumination), even to users with very low vision and that some visual activities of daily living could be significantly augmented, without needing costly vision aids.

Currently, the growing trend is toward designing accessibility applications specifically created for low vision, and mainstream applications with already included adjustable configurations (Zhao et al., 2017, Kinateder et al., 2018). Significant visual function improvements were measured with a new mobile application on a virtual reality device platform that has been presented as an alternative to conventional low-vision devices at affordable cost (Yoon et al., 2019). Such applications will need to be tested and validated systematically as well as clinically.

Development of Efficient Training Programs

In the currently available research literature on head-mounted display use, there are no rehabilitation guidelines for their use and implementation, as there are for the use of traditional low-vision aids in Quebec (2019) and in some European countries (Haute Autorité de Santé, 2012, Institut national d'assurance maladie-invalidité, 2018). Basically, head-mounted display users learn on their own or rely on instructions provided by the device manufacturers. Except for veterans, there is great variability across individual states with regard to access to low-vision rehabilitation services for the provision and training of visual aids in the USA.

Several encouraging benefits of low-vision telerehabilitation emerged in terms of accessibility and acceptability, functional performances, and organizational aspects (Bittner et al., 2018, 2019). A recent prospective study (Lorenzini and Wittich, Under review-b) that explored the use of head-mounted displays via telerehabilitation converged toward a high level of accessibility and acceptability. To help guide evidence-based practice and third-party payer recommendations for head-mounted displays, further studies examining benefits of practice and training are needed. Including users in a living-lab (Niitamo et al., 2006) with an integrated knowledge translation approach to encourage the co-production of intervention programs would be a relevant strategy to optimize the adoption of virtual reality approaches by the population with low vision. Research would be able to better define their needs (e.g., the most appropriate format and content) and would not only be focused on users but also directed by them.

Regarding applications for assessment and training, virtual reality should not replace in-office session or deprive people with visual impairment of the opportunity to navigate real streets but they can extend knowledge and skills with minimal risk. For example, virtual reality training combined with real street training may provide an opportunity to improve the accessibility and affordability of orientation and mobility rehabilitation services. Additionally, such a training approach may facilitate the implementation of proven learning reference frameworks and to provide orientation and mobility telerehabilitation so that individuals with low vision can learn and practice their visual skills at a convenient time and location (Bowman and Liu, 2017).

Predictors Associated with Head-Mounted Display Use

To the extent that psychosocial factors influence the adoption of head-mounted displays, they represent valuable clinical markers and should be systematically evaluated and monitored when a new technology is acquired, prescribed or recommended. Clinicians should consider the use of device-related aspects of quality of life measures in order to identify device users that could benefit from individualized attention during low-vision rehabilitation provision and reduce the probability of device abandonment (Lorenzini et al., 2019). Insofar as the use of such devices involves physical and psychological adjustments, a multidisciplinary team seems important to optimize their proper integration. After head-mounted display acquisition, users should be asked about their potential discomfort, and comparison with other device types should be encouraged. The Simulator Sickness Questionnaire (Kennedy et al., 1993) is easy to administer, making the inclusion of such questionnaires at the acquisition of the device and during the follow-up sessions a feasible recommendation to investigate potential symptoms onset.

Exploring Lived Experience of Virtual Reality in Persons with Low Vision

Overall, the research literature about human factors considering the lived experience of individual with visual impairment receiving virtual reality interventions remains limited. One study (Kinateder et al., 2018) suggested a positive effect of head-mounted displays on the users' visual confidence, referring to their ability to estimate the reliability of their own perception. As with other low-vision aids, training, practice, or calibration is likely to be necessary in order for users to acquire the correct level of visual confidence. Interviews with individuals with visual impairment identified that recognizing faces and reading text were their highest priority functions when using head-mounted displays (Sandnes, 2016). Virtual reality-based technologies enabled people with visual impairment to have visual experiences previously unavailable to them, both in terms of visual performances and visual perceptions (Zolyomi et al., 2017). Computer-assisted sight allows vision to be approached differently, whereby technology-mediated sight refers to a skilled vision, neither entirely human nor entirely digital but more like a combination of human and technological skills.

Further qualitative research and/or a user-centered design approaches (De Vito Dabbs et al., 2009) that consider long-term utilization of such devices is needed to better understand how people with low-vision experience the use of virtual reality. Reports of the lived experiences should enable researchers to better understand the social benefits of these emergent technologies. The ultimate goal of integrating these devices into the lives of persons with low vision is to improve their functional independence and ability to participate in society. This goal needs to remain at the center of the ongoing technological development by making the device user's part of the development and evaluation process.

REFERENCES

Agency for Healthcare Research and Quality. 2004. *Vision Rehabilitation for Elderly Individuals with Low Vision or Blindness*, Rockville (MD), AHRQ.

Aguilar, C. & Castet, E. 2017. Evaluation of a gaze-controlled vision enhancement system for reading in visually impaired people. *PLoS One*, 12, e0174910.

Alabdulkader, B. & Leat, S. J. 2010. Reading in children with low vision. *Journal of Optometry*, 3, 68–73.

Apfelbaum, H., Pelah, A. & Peli, E. 2007. Heading assessment by "tunnel vision" patients and control subjects standing or walking in a virtual reality environment. *ACM Transactions on Applied Perception*, 4, 8.

Arya, S. K., Kalia, A., Pant, K. & Sood, S. 2010. Low vision devices. *Nepalese Journal of Ophthalmology*, 2, 74–77.

Ballinger, R., Lalle, P., Maino, J., Stelmack, J., Tallman, K. & Wacker, R. 2000. Veterans affairs multicenter low vision enhancement system (lves) study: Clinical results. Report 1: Effects of manual-focus LVES on visual acuity and contrast sensitivity. *Optometry*, 71, 764–774.

Binns, A. M., Bunce, C., Dickinson, C., Harper, R., Tudor-Edwards, R., Woodhouse, M., Linck, P., Suttie, A., Jackson, J., Lindsay, J., Wolffsohn, J., Hughes, L. & Margrain, T. H. 2012. How effective is low vision service provision? A systematic review. *Survey of Ophthalmology*, 57, 34–65.

Bittner, A. K., Green, K., Khan, R., Mitesh, M. A., Barnes, M. J. & Ross, N. C. 2019. Changes in reported difficulty with near reading following telerehabilitation for low vision. Poster session presented at: From bench to bedside and back 91st ARVO Annual Meeting, April 28–May 2 2019, Vancouver, Canada. Available: https://iovs.arvojournals.org/article.aspx?articleid=2743712.

Bittner, A. K., Yoshinaga, P., Bowers, A., Shepherd, J. D., Succar, T. & Ross, N. C. 2018. Feasibility of telerehabilitation for low vision: Satisfaction ratings by providers and patients. *Optometry and Vision Science*, 95, 865–872.

Boumenir, Y., Kadri, A., Suire, N., Mury, C. & Klinger, E. 2014. Impact of simulated low vision on perception and action. *International Journal of Child Health and Human Development*, 7, 441–450.

Bowman, E. L. & Liu, L. 2017. Individuals with severely impaired vision can learn useful orientation and mobility skills in virtual streets and can use them to improve real street safety. *PLoS One*, 12, e0176534.

Cakmakci, O. & Rolland, J. 2006. Head-worn displays: A review. *Journal of Display Technology*, 2, 199–216.

Cimarolli, V. R. & Boerner, K. 2005. Social support and well-being in adults who are visually impaired. *Journal of Visual Impairment & Blindness*, 99, 521–534.

Cobb, S., Nichols, S., Ramsey, A. & Wilson, J. 1999. Virtual reality-induced symptoms and effects (VRISE). *Presence: Teleoperators and Virtual Environments*, 8, 169–186.

Coco-Martin, M. B., Pichel-Mouzo, M., Torres, J. C., Vergaz, R., Cuadrado, R., Pinto-Fraga, J. & Coco, R. M. 2019. Development and evaluation of a head-mounted display system based on stereoscopic images and depth algorithms for patients with visual impairment. *Displays*, 56, 49–56.

Colenbrander, A. & Fletcher, D. C. 2006. Contrast sensitivity and ADL performance. *Investigative Ophthalmology & Visual Science*, 47, 5834–5834.

Connors, E. C., Chrastil, E. R., Sanchez, J. & Merabet, L. B. 2014. Virtual environments for the transfer of navigation skills in the blind: A comparison of directed instruction vs. video game based learning approaches. *Frontiers in Human Neuroscience*, 8, 223.

Corn, A. & Lusk, K. E. 2010. Perspectives on low vision. In: Corn, A. & Koenig, A. (eds.) *Foundations of Low Vision: Clinical and Functional Perspectives*, 2nd ed, New York (NY), AFB Press.

Crossland, M. D., Starke, S. D., Imielski, P., Wolffsohn, J. S. & Webster, A. R. 2019. Benefit of an electronic head-mounted low vision aid. *Ophthalmic and Physiological Optics*, 39, 422–431.

Culham, L. E., Chabra, A. & Rubin, G. S. 2004. Clinical performance of electronic, head-mounted, low-vision devices. *Ophthalmic and Physiological Optics*, 24, 281–290.

De Vito Dabbs, A., Myers, B. A., Mc Curry, K. R., Dunbar-Jacob, J., Hawkins, R. P., Begey, A. & Dew, M. A. 2009. User-centered design and interactive health technologies for patients. *Computers, Informatics, Nursing*, 27, 175–183.

Deemer, A. D., Bradley, C. K., Ross, N. C., Natale, D. M., Itthipanichpong, R., Werblin, F. S. & Massof, R. W. 2018. Low vision enhancement with head-mounted video display systems: Are we there yet? *Optometry and Vision Science*, 95, 694–703.

Demer, J. L., Porter, F. I., Goldberg, J., Jenkins, H. A., Schmidt, K. & Ulrich, I. 1989. Predictors of functional success in telescopic spectacle use by low vision patients. *Investigative Ophthalmology & Visual Science*, 30, 1652–1665.

Ehrlich, J. R., Ojeda, L. V., Wicker, D., Day, S., Howson, A., Lakshminarayanan, V. & Moroi, S. E. 2017. Head-mounted display technology for low-vision rehabilitation and vision enhancement. *American Journal of Ophthalmology*, 176, 26–32.

Elliott, D. B., Trukolo-Ilic, M., Strong, J. G., Pace, R., Plotkin, A. & Bevers, P. 1997. Demographic characteristics of the vision-disabled elderly. *Investigative Ophthalmology & Visual Science*, 38, 2566–2575.

Eslambolchilar, P., Hill, K. & Margrain, T. H. 2019. Can assistive digital technologies boost wellbeing in people with sight loss? Poster session presented at: From bench to bedside and back 91st ARVO Annual Meeting, April 28–May 2 2019, Vancouver, Canada. Available: https://iovs.arvojournals.org/article.aspx?articleid=2743718.

Faubert, J. & Overbury, O. 2000. Binocular vision in older people with adventitious visual impairment: Sometimes one eye is better than two. *Journal of the American Geriatrics Society*, 48, 375–380.

Federici, S., Meloni, F. & Borsci, S. 2016. The abandonment of assistive technology in Italy: A survey of national health service users. *European Journal of Physical and Rehabilitation Medicine*, 52, 516–526.

Foster, A., Gilbert, C. & Johnson, G. 2008. Changing patterns in global blindness: 1988-2008. *Community Eye Health*, 21, 37–39.

Geruschat, D., Deremeik, J. & Whited, S. 1999. Head-mounted displays: Are they practical for school-age children? *Journal of Visual Impairment and Blindness*, 93, 485–497.

Gopalakrishnan, S., Jacob, C. E. S., Kumar, M., Karunakaran, V. & Raman, R. 2020. Comparison of visual parameters between normal individuals and people with low vision in a virtual environment. *Cyberpsychology, Behavior, and Social Networking*, 23, 171–178.

Gourlay, D., Lun, K. C., Lee, Y. N. & Tay, J. 2000. Virtual reality for relearning daily living skills. *International Journal of Medical Informatics*, 60, 255–261.

Gupta, A., Mesik, J., Engel, S. A., Smith, R., Schatza, M., Calabrese, A., Van Kuijk, F. J., Erdman, A. G. & Legge, G. E. 2018. Beneficial effects of spatial remapping for reading with simulated central field loss. *Investigative Ophthalmology & Visual Science*, 59, 1105–1112.

Harper, R., Culham, L. & Dickinson, C. 1999. Head mounted video magnification devices for low vision rehabilitation: A comparison with existing technology. *British Journal of Ophthalmology*, 83, 495–500.

Haute Autorité De Santé. 2012. *DégénéRescence Maculaire LiéE À L'âGe: La RééDucation De Basse Vision*, Paris, HAS.

Hooper, P., Jutai, J. W., Strong, G. & Russell-Minda, E. 2008. Age-related macular degeneration and low-vision rehabilitation: A systematic review. *Canadian Journal of Ophthalmology*, 43, 180–187.

Hwang, A. D. & Peli, E. 2014. An augmented-reality edge enhancement application for google glass. *Optometry and Vision Science*, 91, 1021–1030.

Institut National D'assurance Maladie-Invalidité. 2018. Déficience visuelle: Intervention dans le coût de la rééducation par des centres spécialisés. Réglementation d'application jusqu'au 31 décembre 2018 inclus. Bruxelles, INAMI.

Jeganathan, V. S. E., Kumagai, A., Shergill, H., Fetters, M., Gosbee, J., Moroi, S. E., Weiland, J. D. & Ehrlich, J. R. 2019. Design of smart head-mounted display technology: A qualitative study. Poster session presented at: From bench to bedside and back 91st ARVO Annual Meeting, April 28–May 2 2019, Vancouver, Canada. Available: https://iovs.arvojournals.org/article.aspx?articleid=2746959.

Kammer, R., Kim, B., Kuppermann, B. D., Watola, D. A., Tsang, T. & Mehta, M. C. 2019. Performance of an augmented reality device on functional activities. Poster session presented at: From bench to bedside and back 91st ARVO Annual Meeting, April 28–May 2 2019, Vancouver, Canada. Available: https://iovs.arvojournals.org/article.aspx?articleid=2743709.

Kennedy, R. S., Lane, N. E., Berbaum, K. S. & Lilienthal, M. G. 1993. Simulator sickness questionnaire: An enhanced method for quantifying simulator sickness. *International Journal of Aviation Psychology*, 3, 203–220.

Kinateder, M., Gualtieri, J., Dunn, M. J., Jarosz, W., Yang, X. D. & Cooper, E. A. 2018. Using an augmented reality device as a distance-based vision aid-promise and limitations. *Optometry and Vision Science*, 95, 727–737.

Klinger, E., Boumenir, Y., Kadri, A., Mury, C., Suire, N. & Aubin, P. 2013. Perceptual abilities in case of low vision, using a virtual reality environment. 2013 International Conference on Virtual Rehabilitation (ICVR), 26–29 Aug. 2013. 63–69.

Kolasinski, E. M. 1995. *Simulator sickness in virtual environments: Technical report 1027* [Online]. United States Army Research Institute for Behavioral and Social Sciences. Available: https://apps.dtic.mil/dtic/tr/fulltext/u2/a295861.pdf [Accessed].

Laviola, J.J., Jr. 2000. A discussion of cybersickness in virtual environments. *SIGCHI Bulletin*, 32, 47–56.

Lodato, C. & Ribino, P. 2018. A novel vision-enhancing technology for low-vision impairments. *Journal of Medical Systems*, 42, 256.

Lorenzini, M. C., Hämäläinen, A. M. & Wittich, W. 2019. Factors related to the use of a head-mounted display for individuals with low vision. *Disability and Rehabilitation*, 1–15 (in press).

Lorenzini, M. C. & Wittich, W. 2019. Factors related to the use of magnifying low vision aids: A scoping review. *Disability and Rehabilitation*, 1–13 (in press).

Lorenzini, M. C. & Wittich, W. Under revision-a. Head-mounted visual assistive technology-related Quality of Life changes after Telerehabilitation. *Optometry and Vision Science*.

Lorenzini, M. C. & Wittich, W. Under revision-b. Personalized telerehabilitation for a head-mounted low vision aid: A randomized feasibility study. *Optometry and Vision Science*.

Margrain, T. H. 1999. Minimising the impact of visual impairment. Low vision aids are a simple way of alleviating impairment. *BMJ*, 318, 1504.

Markowitz, S. N. 2006. Principles of modern low vision rehabilitation. *Canadian Journal of Ophthalmology*, 41, 289–312.

Massof, R. W., Baker, F. H., Dagnelie, G., Derose, J. L., Alibhai, S., Deremeik, J. T. & Ewart, C. 1995. Low vision enhancement system: Improvements in acuity and contrast sensitivity. *Optometry and Vision Science*, 72, 20.

Massof, R. W. & Rickman, D. L. 1992. Obstacles encountered in the development of the low vision enhancement system. *Optometry and Vision Science*, 69, 32–41.

Niitamo, V. P., Kulkki, S., Eriksson, M. & Hribernik, K. A. 2006. State-of-the-art and good practice in the field of living labs. 12th International Conference on Concurrent Enterprising: Innovative Products and Services through Collaborative Networks, Milan, Italy.

Ortiz, A., Chung, S. T., Legge, G. E. & Jobling, J. T. 1999. Reading with a head-mounted video magnifier. *Optometry and Vision Science*, 76, 755–763.

Owsley, C., Mcgwin, G., Jr., Lee, P. P., Wasserman, N. & Searcey, K. 2009. Characteristics of low-vision rehabilitation services in the United States. *Archives of Ophthalmology*, 127, 681–689.

Papageorgiou, E., Hardiess, G., Schaeffel, F., Wiethoelter, H., Karnath, H. O., Mallot, H., Schoenfisch, B. & Schiefer, U. 2007. Assessment of vision-related quality of life in patients with homonymous visual field defects. *Graefe's Archive for Clinical and Experimental Ophthalmology*, 245, 1749–1758.

Pelaez-Coca, M. D., Vargas-Martin, F., Mota, S., Diaz, J. & Ros-Vidal, E. 2009. A versatile optoelectronic aid for low vision patients. *Ophthalmic and Physiological Optics*, 29, 565–572.

Peli, E. 1996. *Visual Perceptual, and Optometric Issues with Head-Mounted Displays (HMD)*, Playa del Rey (CA), Society for Information Display.

Peli, E., Kim, J., Yitzhaky, Y., Goldstein, R. B. & Woods, R. L. 2004. Wideband enhancement of television images for people with visual impairments. *Journal of the Optical Society of America A: Optics, Image Science & Vision*, 21, 937–950.

Peli, E., Luo, G., Bowers, A. & Rensing, N. 2007. Applications of augmented vision head-mounted systems in vision rehabilitation. *Journal of the Society for Information Display*, 15, 1037–1045.

Peterson, R. C., Wolffsohn, J. S., Rubinstein, M. & Lowe, J. 2003. Benefits of electronic vision enhancement systems (EVES) for the visually impaired. *American Journal of Ophthalmology*, 136, 1129–1135.

Powell, W., Powell, V. & Cook, M. 2020. The accessibility of commercial off-the-shelf virtual reality for low vision users: A macular degeneration case study. *Cyberpsychology, Behavior, and Social Networking*, 23, 185–191.

Quebec. 2019. *Régie d'Assurance Maladie du Québec* [Online]. Québec: RAMQ. Available: http://www.ramq.gouv.qc.ca/fr/citoyens/programmes-aide/aides-visuelles/Pages/aides-visuelles.aspx [Accessed 2019].

Rassi, S. Z., Saint-Amour, D. & Wittich, W. 2016. Drug-induced deficits in color perception: Implications for vision rehabilitation professionals. *Journal of Visual Impairment & Blindness*, 110, 448–453.

Robillard, N. & Overbury, O. 2006. Quebec model for low vision rehabilitation. *Canadian Journal of Ophthalmology*, 41, 362–366.

Rubin, G. S. 2013. Measuring reading performance. *Vision Research*, 90, 43–51.

Ruddle, R. A. & Lessels, S. 2006. For efficient navigational search, humans require full physical movement, but not a rich visual scene. *Psychological Science*, 17, 460–465.

Sandnes, F. E. 2016. What do low-vision users really want from smart glasses? In: Miesenberger, K., Bühler, C. & Penaz, P. (eds.) *Faces, Text and Perhaps No Glasses at All*, Cham, Springer International Publishing, 187–194.

Scott, I. U., Smiddy, W. E., Schiffman, J., Feuer, W. J. & Pappas, C. J. 1999. Quality of life of low-vision patients and the impact of low-vision services. *American Journal of Ophthalmology*, 128, 54–62.

Seki, Y. & Sato, T. 2011. A training system of orientation and mobility for blind people using acoustic virtual reality. *IEEE Transactions on Neural Systems and Rehabilitation Engineering*, 19, 95–104.

Spitzer, C., Ferrell, U. & Ferrell, T. 2015. *Digital Avionics Handbook*, Boca Raton (FL), CRC Press.

Stelmack, J. A., Tang, X. C., Reda, D. J., Rinne, S., Mancil, R. M., Massof, R. W. & Group, L. S. 2008. Outcomes of the veterans affairs low vision intervention trial (LOVIT). *Archives of Ophthalmology*, 126, 608–617.

Tarita-Nistor, L., Gonzalez, E. G., Markowitz, S. N. & Steinbach, M. J. 2006. Binocular interactions in patients with age-related macular degeneration: Acuity summation and rivalry. *Vision Research*, 46, 2487–2498.

Trauzettel-Klosinski, S. 2010. Rehabilitation for visual disorders. *Journal of Neuroophthalmology*, 30, 73–84.

Van Rheede, J. J., Wilson, I. R., Qian, R. I., Downes, S. M., Kennard, C. & Hicks, S. L. 2015. Improving mobility performance in low vision with a distance-based representation of the visual scene. *Investigative Ophthalmology & Visual Science*, 56, 4802–4809.

Virtanen, P. & Laatikainen, L. 1993. Low-vision aids in age-related macular degeneration. *Current Opinion in Ophthalmology*, 4, 33–35.

Vos, T., Allen, C., Arora, M., Barber, R., Bhutta, Z., Brown, A., Carter, A. & Casey, D. 2016. Global, regional, and national incidence, prevalence, and years lived with disability for 310 diseases and injuries, 1990-2015: A systematic analysis for the global burden of disease study 2015. *Lancet*, 388, 1545–1602.

Watson, G. R. 2001. Low vision in the geriatric population: Rehabilitation and management. *Journal of the American Geriatrics Society*, 49, 317–330.

Wessels, R., Dijcks, B., Soede, M., Gelderblom, G. J. & De Witte, L. 2003. Non-use of provided assistive technology devices, a literature overview. *Technology and Disability*, 15, 231–238.

Wittich, W., Lorenzini, M. C., Markowitz, S. N., Tolentino, M., Gartner, S. A., Goldstein, J. E. & Dagnelie, G. 2018. The effect of a head-mounted low vision device on visual function. *Optometry and Vision Science*, 95, 774–784.

Wolffsohn, J. S. & Peterson, R. C. 2003a. A review of current knowledge on electronic vision enhancement systems for the visually impaired. *Ophthalmic and Physiological Optics*, 23, 35–42.

Wolffsohn, J. S. & Peterson, R. C. 2003b. A review of current knowledge on electronic vision enhancement systems for the visually impaired. *Ophthalmic and Physiological Optics*, 23, 35–42.

World Health Organization. 2001. *International Classification of Functioning, Disability and Health (ICF)*, Geneva, WHO.

World Health Organization. 2018. *International Classification of Diseases 11(ICD-11)*, Geneva, WHO.

Yoon, D. Y., Jeon, H. S., Wee, W. R. & Hyon, J. Y. 2019. Low vision aids using virtual reality (VR) headsets and mobile application; preliminary report. Poster session presented at: From bench to bedside and back 91st ARVO Annual Meeting, April 28–May 2 2019, Vancouver. Available: https://iovs.arvojournals.org/article.aspx?articleid=2743720.

Zhao, Y., Hu, M., Hashash, S. & Azenkot, S. 2017. Understanding low vision people's visual perception on commercial augmented reality glasses. *CHI '17 Proceedings of the 2017 CHI Conference On Human Factors in Computing Systems*, May 06–11, 2017, Denver, Colorado, pp. 4170–4181.

Zolyomi, A., Shukla, A. & Snyder, J. L. 2017. Technology-mediated sight: A case study of early adopters of a low vision assistive technology. *ASSETS '17 Proceedings of the 19th International ACM SIGACCESS Conference on Computers and Accessibility* [Online]. Available: https://dl.acm.org/citation.cfm?id=3132552.

20 Response of Women with Mobility Impairments to a Group Weight Management Intervention in the Virtual World of SecondLife©

Margaret A. Nosek *
Baylor College of Medicine
Houston, Texas

Susan Robinson-Whelen
Baylor College of Medicine
TIRR, Memorial Hermann Hospital System
Houston, Texas

Rosemary B. Hughes
The University of Montana
Missoula, Montana

Stephanie L. Silveira *
University of Alabama
Birmingham, Alabama

Rachel Markley *
Houston Methodist Research Institute
Houston, Texas

Tracey A. Ledoux and Daniel P. O'Connor
University of Houston
Houston, Texas

Rebecca E. Lee
Arizona State University
Tempe, Arizona

Thomas M. Nosek
Case Western Reserve University School of Medicine
Cleveland, Ohio

---

*Investigators directly involved in the qualitative analysis.

CONTENTS

INTRODUCTION

Virtual worlds hold the promise of bringing health promotion interventions to the underserved population of people with mobility impairments, a population that faces significant health disparities. Barriers such as lack of transportation, cost, inaccessibility of the built and natural environment, providers lacking disability-related knowledge and sensitivity, unavailability of personal assistants, pain, and fatigue, often limit their participation in clinic- or community-based programs (Mojtahedi et al., 2008, Rimmer et al., 2004, 2007). When programs are available virtually, many of these barriers can disappear. This chapter examines the responses of women with various conditions that limit mobility to a weight management intervention offered in a virtual world.

Population-based studies show compelling evidence that women with disabilities display significant disparities in rates of obesity (Eliason et al., 2017, Froehlich-Grobe et al., 2013); yet, the barriers that they face when attempting to participate in community-based weight management programs are pervasive and discouraging (Nosek et al., 2014a). Women living with physical disability not only have a higher prevalence of obesity than men with disabilities but they also have an obesity prevalence that is nearly double that of women in the general population (Froehlich-Grobe et al., 2013, Nosek et al., 2008). As the impact of the obesity epidemic continues to strengthen, there is an increasing necessity for interventions that address the trends, barriers, and facilitators for weight management in distinct subpopulations (Wang and Beydoun, 2007). The Center for Research on Women with Disabilities developed a weight management intervention tailored to the needs of the severely disadvantaged and neglected population of women with mobility impairments. The medium for delivery was the free, online virtual world of SecondLife® (SL). While the developmental process and results of the pilot test of this intervention have been published previously (Nosek et al., 2014b, 2018), we present here a detailed discussion of how participants in the beta and pilot tests responded to group learning activities in SL.

SECONDLIFE®

In the current study, we sought to circumvent some of the barriers to weight management experienced by women who have more severe physical disabilities, live in more remote areas, and face other barriers to participating in face-to-face programs by offering a small group, weight management intervention, called GoWoman, over the Internet. We chose to use the massive multiuser virtual world of SL (Linden Lab, 2011), which is accessed by downloading a free program at www.Secondlife.com to a hard drive on a computer, tablet, or smartphone. SL has 63 million registered users internationally with approximately 50,000 online at any given time on 1,632 square kilometers of virtual land (Shepherd, 2020). Usage has increased during the COVID-19 pandemic, suggesting the value of virtual worlds for making social connections during a time of social distancing (James, 2020). No data are available on how many SL users have disabilities.

SL enables participants to create and outfit a personal avatar (a three-dimensional animated image) that they can manipulate using a keyboard, mouse, track ball, or on-screen controls. Users communicate with other avatars by text or voice. They have the option of giving their avatar a visible disability or assistive devices (e.g., wheelchair). This supportive virtual environment offers people with significant mobility limitations opportunities for social engagement and movement, such as walking, running, jumping, dancing, and flying, that can only be imagined in real life (Stewart et al., 2010). SL also offers the unique experience of trying alternative forms of embodiment and presence in a social context.

REAL-LIFE BENEFITS FROM ACTIVITIES IN SL

The GoWoman team previously tested a self-esteem enhancement intervention in SL and found a strong sense of community among participants (Nosek et al., 2016, Robinson-Whelen et al., 2020). Women in that study indicated that participating in this virtual intervention was more convenient and enjoyable than having teleconferences or face-to-face sessions. On our secure GoWoman parcel of land in SL, participants can observe a real-time PowerPoint presentation by a facilitator, ask questions, dine together in the kitchen area or cafe, dance, and use numerous simulated types of exercise equipment with attachable animations that make their avatar go through exercise motions. Unlike a videoconference or other meeting platform, these interactions can create an enriched sense of active community. Some research indicates behavioral changes in real life are associated with the creation of an avatar that reflects the user's ideal body characteristics and virtual social interaction styles (Yee et al., 2009). This implies the possibility of virtual exercise affecting real-life exercise behaviors.

General weight loss programs have been tried previously in SL. In one study, Sullivan and colleagues (Sullivan et al., 2013) randomly assigned weight management program participants to face-to-face or SL-based weight maintenance sessions, with both nutritional and exercise components. Although net weight loss was greater with face-to-face sessions, the SL treatment saw greater weight maintenance over time (Sullivan et al., 2013).

The use of an avatar can impact a participant's response to a weight management program. For instance, another study conducted survey interviews within SL (avatar-to-avatar) with questions related to fitness in the virtual world, in comparison to fitness in the real world. Interview responses revealed common themes. A majority of participants not routinely physically active in real life were physically active in SL (Dean et al., 2009). While some claim that virtual fitness alone will not result in the physiological benefits of healthful diet habits and exercise in the real world, avatar-based virtual interventions improve perceived social support for weight loss and increase self-efficacy, the crucial determinant of weight management (Coons et al., 2011, Ruiz et al., 2011). The use of avatars is also associated with changing attitudes related to poor eating habits prior to physical activity performance (Riva et al., 2001).

The avatar has a unique role for SL users with physical disabilities, especially women. Many women with disability value the opportunity to decide how to (or if they will) visibly represent their disability in SL (Stendal et al., 2012). One caveat to this benefit is that the relationship with one's avatar and the ability to build virtual intrapersonal relationships with other avatars requires technological proficiency and experience in virtual worlds (Stendal et al., 2012, 2013). Women may also find themselves uncomfortable with the pre-determined appearance of SL avatars, although proficient users are capable of significantly altering the representation of their avatar (Dumitrica and Gaden, 2009). Women SL-users are more likely to create thin avatars as compared to men. This may serve less as an expression of gender expectations and more as a coping mechanism for body image disturbance (Dunn and Guadagno, 2012).

No studies to date have documented the experiences of women with mobility impairments in structured SL weight management interventions. Qualitative feedback on these experiences may provide insight to both the effective and ineffective aspects of virtual interventions for real-world weight management. The purpose of the current study is to evaluate participant feedback on the GoWoman weight management intervention in SL.

METHODS

Our multidisciplinary research team created an adaptation of the Diabetes Prevention Program (DPP) (Diabetes Prevention Program Research Group, 2002a, 2002b, Knowler et al., 2002, Kramer et al., 2009) that was responsive to the weight management needs of women and those with disabilities. We infused it with settings and activities that are unique to virtual environments, such as installing interactive educational PowerPoint displays and games on the kitchen walls. The research team made adaptations to the DPP based on Social Cognitive Theory (Bandura, 1986) and related self-management strategies shown to be effective in our prior intervention research with this population of women (Hughes et al., 2003, 2006, Robinson-Whelen et al., 2006). We also infused the curriculum with gender- and disability-specific issues by incorporating principles of feminist psychology (Jordan, 2001) and the philosophy of independent living (Nosek and Fuhrer, 1992, Nosek and Hughes, 2004).

The project's five-member Community Advisory Board (CAB) reviewed and beta tested the full GoWoman curriculum in SL. After receiving approval from the Institutional Review Board for human research at the first author's institution and the two collaborating institutions, we recruited women locally for the pilot test by posting flyers at academic institutions, community advocacy organizations, disability service providers, and metropolitan recreational facilities and posting announcements on social media and our website. Staff conducted an initial telephone screening followed by a full in-person screening using wheelchair accessible equipment, including a platform scale and adjustable height table for transferring. Inclusion and exclusion criteria have been published elsewhere (Nosek et al., 2018) including age requirements, disability type and duration, body mass index, weight loss intention, physician's permission, and computer capacity. Examples of ineligibility criteria were current pregnancy, insulin-dependent diabetes, untreated hypertension, pressure ulcers, and weight loss of more than ten pounds in the past three months.

We screened participants for computer capability by guiding them over the phone on how to download the program and use a guest username and password during the screening process. Participants then logged into SL and told a research assistant what they were seeing over the phone as the island loaded. Once the participant could see the island and the guest avatar fully, we asked her to walk around and view a whiteboard with presentation materials.

Immediately after telephone screening, each woman came to the project office to sign consent forms, provide written authorization for staff to contact their physician, complete a questionnaire on a secure computer using SurveyMonkey (SurveyMonkey, 2011), and have her blood pressure, weight, and height measured. We assured everyone that their responses were completely confidential and their only identification would be a unique number assigned to them by the research staff.

Before beginning the GoWoman intervention, participants new to SL received up to three training sessions as needed. The first two sessions involved learning basic navigation and communication skills and exploring the layout of the GoWoman venue. In the third session, participants met in groups of three or four to practice communication etiquette and interaction, play educational games, and use exercise animations.

Beta testers attended two-hour long sessions twice weekly for eight weeks. Since we mainly wanted their feedback on intervention content, they were not required to give informed consent and did not complete the pre- or post-tests. Pilot testers were asked to attend 16 weekly meetings in SL for two hours with a brief break. Sessions consisted of facilitators presenting information outlined in PowerPoint displays projected onto whiteboards in SL, verbal exercises, discussions, and optional excursions to different parts SL. The venue featured a lily pond, a pool, a beach, and an exercise area that offered animations for workouts mats, stationary bikes, trampolines, and rock climbing. Participants created their weekly action plans at the end of each session. At the beginning of the next session, each woman reported her progress including successes and barriers and received feedback from the group. Participants completed a weekly feedback survey in SurveyMonkey to report on their satisfaction with each session. They returned for a post-test appointment for weight and other measurements within two weeks of the final session in SL. Facilitators recorded field notes after each session.

The post-test contained summative evaluation items. Among the topics were overall program satisfaction, satisfaction with group leaders and the setting, preferences for future programing, satisfaction with training on how to use SL, and experiences with SL. It also asked open-ended questions concerning likes and dislikes of the program, feelings about using an avatar, and additional comments or suggestions.

Another source of feedback was from CAB members who wrote blog posts on a topic of their choosing related to weight loss or SL. These appeared on the GoWoman blog on the Baylor College of Medicine website.

We drew content for this qualitative analysis from feedback forms that were part of the beta test with CAB members, qualitative items in the post-intervention evaluations, facilitators' notes, minutes of CAB meetings, and CAB blog posts. Techniques of qualitative analysis were applied, including constant comparison and thematic extraction (Lincoln and Guba, 1985). Each data element was categorized by topic. Themes emerged from this categorization process.

RESULTS

SAMPLE CHARACTERISTICS

There were 5 beta testers and 13 participants with complete data in the pilot test. The sample of 18 women was predominantly white, middle-aged, and well educated. The majority had some type of spinal cord injury or dysfunction, were married, and worked full- or part-time. See Table 20.1 for more details.

THEMES

Analysis revealed seven themes from comments about SL made by the community advisers and intervention participants as they offered feedback on their experiences after the beta test and pilot test of the GoWoman intervention.

Theme #1: It Is Difficult to Balance Replicating Disability and Accessibility from Real Life and Allowing Fantasy in the Gowoman Environment

Most participants initially felt the need to replicate their image, including their disability and mobility patterns, in SL. This is difficult logistically because default images that are most easily available

TABLE 20.1
Sample Characteristics (*N* = 18*)

Variable	*M* (SD) or *n* (%)
Age	49.62 (8.55)
	Range = 28–59 years
Level of education	5 (28%)
High school (HS) diploma or GED	5 (28%)
HS plus some college or technical school	6 (33%)
College degree	2 (11%)
Graduate degree	
Race	13 (72%)
White	4 (22%)
Black	1 (6%)
Other	
Ethnicity	2 (11%)
Hispanic	14 (78%)
Non-hispanic	2 (11%)
Not known	
Marital status	10 (56%)
Married	
Employment, occupation	9 (50%)
Full-time	5 (28%)
Part-time	2 (11%)
Student	2 (11%)
Unemployed	
Disability/health condition	6 (33%)
Multiple sclerosis	3 (17%)
Spinal cord dysfunction (spinal cord injury, spina bifida)	5 (28%)
Neuromuscular (cerebral palsy, post-polio, muscular dystrophy)	2 (11%)
Back injury/back pain	2 (11%)
Joint and connective tissue diseases (arthritis, fibromyalgia)	

* This number is the aggregate of 5 beta testers and 13 research participants who completed the pilot test of the GoWoman intervention.

in SL conform to the nondisabled norm. It takes considerable effort to modify an avatar's body to look asymmetrical or to have a missing limb, for example. Some may interpret this as unequal treatment or even discrimination against people with disabilities in SL. Users of the program have reported going years before discovering that it is possible to obtain a wheelchair as an object in SL. When you click on it, you do not get the option to "sit," as if it were a regular chair. Instead, you get the option to "ride," which is an animation, as if it were a motorcycle or a surfboard. The wheelchair remains attached to the avatar until the user clicks on "remove."

Unlike real life, SL gives users the power to make the decision about using a mobility device. Some women initially protested the lack of disability options when creating their avatar. Embodiment as a walking person was uncomfortable and foreign to some of them. One beta tester said, "It's difficult to decide whether or not to leave your disability behind." Others had no trouble at all. One facilitator said, "I've had a disability all my life. Here, I'm liberated!" A participant did not want a wheelchair in SL as she preferred to focus more on an ideal self in the virtual world. One CAB member wrote a blog post about her experience in having an avatar. She initially disliked the idea of having a nondisabled, thin avatar. Later, however, she adapted and felt liberated by all the capabilities and actions that were possible vicariously through her avatar in SL. She wrote:

"Being part of Second Life was a joy in many different ways. One that stands out most to me is the aspect of not needing my wheelchair. I have been in a motorized wheelchair for nearly 16 years and for the first time I could … casually be able to walk, run, or fly to our destinations. I did not need to worry about if there was a ramp to allow me access, if the door would be too heavy to open, if there was an accessible table for me to sit. Instead, I could enjoy all aspects of any of the discussions … in our group and Second Life without worry of inability of access. I found this to be very empowering and relaxing, as I was not worried about the logistics of moving from location to location, instead I was able to focus on the group dynamics of communication."

It took time for some participants to adopt a fantasy mindset and become more adventurous in exploring options for their avatar. There were complaints that the available clothes and accessories were too sexy for their taste, but as the intervention progressed, we found that some came to sessions wearing new items, such as sparkling shoes, scarves, earrings, and new hairstyles, that they found while exploring other sites or received from friends in SL. One woman who was lesbian found that it was easier for her to start with a male avatar and adjust the dimensions and accessories to achieve the look she wanted. Participants appreciated the variety of choices available for creating an avatar that met their emotional and creative needs in how they portrayed themselves.

Because avatars have the capability of walking, running, and flying, accessibility is not the same issue as it is in real life. Barriers in SL take other forms, such as the lack of navigational cues, so avatars can get trapped in a room, in a grove of trees, at the bottom of a river, or in some environmental feature where the exit path is difficult to find. Another common problem is when a person experiments with flying, sometimes their avatar can get locked in that mode and fly off to infinity. To solve this problem, avatars can either stop flying and return home or completely exit the SL program. Technological glitches like this frustrated novice SL users, just as managing the lack of curb cuts frustrates real-life wheelchair users.

The beta group debated extensively whether to replicate real-life accessibility modifications, such as ramps, in SL, even though they were not physically necessary. Avatars with a wheelchair animation can easily mount a flight of stairs or jump into a boat, wheelchair, and all. The final decision was to include representations of real-life accessibility in the virtual environment for their symbolic value.

Participants felt the need to clarify their identity as a woman with a disability, especially if their avatar was not using a wheelchair. One woman stated, "Just because I don't have a disability in SL doesn't mean I am ashamed of my disability." A facilitator felt awkwardness going into a crowded virtual room and striking up a conversation with someone new. She felt obliged to tell them, "Well, you know, I'm really disabled."

Some women seemed to feel more connected with their able-bodied avatar as the program progressed. One woman stated, "Towards the end of the program, I had kinda gotten used to relating to my able-bodied avatar." Other women were excited by the opportunity to exist in a world without their disability, even if only virtually. One participant said, "It was exuberating to be able to run and jump and fly. I felt free." Still other women saw the avatar as simply a means of participating in the program and receiving the information, without investing any time or energy in the avatar itself.

Theme #2: SL Poses New Kinds of Challenges to Movement

As women with mobility impairments, our advisers were familiar with learning alternative strategies for navigating the environment to accommodate their limitations. It was an "aha" moment when they realized that navigating in SL poses a completely new set of challenges to mobility. Moving from one location on the island to another required using arrow keys to manipulate the avatar like pulling the strings on a marionette. Those with slow Internet connections had to deal with the problem of delayed responses, such that hitting an arrow key did not necessarily result in an immediate effect. Avatars sometimes moved in a discontinuous series of lurches, crashing into furniture and each other. This problem was worse for participants whose disability caused hand weakness or incoordination.

We installed transport boards in various locations that had a button for each location on the island. Clicking on the desired destination resulted in a "whooshing" sound and magically the avatar appears in the new location without taking a step. Many women, especially those with visual impairments, found it easier to make their way to the transport board and simply touch the button for their destination, instead of trying to navigate their way across the island. One CAB member recommended that we provide training to participants about how to navigate around virtual environmental barriers.

It is challenging to move spontaneously in SL. This difficulty inhibits movement as a form of expression. Each component of motion requires using a series of keystrokes and menu-driven commands. As animations are continuously evolving, eventually avatars will be able to move more smoothly and expressively. Until then, however, movement as a form of expression is still rather crude. Preprogrammed dance animations are available for download. They attach to an avatar like a piece of clothing, making the avatar move in a repeating sequence of dance steps until the user removes the animation. We installed a "dance ball" over the deck of the pool area and used it for a party at the conclusion of the intervention. By touching or clicking on it, participants could activate a series of rock 'n roll and other dance animations. Participants found it entertaining and both physically and emotionally invigorating.

Theme #3: Disability Affects Group Dynamics in SL as it Does in Real Life

Whether participants chose to represent disability as a visible part of their avatar, it presented itself in their conversation, nonetheless. It seemed some women felt apologetic that their avatar did not have a disability, so they had to describe their limitations to other participants as an explanation for their comments and feelings. Occasionally, they had to explain how their mobility limitations contributed to their weight status. This type of explanation may not have been necessary in a real-life weight-loss program in which a participant's physical disability is readily apparent to other participants. Some expressed wanting to know more about each other's disabilities, commenting that SL limited their understanding of each other's disabilities and disability severity.

Some women preferred or needed to participate by typing their comments instead of using their voice and the audio capacity of the SL program. For example, one woman with significant communication limitations found typing to be more accessible. This was an essential accommodation requiring the facilitators to read the typed comments to make sure everyone noticed and for the benefit of participants with visual impairments. On the other hand, one participant stated, "I like using my voice to communicate because it brings more closeness to the group; it makes us feel like we are actually together."

Sharing progress on action plans at the beginning of each session was rated as a helpful component of the program. Some women indicated that it added accountability and minimized their sense of being the only ones facing barriers to implementing their plans. As they received helpful suggestions from others in the group, they felt reinforced and validated. Facilitators also observed the strong, positive energy after these discussions, even when participants expressed discouragement about not being able to carry out their action plans. The women repeatedly mentioned the value of peer mentoring.

Some women talked about feeling anxious when encountering strangers in SL. In off-island situations, some felt uncomfortable and distrustful not knowing who was behind the other person's avatar. They appreciated the secure nature of the GoWoman venue and the knowledge that everyone in the group was there for the same purpose.

The lack of social cues and body language can hamper group dynamics in SL. Avatars have no variation in facial expressions or gestures unless the user specifically engages or attaches them. This prevents others from perceiving nonverbal emotional responses to conversations. On one hand, this eliminates one level of data for social and emotional interpretation. On the other, it puts everyone on equal standing with participants who have visual impairments and must constantly rely on verbal cues or tips from conversation dynamics to determine social and emotional subtleties.

Theme #4: Exercise in SL has Effects in Real Life

A few of the women commented on the real-life physical effects they felt from their activities in SL. When their avatar was involved in vigorous exercise, such as doing jumping jacks, riding a stationary bike, or dancing, they reported that their real-life breathing rate increased. In addition to the cognitive benefit of improving awareness about different types of physical activity, exercise, and relaxation techniques, they reported a physiologic effect at the end of each session. They also found the exercise and dance portions of the sessions highly motivational. One participant commented, "If I can feel this good by just imagining I'm dancing through my avatar in SL, who knows how good I'll feel if I really dance in real life!" Another said, "it almost feels like I've done some exercise."

One participant chose to write her blog on the effect exercising in SL had on her body in real life.

"[GoWoman] was a space with no intimidation. There were no problems due to inaccessible equipment, no sweaty locker rooms and no long drive across town to the gym. In other words, it was perfect! I especially liked seeing myself (meaning my 'avatar') high kicking a punching bag over and over with no achy muscles or loss of balance. Talk about therapeutic! ... You may be wondering how useful a virtual workout is in a digital world that isn't real. I felt it created a strong message through my neural pathways in the brain when watching "me" perform physical feats I couldn't do IRL (In Real Life) ... Maybe I didn't lose 20 pounds as my avatar jumped up and down hitting that bag, but to see "myself" this physically active and able, works on a deeper level. I feel [it] creates possibility within me.

Theme #5: Privacy and Anonymity in SL Facilitate Open Communication

SL tends to be an anonymous environment. There is no requirement to use real names. Participants know they are in an intervention for women with mobility impairments, but with no visual cues associated with their disability, they are able to communicate as much or as little information about their unique disability as they feel is necessary. Participants said "Second Life allowed you to have some privacy but yet have a real feeling of communicating with others" and "Second life was a good way to interact with the other ladies and maintain anonymity."

The physical awkwardness of human contact and conversation with someone who uses a wheelchair can disappear in SL. In real life, it takes a degree of comfort to come to the same eye level as an individual in a wheelchair or casually touch them on the shoulder. Many people understand that touching a person's wheelchair can violate their space. Some participants felt relief at not having to deal with these barriers to conversation in SL.

Participants rated the program highly due to ease of candid communication through an avatar on the sensitive topic of weight loss. One participant said, "I felt like we were able to share more freely than we might have done without the avatar or SL." Weight loss can be an extremely difficult topic for individuals to discuss and the more honest and open participants are, the richer the discussions will be. Several participants indicated that removing the stigma of disability and obesity in communicating through avatars was a strength of this program.

Theme #6: Balance of Pros and Cons of a Virtual Intervention

Participants were asked which format they would prefer if they were to complete the program again–face-to-face and in-person, telephone-based, internet-based with audio and video, or in a virtual world like SL. The average rating for face-to-face and a virtual reality program were very close with ratings of 1.94 and 2.13, respectively, on a scale of 1 to 5 with 1 being the highest. Participants then explained their preference.

Many women stated that although they would have preferred a traditional face-to-face format, they liked the ease of participating from home on the computer. One woman appreciated using SL because "a virtual environment allowed me to participate more often when ill health would have made it that much more difficult otherwise."

Fatigue was also a reason some preferred a virtual format. One woman shared that "There was the convenience of being in my own home; therefore, if any fatigue began to set in I did not have to worry about any transportation endeavors after the session. ... Fatigue is one of the key aspects that often times keeps me from joining other groups, hence SL proved to be a key component to decrease fatigue while enjoying updated information on the topic at hand, as well as socially being able to communicate with others." Although she felt she might have missed the emotional aspect of participating face-to-face, the physical benefits and convenience of participating at home outweighed this negative aspect.

As discussed above, some women felt the virtual world format allowed for a greater sense of privacy and anonymity than other formats and this allowed them to speak more freely. Several beta testers and pilot test participants said that the virtual world demanded time for adjustment. It was more comfortable for them when all participants were using the voice instead of text as it felt more like a face-to-face format. We appreciate their feedback but at the same time, we also appreciate the need to make our program accessible to women with significant communication limitations who may encounter serious barriers to participating in community-based weight loss programs. Nevertheless, all who expressed initial discomfort reported adjusting over time.

One of the most commonly reported barrier to the virtual reality format was accountability. A large part of the program involved participants encouraging other group members to fulfill their action plans and assisting each other with problem solving to overcome barriers. One participant reported a preference for a face-to-face format as she indicated feeling more accountable when she knows she will be facing the group and giving updates each week. Another said it was "too easy to backslide" when not seeing everyone in person.

The technological demands of using SL are substantial for some new users. Problems like slow resolution time causing avatars to appear as a cloud on opening the program, difficulty establishing an audio connection, and freezing of the program caused some to be discouraged at first. Nevertheless, only a few comments related to this were among responses to the question of pros and cons of participating in an intervention offered in SL. This could be attributed to the fact that women who were beset with technological problems tended to drop out early in the intervention.

DISCUSSION

Qualitative methodology was useful in helping us gain insights from the women who participated in the GoWoman intervention. The seven themes we identified yielded some lessons learned that represent the intersection of disability and virtual worlds. Table 20.2 lists these categorizations in relation to each other.

Orientation is the process of adjusting to new surroundings, employment, activity, or the like (Dictionary.com, 2020). Health promotion interventions in SL must allow time for training new users on how to become oriented to the virtual environment (Theme #1 and #6). Teaching basic navigation skills is easier if it is a one-on-one exercise and if the user has some familiarity with video games. For inexperienced users, however, orientation can take longer, especially if they have a visual or cognitive impairment. Some participants never reached a level of comfort in the virtual environment and were more likely to drop out. For others, the orientation process struck excitement for exploring a new way of moving and interacting with others. These participants were more likely to adorn their avatar with accessories and try out the exercise equipment. Careful attention to a participant's readiness to embrace SL can increase her capacity to enjoy the intervention.

Liberation is the act of gaining equal rights or full social or economic opportunities for a particular group (Dictionary.com, 2020). People with disabilities live day in and day out within the limitations that are part of their disabling condition. Running, jumping, dancing, and kicking a punching bag are stress relieving experiences many can only imagine. In SL, these are all possible with only a mouse click. Even flying is possible–the ultimate in liberating experiences.

Theme #1 speaks about replicating disability and accessibility. In addition to feeling liberated from physical constraints, participants reported feeling freed of the stereotypes, misunderstandings, and biases others have in relating to women with disabilities and women who are overweight or obese. This also affected group dynamics (Theme #3). Some participants claimed exhilaration

TABLE 20.2
Lessons, Themes, and Comments

Lessons	Themes	Comments
Orientation	Theme #1: Replicating disability and accessibility Theme #6: Pros and cons	Participants must become familiar and comfortable with the virtual environment and their avatar if they are to attain maximum benefit from virtual interventions.
Liberation	Theme #1: Replicating disability and accessibility Theme #3: Group dynamics Theme #5: Privacy and anonymity	Some women expressed feelings of liberation at having an able-bodied avatar; others were bothered by it. With no pressure to disclose their real-life disability, this was a new experience in relating with others in the group.
Embodiment	Theme #1: Replicating disability and accessibility Theme #3: Group dynamics	Choices participants make in manifesting their avatar reflect their personality and how they wish to interact with others in the virtual world.
Experimentation	Theme #2: Movement	The virtual world allows participants to experiment with new types of movement that they can only imagine in real life.
Transference	Theme #4: Exercise	Exercise in the virtual world may stimulate a physical response in real life. Attitudinal and behavioral changes gained from virtual interventions may transfer to real life.

at the release of responsibility to counter the burden of other people's attitudes. Some women find it comforting to talk about weight in an anonymous setting so that others cannot define them by their weight status (Theme #5).

On the other side of liberation were feelings of guilt, of being an imposter. By "abandoning" their disability, some felt they had betrayed loyalty to their disability and the disability rights movement. SL challenges disability identity (Theme #1) and membership in peer groups by forcing the choice of whether to adopt the outward signs of disability. Most women who "rode" on wheelchairs did so as an experiment and quickly reverted to the avatar's natural, able-bodied state, mostly as a matter of convenience and not as a statement of identity or allegiance. There were comments about feeling uncomfortable with a thin avatar, but no expressions of guilt.

The debate over installing ramps as symbols of accessibility (Theme #1) even though it was unnecessary reflects the push to carry the banner of disability rights. It is possible that participants who considered disability as an integral part of their being and valued their role as a disability rights advocate struggled the most with these decisions.

Embodiment is synonymous with avatar, epitome, exemplar, expression, and personification (Thesaurus.com, 2020). SL allows participants unlimited choices in creating an avatar. Variations are possible in bodily characteristics, such as sex, skin tone, size, shape, and hair, or replication of a character from history or the future, or even a fantasy creature. These choices reflect the participant's sense of adventure, enjoyment of experimentation, desired or ideal self, and even personality. SL allows users to try out new looks and test their level of comfort. This level of self-discovery is not available in traditional interventions.

Physical disability no longer needs to be part of human interactions in virtual worlds (Theme #1). No disability-related cues have to be present in SL. Therefore, for the first time in many years or even in their lives, participants were able to hold conversations without being identified or defined by their disability (Theme #3). These kinds of interactions fostered new social relationships based more strongly on content than appearance.

Experimentation is a trial and error process. With the right training, participants can learn how to pursue the many opportunities for exploration and experimentation in SL. Using the new types of movement (Theme #2) can be thrilling or frustrating. Luckily, if a participant finds herself in an undesirable or inescapable situation, all she has to do is exit the program. It takes courage and self-confidence to engage in virtually risky activities. There are no known studies documenting how interest in experimentation in virtual worlds transfers to better intervention outcomes and long-term behavior change in real life.

Transference is the movement of a characteristic or behavior from one state of being to another. It implies agency on the part of the individual who activates the movement. SL offers new opportunities for developing this sense of agency. One of the biggest gaps in knowledge about virtual health promotion interventions is how much virtual learning and interactions transfer into real world attitudinal, physical, and behavioral change (Theme #4). Previous literature and evidence from our data suggest that individuals believe their experience in SL affects their behavior in the real world.

Several of our beta testers described physiological changes, such as increased heart rate, from watching their avatars exercising. By witnessing the avatar exercise, participants can imagine new behaviors and possibilities for their real life. These experiences reflect the Proteus Effect, the process by which an individual's behavior matches their online self-representation. Scientists at Stanford University's Virtual Human Interaction Lab (https://vhil.stanford.edu/) (Yee et al., 2009) are examining the mechanisms that underlie this effect.

IMPLICATIONS FOR FUTURE RESEARCH

These findings have important implications for future research. Scant literature addresses the responses of women with disabilities to activities in SL. Future research should examine more closely disability identity and its relation to virtual embodiment, characteristics of women who engage in exploration and experimentation, strategies for increasing comfort in virtual environments, and transference of learnings in SL to real-life change in attitudes and behaviors. Research on physical responses to virtual exercise would advance our understanding of this as a biofeedback mechanism, a visualization technique, or the Proteus Effect. Choice about adding disability characteristics or devices to the avatar may be associated with the participant's real-life disability. The study of embodiment needs much more attention, especially as it relates to intervention outcomes. Closer examination of the Proteus Effect in people with disabilities could add to the strategies for achieving rehabilitation outcomes.

CONCLUSION

A weight management intervention in SL can transcend disability-related barriers to participation and yield real-life attitudinal and behavioral changes. Future research should examine how women with mobility impairments in SL make decisions in creating their avatars, move around the virtual environment, engage with interactive displays, and transfer learnings from virtual interventions to real life.

ACKNOWLEDGEMENTS

We thank Case Western Reserve University for their generosity and technical assistance provided by Sue Simonson Shick in helping us construct and maintain the GoWoman intervention venue on their Second Life island, ClevelandPlus.

Grateful appreciation goes to the dedicated members of the project's community advisory board, Titilayo Awoniyi, Leanne Beers, Anita Cameron, Terri O'Hare, and Alejandra Ospina, for their involvement in all phases of this study.

DISCLOSURE STATEMENT

Authors have disclosed no financial conflicts of interest due to commercial associations, consultancies, equity interests, or patent licensing arrangements in connection with this work.

FUNDING

Funding for this study was a field-initiated development grant from the National Institute on Disability, Independent Living, and Rehabilitation Research in the US Department of Health and Human Services (#90IF0036).

REFERENCES

Bandura, A. 1986. *Social Foundations of Thought and Action: A Social Cognitive Theory*, Englewood Cliffs, NJ, Prentice-Hall.

Coons, M. J., Roehrig, M. & Spring, B. 2011 The potential of virtual reality technologies to improve adherence to weight loss behaviors. *J Diabetes Sci Technol*, 5, 340–4.

Dean, E., Cook, S., Keating, M. & Murphy, J. 2009. Does this avatar make me look fat? Obesity and interviewing in Second Life. *J Virtual Worlds Res*, 2(2), 3–11.

Diabetes Prevention Program Research Group. 2002a. The diabetes prevention program (DPP): description of lifestyle intervention. *Diabetes Care*, 25, 2165–71.

Diabetes Prevention Program Research Group. 2002b. Reduction in the incidence of type 2 diabetes with lifestyle intervention or metformin. *N Engl J Med*, 346, 393–403.

Dictionary.Com. 2020. *Dictionary.com* [Online]. Dictionary.com LLC. Available: www.Dictionary.com [Accessed May 22, 2020].

Dumitrica, D. & Gaden, G. 2009. Knee-high boots and six-pack abs: autoethnographic reflections on gender and technology in Second Life. *J Virt Worlds Res*, 1(3), 422–45.

Dunn, R. A. & Guadagno, R. E. 2012. My avatar and me - gender and personality predictors of avatar-self discrepancy. *Comput Human Behav*, 28, 97–106.

Eliason, M. J., Mcelroy, J. A., Garbers, S., Radix, A. & Barker, L. 2017. Comparing women with and without disabilities in five-site "Healthy Weight" interventions for lesbian/bisexual women over 40. *Disabil Health J*, 10, 271–8.

Froehlich-Grobe, K., Lee, J. & Washburn, R. A. 2013. Disparities in obesity and related conditions among Americans with disabilities. *Am J Prev Med*, 45, 83–90.

Hughes, R. B., Nosek, M. A., Howland, C. A., Groff, J. & Mullen, P. D. 2003. Health promotion workshop for women with physical disabilities: a pilot study. *Rehabil Psychol*, 48, 182–8.

Hughes, R. B., Robinson-Whelen, S., Taylor, H. B. & Hall, J. W. 2006. Stress self-management: an intervention for women with physical disabilities. *Womens Health Issues*, 16, 389–99.

James, W. 2020. Second Life sees significant user concurrency growth of old & returning players, post-pandemic declaration. *New World Notes* [Online]. Available from: https://nwn.blogs.com/nwn/2020/03/second-life-coronavirus-concurrency.html [Accessed 3/18/2020].

Jordan, J. V. 2001. A relational-cultural model: healing through mutual empathy. *Bull Menninger Clin: Treatment Approach N Millennium*, 65, 92–103.

Knowler, W. C., Barrett-Connor, E., Fowler, S. E., Hamman, R. F., Lachin, J. M., Walker, E. A. & et al. 2002. Reduction in the incidence of type 2 diabetes with lifestyle intervention or metformin. *N Engl J Med*, 346, 393–403.

Kramer, M. K., Kriska, A. M., Venditti, E. M., Miller, R. G., Brooks, M. M., Burke, L. E., Siminerio, L. M., Solano, F. X. & Orchard, T. J. 2009. Translating the diabetes prevention program: a comprehensive model for prevention training and program delivery. *Am J Prev Med*, 37, 505–11.

Lincoln, Y. S. & Guba, E. G. 1985. *Naturalistic Inquiry*, Newbury Park, CA, Sage.

Linden Lab. 2011. *Second Life* [Online]. Linden Lab, Inc. Available: http://secondlife.com/ [Accessed June 15, 2011].

Mojtahedi, M. C., Boblick, P., Rimmer, J. H., Rowland, J. L., Jones, R. A. & Braunschweig, C. L. 2008. Environmental barriers to and availability of healthy foods for people with mobility disabilities living in urban and suburban neighborhoods. *Arch Phys Med Rehabil*, 89, 2174–9.

Nosek, M., Lee, R. E., Robinson-Whelen, S., Ledoux, T. A., Hughes, R. B., O'Connor, D. P., Goe, R. & Nosek, T. M. 2014a. The inclusion challenge: how can general weight loss programs accommodate the needs of women with mobility impairments? *Ann Behav Med*, 47, s64.

Nosek, M. A. & Fuhrer, M. J. 1992. Independence among people with disabilities: I. A heuristic model. *Rehabil Counsel Bull*, 36, 6–20.

Nosek, M. A. & Hughes, R. B. 2004. Navigating the road to independent living. In: Atkinson, D. R. & Hackett, G. (eds.) *Counseling Diverse Populations*. Boston, MA, McGraw-Hill.

Nosek, M. A., Robinson-Whelen, S., Hughes, R. B., Ledoux, T. A., Lee, R. E., O'connor, D. P., Nosek, T. M., Goe, R. & Silveira, S. L. 2014b. GoWoman: the development of a virtual world weight management intervention for women with mobility impairments. *Annual Meeting of the American Public Health Association*. New Orleans, LA, American Public Health Association.

Nosek, M. A., Robinson-Whelen, S., Hughes, R. B. & Nosek, T. M. 2016. An internet-based virtual reality intervention for enhancing self-esteem in women with disabilities: results of a feasibility study. *Rehabil Psychol*, 61(4), 358–370

Nosek, M. A., Robinson-Whelen, S., Hughes, R. B., Petersen, N. J., Taylor, H. B., Byrne, M. M. & Morgan, R. 2008. Overweight and obesity in women with physical disabilities: associations with demographic and disability characteristics and secondary conditions. *Disabil Health J*, 1, 89–98.

Nosek, M. A., Robinson-Whelen, S., Ledoux, T. A., Hughes, R. B., O'Connor, D. P., Lee, R. E., Goe, R., Silveira, S. L., Markley, R., Nosek, T. M. & The GoWoman Consortium. 2019. A pilot test of the GoWoman weight management intervention for women with mobility impairments in the online virtual world of second life. *Disabil Rehabil*, 41(22), 2718–2729.

Rimmer, J. H., Riley, B., Wang, E., Rauworth, A. & Jurkowski, J. 2004. Physical activity participation among persons with disabilities: barriers and facilitators. *Am J Prev Med*, 26, 419–25.

Rimmer, J. H., Wolf, L. A., Armour, B. S. & et al. 2007. Physical activity among adults with a disability - United States, 2005. *Morbid Mortal Wkly Rep*, 56, 1021–4.

Riva, G., Bacchetta, M., Baruffi, M. & Molinari, E. 2001. Virtual reality-based multidimensional therapy for the treatment of body image disturbances in obesity: a controlled study. *Cyberpsychol Behav*, 4, 511–26.

Robinson-Whelen, S., Hughes, R. B., Taylor, H. B., Colvard, M., Mastel-Smith, B. & Nosek, M. A. 2006. Improving the health and health behaviors of women aging with physical disabilities: a peer-led health promotion program. *Women's Health Issues*, 16, 334–45.

Robinson-Whelen, S., Hughes, R. B., Taylor, H. B., Markley, R., Vega, J. C., Nosek, T. M. & Nosek., M. A. 2020. Promoting psychological health in women with SCI: development of an online self-esteem intervention. *Disabil Health J*, 13, 100867.

Ruiz, J. G., Andrade, A. D., Anam, R., Aguiar, R., Sun, H. & Roos, B. A. 2011. Using anthropomorphic avatars resembling sedentary older individuals as models to enhance self-efficacy and adherence to physical activity: psychophysiological correlates. *Stud Health Technol Inform*, 173, 405–11.

Shepherd, T. 2020. Second Life Grid Survey - Region Database. Retrieved from http://www.gridsurvey.com/ [May 10, 2020].

Stendal, K., Molka-Danielsen, J., Munkvold, B. E. & Balandin, S. 2012. Virtual worlds and people with lifelong disability: exploring the relationship with virtual self and others. ECIS 2012.

Stendal, K., Molka-Danielsen, J., Munkvold, B. E. & Balandin, S. 2013. Social affordances for people with lifelong disability through using virtual worlds. 2013 46th Hawaii International Conference on System Sciences, January 2013. IEEE, 873–882.

Stewart, S., Hansen, T. S. & Carey, T. A. 2010. Opportunities for people with disabilities in the virtual world of Second Life. *Rehabil Nurs*, 35, 254–59.

Sullivan, D. K., Goetz, J. R., Gibson, C. A., Washburn, R. A., Smith, B. K., Lee, J., … & Donnelly, J. E. 2013. Improving weight maintenance using virtual reality (second life). *J Nutr Educ Behav*, 45, 264–8.

Surveymonkey. 2011. *Survey Monkey* [Online]. Available: http://www.surveymonkey.com [Accessed June 15, 2011].

Thesaurus.Com. 2020. *Thesaurus.com* [Online]. Dictionary.com LLC. [Accessed May 22, 2020].

Wang, Y. & Beydoun, M. A. 2007. The obesity epidemic in the United States–gender, age, socioeconomic, racial/ethnic, and geographic characteristics: a systematic review and meta-regression analysis. *Epidemiol Rev*, 29, 6–28.

Yee, N., Bailenson, J. N. & Ducheneaut, N. 2009. The proteus effect: implications of transformed digital self-representation on online and offline behavior. *Commun Res*, 36, 285–312.

Index

Printed in the United States
By Bookmasters